中国外来入侵物种图鉴

Pictorial Alien Invasive Species in China

鄢 建 主编

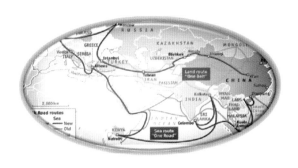

中国农业科学技术出版社

图书在版编目（CIP）数据

中国外来入侵物种图鉴/鄢建主编．—北京：中国农业科学技术出版社，2018.1
　ISBN 978-7-5116-3452-8

Ⅰ.①中… Ⅱ.①鄢… Ⅲ.①外来种—侵入种—中国—图集 Ⅳ.①Q16-64

中国版本图书馆 CIP 数据核字（2017）第 321600 号

责任编辑　贺可香
责任校对　贾海霞

出 版 者	中国农业科学技术出版社
	北京市中关村南大街12号　　邮编：100081
电　　话	（010）82106638（编辑室）（010）82109702（发行部）
	（010）82109709（读者服务部）
传　　真	（010）82106650
网　　址	http://www.castp.cn
经 销 者	全国各地新华书店
印 刷 者	北京科信印刷有限公司
开　　本	880mm×1 230mm　1/16
印　　张	28.5
字　　数	900千字
版　　次	2018年1月第1版　2018年1月第1次印刷
定　　价	398.00元

◆ 版权所有·翻印必究 ◆

《中国外来入侵物种图鉴》
编委会

主　任　高玉潮
副主任　李建伟　毕克新　张宝峰　陈洪俊　唐光江　赵增连　郭雪艳
　　　　　张柏青　石　勇　孙江功　刘心同　陈洪波
委　员　张铁军　孙明钊　蒋小龙　安治国　印丽萍　邓学农　乔华峰
　　　　　宋振乾　马　昕　陈彦兵　李建军　吴杏霞　黄庆林　张成标
　　　　　黄从平　马晓光　赵文胜

主　编　鄢　建
副主编　王洪兵　张艺兵　焦宏强　徐海涛　李英强
编　委（按姓氏笔画排序）
　　　　　于长友　于立欣　于红光　门爱军　王　丹　王军工　王守国
　　　　　王　岩　王辉波　冯春光　白章红　吕　朋　刘文鹏　刘学科
　　　　　刘　涛　江兴培　孙双燕　孙　凌　苏保乐　沐咏民　汤德良
　　　　　吴　昊　何金英　张　明　陆福军　陈德明　郑大勇　范京安
　　　　　赵怡芳　赵　莎　姜晓菲　袁　平　袁　涛　原志伟　党国瑞
　　　　　徐　卫　徐　亮　徐　强　黄法余　葛　岚　鄢　建　蒲　民
　　　　　赖　凡　谭青安　熊先军

前　言

国家生态安全，就如同国家政治安全、经济安全、文化安全、社会安全和国防安全一样，是构成国家安全体系的重要基石。在国家生态安全体系中，国门生物安全则是其最主要的组成部分之一。外来入侵物种不仅直接影响着国门生物安全，更是威胁到国家生态安全乃至国家安全。

在维护国家生态安全和国门生物安全方面，国家口岸管理部门，尤其是出入境检验检疫部门，更应该担负起国门卫士的庄严使命，不负国家和人民的重托。

"外来入侵物种"（Alien Invasive Species，IAS），是指"一个定殖在其过去或现在的自然分布区之外，但一旦进入和/或扩散，将威胁生物多样性的物种"（《生物多样性公约》（*Convention on Biological Diversity*，CBD））。

生物入侵尤其是外来物种入侵问题，已成为21世纪的世界性难题。目前虽然还没有一个准确的权威统计数据，但根据国际自然保护联盟（International Union for Conservation of Nature，IUCN）的报告，外来入侵物种每年给全球造成的直接经济损失至少超过4 000亿美元。在美国，外来入侵物种每年造成的直接损失高达1 200亿美元，如果加上防治费用，则更是超过1 380亿美元（Pimental等，2005）。

我国是遭受外来入侵物种危害最严重的国家之一。从2013年10月24日在青岛召开的"第二届国际生物入侵大会"上获悉，当时确认的中国外来入侵物种就有544种。这些外来入侵物种每年给国家造成的经济损失已超过2 000亿元。在IUCN公布的世界最严重的100种入侵物种中，我国占50余种。

外来入侵物种除了造成直接经济损失外，其所带来的全球性生态灾难更为突出和危险。这充分表现在，当外来入侵物种进入后，如果一旦定殖成功，种群就会迅速增殖并广泛扩散，导致当地物种种群急剧减少、濒危乃至灭绝，从而破坏生物多样性，打破生态平衡，最终使得生态系统崩溃，引发生态灾难。生态灾难可能诱发社会危机，甚至是造成一些人类灿烂文明消亡的直接原因之一。目前，外来入侵物种已经扩散到我国34个省、直辖市、自治区的部分地区，涉及森林、水域、湿地、草原、农牧区和城市居民区等几乎所有的生态系统，并已经对其构成了严重威胁。

生物多样性（Biological Diversity），是指各种活体生物间的变异性，包括陆生、海生、其他水生生态系统以及由它们组成的复合生态系统的多样性，因此也包括生物种内多样性、种间多样性和生态系统多样性。生物多样性不仅具有极高的经济价值，而且是人类赖以生存的基本条件。据美国生态经济学家Robert Constanza等（1997）评估，由生物多样性构建的全球生态系统，每年可创造价值16万亿～54万亿美元，平均33万亿美元的经济效益，大大超过当时全球国民生产总值（GDP）的总和（18万亿美元）。而生物多样性为人类所提供的基本生存条件，在很大程度上是其他方法根本无法替代的。因此，从这个意义上讲，保护生物多样性，就是保护生物赖以生存的生态系统，就是保护人类的基本生存环境。我国是世界生物多样性最丰富的国家之一，占全球生物多样性的10%以上。所以说，保护好我国的生物多样性，不仅是为中国，也是在为全人类的生存和发展做千秋伟业。

随着"一带一路"倡议和建设的不断落实与推进，我国必将成为全球最重要的政治、经济、文化、旅游、商贸和交通运输中心。因此，外来入侵物种对我国的入侵风险，将与日俱增。如何平衡外来入侵物种管理和保护生物多样性之间的关系，控制风险，趋利避害，为"一带一路"建设保驾护航，已经成为摆在我们面前的迫切问题。我们没有退路，只能未雨绸缪，积极应对，早作安排和部署。

我们今天做这项工作，无疑是中国对构建"人类命运共同体"、解决人类可持续发展提供"中国方案"。但这将涉及"一带一路"沿线众多国家和地区，甚至是整个世界，单靠中国自身的力量是远远不够的。所以，相关国家、国际组织乃至个人，都有义务通力合作，惠利互享，全球共治，做好外来入侵物种的管理工作。编写本书的主要目的之一，也正是因为这个新形势和新要求。

本书以国家环保部（总局）、中国科学院联合对外正式公布的四批71种中国外来入侵物种名录为基础，参考国内外大量文献资料，且按生物类别编写，并配以精美图片，图文并茂，使得本书更有可读性，查阅亦很方便。本书除收集了我国对外公布的所有外来入侵物种外，还增补了部分近似物种（21种），既是目前我国第一部最新最全的外来入侵物种图鉴，也是研究生物入侵的基础工具书，尤其适合从事国门生物安全管理的口岸管理部门如出入境检验检疫、海关、港口管理部门，也适合国内环保、农林牧渔部门、法制部门、外经贸企业使用。同时，适合生物入侵相关的教学、科研等大中专院校和科研院所参考使用。当然也适合那些对此领域感兴趣的读者朋友，以为科普之用。在书中，我们还增加了外来入侵物种的英文参考资料，以及最新《中华人民共和国进境植物检疫性有害生物名录》等附录文件，这是为了方便读者进一步学习、研究和开展对外交流之用。

在本书的编写过程中，得到了山东检验检疫局领导的大力支持，得到了食品处同事们的无私帮助。同时，还得到了国内外同行的热情鼓励和中肯建议。在此一并表示衷心感谢！

由于本书所涉及内容非常广，专业性极强，我们虽然几易其稿，力求做得更好一些，但由于编者专业知识、科学水平有限，加之时间仓促，书中一定会存在不少缺点甚至错误。敬请读者批评指正，并提出建设性意见和建议。再次表示衷心感谢！

建设"一带一路"，是中华民族更快实现"中国梦"的必由之路。在铺就伟大"中国梦"的道路上，我们愿作一粒"铺路石"！

编 者

2017年10月

内容简介

本书收录了国家环保部（总局）、中国科学院联合对外正式公布的所有四批71种中国外来入侵物种，还增补了部分外来入侵物种的近似种21种，是目前我国第一部最新最全的外来入侵物种图鉴，也是研究生物入侵的基础工具书。

本书介绍了每个外来入侵物种的信息，包括学名、异名、别名、英文名、形态特征、生物危害、地理分布、传播途径和管理措施等内容，还尽可能地配上在入侵物种生长发育过程中，最具有代表性特征的精美图片，方便读者查阅。

读者对象

◆ 适合国家安全管理部门

◆ 适合从事国门生物安全管理的口岸部门，如出入境检验检疫、海关、港口等管理部门

◆ 适合国内环保、法制、农林牧渔业部门

◆ 适合与生物入侵相关的教学、科研等大中专院校和科研院所

◆ 适合外经贸企业

◆ 科普读物

◆ 对外交流

目　录

外来入侵植物

紫茎泽兰	3
飞机草	12
藿香蓟	19
薇甘菊	26
空心莲子草	35
豚　草	41
三裂叶豚草	48
毒　麦	55
互花米草	61
凤眼莲	70
假高粱	75
马缨丹	87
大　薸	93
加拿大一枝黄花	99
蒺藜草	106
长刺蒺藜草	110
银胶菊	114
黄顶菊	121
土荆芥	126
刺　苋	130
反枝苋	136
长芒苋	142
落葵薯	150
钻形紫菀	155
三叶鬼针草	161
大狼杷草	169

小蓬草 ··· 182

苏门白酒草 ··· 188

一年蓬 ··· 198

假臭草 ··· 204

刺苍耳 ··· 209

圆叶牵牛 ·· 215

垂序商陆 ·· 224

光荚含羞木 ··· 234

五爪金龙 ·· 245

喀西茄 ··· 251

黄花刺茄 ·· 255

刺果瓜 ··· 270

野燕麦 ··· 276

水盾草 ··· 290

外来入侵动物

昆 虫

蔗扁蛾 ··· 297

湿地松粉蚧 ··· 301

强大小蠹 ·· 304

美国白蛾 ·· 308

桉树枝瘿姬小蜂 ··· 313

稻水象甲 ·· 317

入侵红火蚁 ··· 321

苹果蠹蛾 ·· 325

三叶斑潜蝇 ··· 330

松突圆蚧 ·· 335

椰心叶甲 ·· 339

红棕象甲 ·· 343

铃木方翅网蝽 ·· 348

扶桑绵粉蚧 ··· 351

刺桐姬小蜂 ··· 354

美洲大蠊 ·· 357

德国小蠊 ·· 360

无花果蜡蚧 ··· 364

　　枣实蝇 ··· 368

　　椰子织蛾 ··· 373

　　松树蜂 ··· 376

线　虫

　　松材线虫 ··· 380

软体动物

　　非洲大蜗牛 ··· 384

　　福寿螺 ··· 388

两栖类

　　牛　蛙 ··· 393

甲壳类

　　克氏原螯虾 ··· 396

爬行类

　　巴西红耳龟 ··· 399

鱼　类

　　豹纹脂身鲇 ··· 403

　　腹锯鲑脂鲤 ··· 407

　　尼罗罗非鱼 ··· 410

　　食蚊鱼 ··· 413

主要参考文献 ··· 420

主要网站 ··· 421

附　录 ··· 427

　　Ⅰ 世界最严重的100种入侵物种 ································ 427

　　Ⅱ 世界十大外来入侵物种 ···································· 432

　　Ⅲ 中华人民共和国进境植物检疫性有害生物名录 ················ 433

Part 1

外来入侵植物

紫茎泽兰

学　　名：***Ageratina adenophora*** (Spreng.) King & H.Rob.
异　　名：*Eupatorium adenophorum* Spreng.，*Eupatorium glandulosum* Michx.，*Eupatorium glandulosum* Hort. ex Kunth，*Eupatorium pasadenense* Parish
别　　名：解放草、马鹿草、破坏草、黑头草、大泽兰、飞机草、臭草、腺泽兰
英文名：Crofton weed，sticky eupatorium

形态特征

紫茎泽兰为多年生草本植物或亚灌木状植物。

植物根茎粗壮发达，直立。植株高10～200cm。

植物茎紫色，被白色或锈色短柔毛。分枝对生，斜上。

植物叶对生。叶片质薄，卵形、三角形或菱状卵形。叶腹面绿色，背面色浅；两面被稀疏的短柔毛，在叶背面及沿叶脉处毛稍密；基部平截或稍呈心形，顶端急尖，基出三脉；边缘有稀疏粗大而不规则的锯齿。在花序下方，叶边缘则为波状浅锯齿或近全缘。

头状花序小，直径可达6mm。在枝端排列成复伞房或伞房花序。总苞片3～4层，含40～50朵小花。管状花两性，白色，花药基部钝。

子实为瘦果，黑褐色。每株可年产种子1万粒左右，瘦果藉冠毛随风传播。

花期为11月至翌年4月。果期为3～4月。

生物危害

紫茎泽兰属于我国检疫性有害生物。其入侵农田、林地、牧场后，与农作物、牧草和林木争夺肥、水、阳光和空间；还可分泌克生性物质，抑制周围其他植物生长，形成单优种群，破坏生物多样性；对农作物和经济作物产量、草地维护、森林更新有极大影响。亦破坏园林景观。

紫茎泽兰植株体内含有的芳香物质和辛辣化学物质均为有毒物质，家畜误食后，可致其中毒甚至死亡。

紫茎泽兰的花粉为过敏原，可引起人类过敏性疾病。

地理分布

紫茎泽兰原产于美洲墨西哥至哥斯达黎加一带，分布在北纬37°至南纬35°范围内。现分布于美国、澳大利亚、新西兰、南非、西班牙、印度、菲律宾、马来西亚、新加坡、印度尼西亚、巴布亚新几内亚、泰国、缅甸、越南、尼泊尔、巴基斯坦以及太平洋岛屿等30多个国家和地区。

大约在20世纪40年代，紫茎泽兰由中缅边境传入我国云南南部。目前在我国云南、贵州、四川、广西壮族自治区（以下简称广西）、西藏自治区（以下简称西藏）等地都有分布，并以每年10～30km的速度向北和向东扩散。

| 传播途径 |

紫茎泽兰除种子传播外，其根、茎也能进行无性繁殖。因此，这些繁殖材料均可以人为传播或自然扩散。

| 管理措施 |

可以采取有害生物综合治理（Integrated Pest Management，IPM）（简称综合治理，下同）或有害生物综合控制（Integrated Pest Control，IPC）（简称综合控制，下同）措施，包括植物卫生措施（Phytosanitary Measures）。

 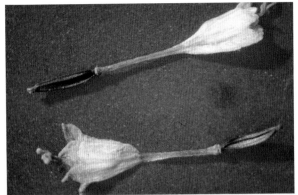
 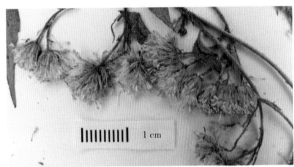
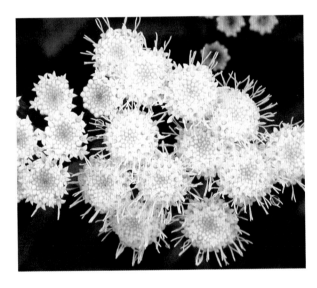 
 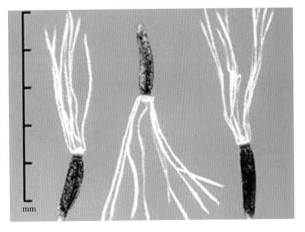

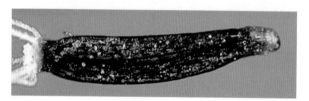

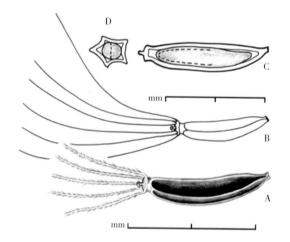

泽兰实蝇（*Procecidochares utilis*）为害后形成虫瘿

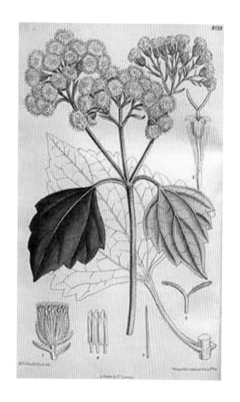

（图片选自Wikipedia、pic.sogou.com，tupian.baike.com，idtools.org，upload.wikimedia.org，invasoras.pt，bugwoodcloud.org，nathistoc.bio.uci.edu，lookformedical.com，hasbrouck.asu.edu，agpest.co.nz，plantgenera.org，keyserver.lucidcentral.org）

紫茎泽兰（*Ageratina adenophora*）

【www.efloras.org】*Ageratina adenophora* (Sprengel) R. M. King & H. Robinson, Phytologia. 19: 211. 1970.

Eupatorium adenophorum Sprengel, Syst. Veg., ed. 16, 3: 420. 1826, based on *E. glandulosum* Kunth in Humboldt et al., Nov. Gen. Sp. 4, ed. f°: 96. 1818, not Michaux (1803).

Shrubs or perennial herbs, 30-90 (-200) cm tall. Stems erect; branches opposite, obliquely ascending, white or ferruginous puberulent, upper part and peduncles more densely so, glabrescent or glabrous in lower part by anthesis. Leaves opposite, long petiolate; blade abaxially pale, adaxially green, ovate, triangular-ovate, or rhombic-ovate, 3.5-7.5 cm × 1.5-3 cm, thin, both surfaces sparsely puberulent, more densely so abaxially and on veins, basally 3-veined, base truncate or slightly cordate, margin coarsely crenate, apex acute. Synflorescences terminal, corymbose or compound-corymbose, to 12 cm in diam. Capitula numerous, 2-4 cm, 40-50-flowered; involucre broadly campanulate, ca. 3 × 4 mm; phyllaries 2-seriate, linear or linear-lanceolate, 3.5-5 mm, apex acute or acuminate; receptacle convex to conical; corollas purplish, tubular, ca. 3.5 mm. Achenes black-brown, narrowly elliptic, 1-1.5 mm, 5-angled, without hairs and glands; pappus setae 10, basally connate, white, fine, equal to corolla. Fl. and fr. Apr-Oct. $2n = 51$.

【www.efloras.org】*Ageratina adenophora* (Sprengel) R. M. King & H. Robinson, Phytologia. 19: 211. 1970.

Crofton weed, sticky snakeroot

Eupatorium adenophorum Sprengel, Syst. Veg. 3: 420. 1826, based on *E. glandulosum* Kunth in A. von Humboldt et al., Nov. Gen. Sp. 4 (fol.): 96, plate 346. 1818; 4 (qto.): 122, plate 346. 1820, not Michaux 1803

Subshrubs, 50-220 cm.

Stems (usually purplish when young) erect, stipitate-glandular.

Leaves opposite; petioles 10-25 mm; blades (abaxially purple) ovate-lanceolate or ovate-deltate to lanceolate-ovate, (1.5-) 2.5-5.5 (-8) cm × 1.5-4 (-6) cm, bases cuneate to obtuse or nearly truncate, margins serrate, apices acute to acuminate, abaxial faces stipitate- to sessile-glandular.

Heads clustered.

Peduncles 5-12 mm, densely stipitate-glandular and sometimes also sparsely viscid-puberulent.

Involucres 3.5-4 mm.

Phyllaries: apices acute, abaxial faces stipitate-glandular

.**Corollas** white, pink-tinged, lobes sparsely hispidulous.

Cypselae glabrous.

$2n = 51$.

【www.cabi.org】*Ageratina adenophora* (Croftonweed)

Broadleaved, Herbaceous, Perennial, Seed propagated Shrub, Vegetatively propagated plant.

A many stemmed, perennial herb 1-2 m high, reproducing by seed and vegetatively from a short, pale, yellow rootstock. The following description is taken from Parsons and Cuthbertson (1992). Stems are purplish, numerous, erect, smooth, cylindrical; shortly branched towards the apex, 1-2 m long, occasionally longer; glandular, hairy at first but becoming woody with age and rooting at the nodes if damaged. Stems arise from a short, thick, pale-yellow rootstock with a carrot-like odour when broken, giving rise to numerous branching secondary roots extending laterally to a radius of 1 m and downwards to 40 cm; adventitious roots may form on the first 3 cm of stem. Leaves dark green; opposite, broadly trowel-shaped, 5-8 cm long, (2.5-) 3-7.5 cm wide, with serrated edges, tapering towards the apex and narrowing abruptly at the base into a slender stalk 2-4 cm long; 3-nerved, glabrous or slightly pubescent, toothed along the apical margins. Petioles are brown. Flowers comprise 50 to 70 white, tubular florets about 3.5 mm long; grouped into heads 5-6 mm diameter within a row of green bracts and arranged in flat clusters up to 10 cm across at the end of the branches. Seeds are dark brown to black, slender, angular, 1.5-2 mm long; topped by a pappus of 5 to 10 fine white hairs approximately 4 mm long.

【keyserver.lucidcentral.org】 *Ageratina adenophora* (Spreng.) R.M. King & H. Rob.

Distinguishing Features

a long-lived herbaceous plant or small shrub growing up to 2 m tall.

it produces upright, branched stems from a woody rootstock.

its stems are covered in sticky hairs when young and bear pairs of oppositely arranged leaves.

its leaves are relatively broad, diamond-shaped or almost triangular in shape (4-15 cm long and 3-9 cm wide), and have toothed margins.

its small white flower-heads are borne in dense clusters at the tips of the branches.

its roots are yellowish in colour and give off a distinct carrot-like smell when broken or damaged.

Stems and Leaves

The branched stems are densely covered in sticky (i.e. glandular) hairs when young and may be green, reddish or purplish in colour. They become slightly woody and turn brownish-green or brown in colour when mature.

The leaves are oppositely arranged along the stems and are borne on stalks (i.e. petioles) 1-6 cm long. The leaf blades (4-15 cm long and 3-9 cm wide) are trowel-shaped, diamond-shaped (i.e. rhomboid), or triangular with bluntly or sharply toothed (i.e. crenate or serrate) margins. These leaves have sharply pointed tips (i.e. acute apices) and are mostly hairless (i.e. glabrous), but their stalks (i.e. petioles) are often covered in sticky hairs (i.e. they are glandular pubescent).

Flowers and Fruit

The small flower-heads (i.e. capitula) lack large 'petals' (i.e. ray florets) and consist of several tiny flowers (i.e. tubular florets) surrounded by two rows of greenish bracts (i.e. an involucre) 3-5 mm long. These flower-heads (5-8 mm across) are borne in large numbers and arranged in clusters at the tips of the branches (i.e. in terminal corymbose inflorescences). The tiny tubular florets (3-5 mm long) are white and contain both male and female flower parts (i.e. they are bisexual). Flowering occurs mostly during spring and early summer in northern regions, and during late summer and autumn in southern regions.

The 'seeds' (i.e. achenes) are slender, reddish-brown or blackish-brown in colour, and slightly curved. These 'seeds' (1-2 mm long and 0.3-0.5 mm wide) have four or five slight ribs which run lengthwise (i.e. longitudinally) and their bodies are hairless (i.e. glabrous). However, they are topped with a ring (i.e. pappus) of numerous whitish hairs (3-4 mm long), which are readily shed.

【近似种】小紫茎泽兰

学　名：*Ageratina riparia*（Regel）R.M. King & H. Rob.

异　名：*Eupatorium cannabinum* L.（misapplied），*Eupatorium riparium* Regel

英文名：cat's paw，catspaw，creeping crofton weed，creeping croftonweed，mist flower，mistflower，mistweed，river eupatorium，small crofton weed，white weed，William Taylor

Distinguishing Features

a long-lived scrambling herbaceous plant with numerous spreading stems.

its oppositely arranged leaves are relatively narrow with toothed margins.

its numerous small white flower-heads are clustered together at the tips of the branches.

its small blackish-brown seeds（1-2 mm long）are topped with a ring of whitish coloured hairs（3-4 mm long）.

Stems and Leaves

The numerous stems are relatively weak and the lower branches readily produce roots at their joints（i.e. nodes）if they come into contact with the soil. Branches tend to grow obliquely at first, then upwards（i.e. they are decumbent or ascending）. These stems are often reddish or purplish in colour and have a sparse covering of fine hairs（i.e. they are sparsely pubescent）.

The leaves are oppositely arranged along the stems and borne on short stalks（i.e. petioles）5-15 mm long. They are relatively narrow（i.e. narrowly-ovate or lanceolate）and prominently veined. These leaves（3-11 cm long and 8-30 mm wide）are mostly hairless（i.e. glabrous）and taper to a point at their tips（i.e. they have acute apices）. Their margins are sharply toothed（i.e. serrated）, especially toward their tips.

Flowers and Fruit

The small white flower-heads（i.e. capitula）do not have any obvious 'petals'（i.e. ray florets）and consist of several tiny flowers（i.e. tubular florets）surrounded by two rows of greenish bracts（i.e. an involucre）about 4-5 mm long. The tiny tubular florets are white in colour and also about 5 mm long. Several of these flower-heads are clustered together at the tips of the branches（i.e. into terminal corymbose inflorescences）. Flowering occurs from late winter through to late spring, but is most abundant during mid-spring.

The small 'seeds'（i.e. achenes）are slender, blackish-brown in colour and slightly curved. These 'seeds'（1-2 mm long）have four or five hairy ridges which run lengthwise（i.e. longitudinally）and they are topped with a ring（i.e. pappus）of several larger whitish coloured hairs（3-4 mm long）.

【www.hear.org】*Ageratina riparia*（Regel）R.M.King & H.Rob.

"A herb of 40-60 cm height, sometimes exceeding 1 m, with numerous spreading or ascending stems. Leaves have long petioles, are opposite and simple, lanceolate to elliptic, and toothed at the margins of the upper half. The leaf blades are 3-12 cm long and 0.8-3 cm wide. Flower-heads are 5-6 mm in diameter, subtended by green bracts, and contain 15-30 white or cream florets and are arranged in terminal clusters at the ends of branches. The corolla is 3-3.5 mm long. Fruits are dark brown achenes of c. 2 mm length and have a pappus of white hairs. The plant has a short and thick root stock"（Weber, 2003; p. 29）.

"Spreading subshrubs from a creeping rootstock; stems sprawling, 5-10 dm long, sparsely puberulent. Leaves lanceolate to elliptic, 4-8（-12）cm long, 1-1.5 cm wide, margins sharply serrate, petioles 0.5-1.3 cm long. Inflorescences lax; involucral bracts 4-5 mm long, puberulent; corollas white, 3-3.5 mm long. Achenes black, 1.5-2 mm long, puberulent on the angles"（Wagner et al., 1999; p. 255）.

外来入侵植物

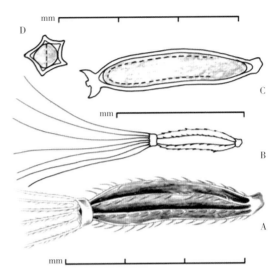

（图片选自keyserver.lucidcentral.org，www.southeastweeds.org.au，a.share.photo.xuite.net，upload.wikimedia.org，idtools.org）

小紫茎泽兰（*Ageratina riparia*）

飞机草

学　　名：**_Chromolaene odorata_** L.

异　　名：_Chrysocoma maculata_ Vell.，_Chrysocoma maculata_ Vell. Conc.，_Chrysocoma volubilis_ Vell. Conc.，_Eupatorium brachiatum_ Sw. ex Wikstr.，_Eupatorium clematitis_ DC.，_Eupatorium conyzoides_ Mill.，_Eupatorium dichotomum_ Sch. Bip.，_Eupatorium divergens_ Less.，_Eupatorium floribundum_ Kunth，_Eupatorium graciliflorum_ DC.，_Eupatorium klattii_ Millsp.，_Eupatorium odoratum_ L.，_Eupatorium sabeanum_ Buckley，_Eupatorium stigmatosum_ Meyen & Walp.，_Osmia atriplicifolia_（Vahl）Sch. Bip.，_Osmia clematitis_（DC.）Sch. Bip.，_Osmia divergens_（Less.）Sch. Bip.，_Osmia floribunda_（Kunth）Sch. Bip.，_Osmia graciliflora_（DC.）Sch. Bip.，_Osmia graciliflorum_（DC.）Sch. Bip.，_Osmia odorata_（L.）Sch. Bip.

别　　名：香泽兰、解放草、马鹿草、破坏草、黑头草、大泽兰

英文名：Chromolaena，Armstrong's weed，baby tea，bitter bush，butterfly weed，Christmas bush，chromolaena，devil weed，eupatorium，Jack in the bush，Jack-in-the-bush，kingweed，paraffinbush，paraffinweed，Siam weed，turpentine weed，triffid weed.

| 形态特征 |

飞机草为多年生草本植物。植物根茎粗壮，横走。

植物茎直立，高1~3m，苍白色，有细条纹。分枝粗壮，常对生，水平射出，与主茎成直角，少有分枝互生而与主茎成锐角的。全部茎枝被稠密黄色茸毛或短柔毛。

植物叶对生，卵形、三角形或卵状三角形。长4~10cm，宽1.5~5cm。上面绿色，下面色淡，两面粗涩，被长柔毛及红棕色腺点；下面及沿脉的毛和腺点稠密；基部平截或浅心形或宽楔形，顶端急尖；基出3脉；侧面纤细，在叶下面稍突起；边缘有稀疏的粗大而不规则的圆锯齿，或全缘，或仅一侧有锯齿，或每侧各有一个粗大的圆齿，或三浅裂状。叶质地稍厚，有叶柄，叶柄长1~2cm。花序下部的叶小，常全缘。

头状花序多数或少数，在茎顶或枝端排成伞房状或复伞房状花序。花序径常为3~6cm，少有13cm的。花序梗粗壮，密被稠密的短柔毛。总苞圆柱形，长1cm，宽4~5mm，约含20个小花；总苞片3~4层，覆瓦状排列，外层苞片卵形，长2mm，外面被短柔毛，顶端钝，向内渐长，中层及内层苞片长圆形，长7~8mm，顶端渐尖。全部苞片有三条宽中脉。花白色或粉红色。花冠长5mm。

瘦果黑褐色，长4mm，5棱，无腺点，沿棱有稀疏的顺向贴紧的白色短柔毛。

花果期为4~12月。

| 生物危害 |

飞机草为我国检疫性有害生物,也是世界公认的恶性有毒杂草。它能分泌感化物质,排挤本地植物,可使草场失去利用价值,也影响林木生长和更新。

飞机草叶中含有香豆素类有毒化合物,能够引起人类皮炎和过敏性疾病。人误食嫩叶会引起头晕、呕吐等中毒症状。家禽、家畜和鱼类误食,也会引起中毒。

| 地理分布 |

飞机草原产于中美洲。现在南美、非洲、亚洲热带地区广泛分布。1934年在我国云南南部首次发现。目前已侵入我国海南、广东、台湾、广西、云南、贵州、香港、澳门等地,并向亚热带地区扩散。为全球性入侵物种。

| 传播途径 |

飞机草的瘦果能借冠毛随风自然扩散,也可人为传播。

| 管理措施 |

综合治理(Integrated Pest Management,IPM)或综合控制(Integrated Pest Control,IPC),包括植物卫生措施(Phytosanitary Measures)。

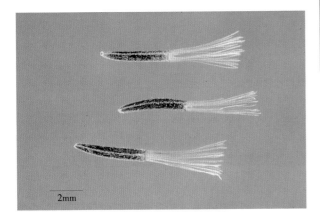

（图片选自www.worldseedsupply.com，baike.baidu.com，pic.baike.soso.com，www.discoverlife.org，www.backyardnature.ne，plantbook.org，keys.lucidcentral.org，www.hear.org，en.wikipedia.org）

飞机草（*Chromolaene odorata*）

紫茎泽兰　飞机草

	紫茎泽兰（*Ageratina adenophorum*）	飞机草（*Chromolaene odoratum*）
茎	呈暗紫色、毛被暗紫，触之有黏手感，异味明显	呈绿色、毛被白色
叶片	呈菱形，厚纸质；叶柄紫褐色，较长	呈三角形，薄纸质，两面都有灰白色绒毛；叶柄绿色，较短
头状花序	呈白色且近圆球形	呈淡绿黄色、绿白色或略淡紫色且为长圆柱形
繁殖方式	有性（瘦果）繁殖+无性繁殖（不定根）	有性（瘦果）繁殖

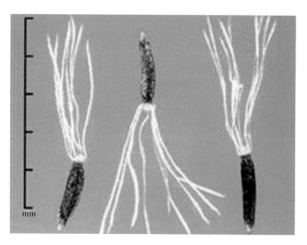
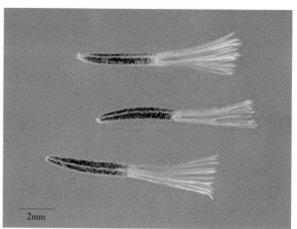

紫茎泽兰（*Ageratina adenophorum*）与飞机草（*Chromolaene odoratum*）比较

（图片选自en.wikipedia.org，www.naturalmedicinefacts.info，keys.lucidcentral.org，luirig.altervista.org，cn.bing.com） 【www.efloras.org】*Chromolaena odorata* (Linnaeus) R. M. King & H. Robinson, Phytologia. 20: 204. 1970.

Eupatorium odoratum Linnaeus, Syst. Nat., ed. 10, 2: 1205. 1759.

Herbs, perennial. Rhizomes robust, procumbent. Stems erect, 1-3 m tall, striate; branches robust, often opposite, spreading and horizontal, rarely alternate forming an acute angle with stem; stems and branches densely fulvous tomentose or shortly pubescent. Leaves opposite; petiole 1-2 cm; blade abaxially pale, adaxially green, ovate, triangular, or ovate-triangular, 4-10 cm × 1.5-5 cm, rather thick, both surfaces scabrid, villous with red-brown glands, abaxially and on veins more densely so, basally 3-veined, lateral veins fine, abaxially slightly raised, base truncate or shallowly cordate, margin sparsely coarsely and irregularly crenate or entire, or serrate on one side, or with one coarse tooth or 3-fid on each side, apex acute; leaves below synflorescence small, often entire. Synflorescence of numerous or few capitula in corymbs or compound corymbs; peduncle thick, densely shortly pubescent. Capitula ca. 20-flowered; involucre cylindric, ca. 10×4-5 mm; phyllaries 3- or 4-seriate, imbricate, outer phyllaries ovate, ca. 2 mm, puberulent, apex obtuse, median and inner phyllaries straw-colored, oblong, 7-8 mm, broadly 3-veined, eglandular, apex acuminate; corollas white or pink, ca. 5 mm. Achenes black-brown, ca. 4 mm, 5-ribbed, eglandular, sparsely white adpressed setuliferous along ribs. Fl. and fr. Apr-Dec. $2n = 58, 60$.

【www.efloras.org】*Chromolaena odorata* (Linnaeus) R. M. King & H. Robinson, Phytologia. 20: 204. 1970.

Eupatorium odoratum Linnaeus, Syst. Nat. ed. 10, 2: 1205. 1759; *Osmia odorata* (Linnaeus) Schultz-Bipontinus

Perennials or subshrubs, mostly 80-250 cm.

Stems erect or sprawling to subscandent, hispidulous to coarsely short-pilose.

Petioles 5-20 mm.

Leaf blades (3-nerved) narrowly lanceolate to deltate-lanceolate or ovate-lanceolate, 3-10 cm × 1-4 cm, margins coarsely dentate to subentire.

Heads usually 5-50+ in (terminal or lateral) corymbiform arrays.

Involucres cylindric, (7-) 8-10 mm.

Phyllaries in 4-6 (-8) series, apices of the inner appressed, rounded to truncate (sometimes slightly white-petaloid or expanded).

Corollas purplish to light blue to nearly white or slightly pinkish.

$2n = 40, 60, 70$.

【www.cabi.org】*Chromolaena odorata* (Siam weed)

Herbaceous, Perennial, Seed propagated Shrub, Woody plant

C. odorata is a herbaceous to woody perennial with a bushy habit which forms a very dense thicket about 2 m high, in almost pure stands. This many-branched plant becomes lianescent when it has the opportunity to climb on a support. Isolated individuals start to branch when they reach a height of about 120 cm. After the first year of growth, the plant develops a strong, woody underground storage organ, which can reach a diameter of 20 cm. Stems are terete and become woody. Twigs are slightly striolate longitudinally, pubescent, opposite-decussate. Leaves are simple, opposite-decussate and without stipules. They are rhomboid-ovate to ovate with an acute apex and a cuneate base. The blades are trinerved a few millimetres after the base, roughly crenate-serrate beyond their maximum breadth, slightly pubescent above and pubescent with numerous small yellow dots below (a lens is needed to see this). The petiole is 1-3 cm long, and the blade 5-14 cm long and 2.5-8 cm broad. Leaves and twigs produce a characteristic smell when crushed (Gautier, 1992a). Capitula are grouped in one, three or five convex trichotomic corymbs 5-10 cm in diameter, at the end of the twigs. The involucre is cylindrical, 8-10 mm long by 3-4 mm broad. It is made of a series of four or five oblong bracts, the external being the shorter. These bracts are obtuse, chartaceous, pale in colour with three or five nerves. The receptacle is convex, without scales. There are 15-35 florets per capitulum. The corolla is 5 mm long and has five lobes. Its colour ranges from pale-lilac to white. Styles are of the same colour, exserted and flexuous. Cypselae are composed of a 3- to 4-mm-long fusiform blackish achene, with five beige barbelate ribs, overtopped by a pappus of about 30 barbelate beige capillary bristles which are 4-5 mm long (Gautier, 1992a).

【www.iucngisd.org】 *Chromolaena odorata*

Chromolaena odorata is an herbaceous perennial that forms dense tangled bushes 1.5-2.0 m in height. It occasionally reaches its maximum height of 6m (as a climber on other plants). Its stems branch freely, with lateral branches developing in pairs from the axillary buds. The older stems are brown and woody near the base; tips and young shoots are green and succulent. The root system is fibrous and does not penetrate beyond 20-30cm in most soils. The flowerheads are borne in terminal corymbs of 20 to 60 heads on all stems and branches. The flowers are white or pale bluish-lilac, and form masses covering the whole surface of the bush (Cruttwell and McFadyen, 1989).

C. odorata is a big bushy herb with long rambling (but not twining) branches; stems terete, pubescent; leaves opposite, flaccid-membranous, velvety-pubescent, deltoid-ovate, acute, 3-nerved, very coarsely toothed, each margin with 1-5 teeth, or entire in youngest leaves; base obtuse or subtruncate but shortly decurrent; petiole slender, 1-1.5cm long; blade mostly 5-12cm long, 3-6cm wide, capitula in sub-corymbose axillary and terminal clusters; peduncles 1-3cm long, bracteate; bracts slender, 10-12mm long; involucre of about 4-5 series of bracts, pale with green nerves, acute, the lowest ones about 2mm long, upper ones 8-9mm long, all acute, distally ciliate, flat, appressed except the extreme divergent tip; florets all alike (disc-florets), pale purple to dull off-white, the styles extending about 4mm beyond the apex of the involucre, spreading radiately; receptacle very narrow; florets about 20-30 or a few more, 10-12mm long; ovarian portion 4mm long; corolla slender trumpet form; pappus of dull white hairs 5mm long; achenes glabrous or nearly so (Stone, 1970). The seeds of Siam weed are small (3-5mm long, ~1mm wide, and weigh about 2.5mg seed-1 (Vanderwoude et al., 2005).

藿香蓟

学　名：***Ageratum conyzoides*** L.

异　名：*Ageratum arsenei* B. L. Rob.，*Ageratum ciliare* L.，*Ageratum conycoides* L.，*Ageratum conyzoides* f. *obtusifolia*（Lam.）Miq.，*Ageratum conyzoides* var. *pilosum* Blume，*Ageratum cordifolium* Roxb.，*Ageratum hirsutum* Poir.，*Ageratum hirtum* Lam. *Ageratum latifolium* var. *galapageium* B. L. Rob.，*Ageratum meridanum* V. M. Badillo，*Ageratum microcarpum*（Benth. ex Benth.）Hemsl.，*Ageratum nanum* Hort. ex Sch. Bip.，*Ageratum obtusifolium* Lam.，*Ageratum pinetorum*（L. O. Williams）R.M.King & H. Rob.，*Ageratum suffruticosum* Regel，*Alomia microcarpa*（Benth. ex Benth.）B. L. Rob.，*Alomia pinetorum* L. O. Williams，*Cacalia mentrasto* Vell.，*Carelia conyzoides*（L.）Kuntze，*Chrysocoma maculata* Vell.，*Eupatorium paleaceum* Sessé & Moc.

别　名：胜红蓟、山羊草、比利山羊草、白花草、脓泡草、绿升麻、白毛苦、毛射香、白花臭草、消炎草、胜红药、水丁药、鱼腥眼、紫红毛草、广马草

英文名：billy goat weed，blue flowered groundsel，blue top，goat weed，mother brinkley，tropical ageratum，white weed，winter weed

| 形态特征 |

藿香蓟为一年生（或多年生）草本（或亚灌木）植物。稍有香味，被粗毛。

植物茎直立，植株高30~100cm。

植物单叶对生，有时上部互生。叶片卵形、菱状卵形或卵状长圆形，长3~8cm，宽2~5cm，茎上部叶较小，变长圆形，先端急尖，基部圆钝或宽楔形，边缘具圆锯齿，两面被白色稀疏柔毛和黄色腺点，基部具3（5）出脉。叶具叶柄，叶柄长1~3cm。

头状花序在枝端排成伞房状。总苞片2~3层，长圆形或披针状长圆形，长3~4mm，边缘撕裂状，具刺状尖头，外面无毛。花冠浅蓝色或白色，长1.5~2.5mm，顶端5浅裂。

瘦果黑褐色，具5棱，长1.2~1.7mm。冠毛膜片状，上部渐狭呈芒状，长1.5~3mm。

花果期5~10月，但在热带地区花果期几乎全年。

| 生物危害 |

藿香蓟常见于山谷、林缘、河边、茶园、农田、草地和荒地等生境，常侵入作物地，如在玉米、甘蔗和甘薯田中发生量大，危害严重。

能产生和释放多种化感物质，抑制本土植物生长。常在入侵地形成单优群落，对入侵地生物多样性造成威胁。目前已入侵到一些自然保护区。

地理分布

藿香蓟原产于热带美洲。现已广泛分布于非洲全境、印度、印度尼西亚、老挝、柬埔寨、越南等地。19世纪在我国香港记录。现主要分布于北京、天津、河北、辽宁、吉林、黑龙江、上海、江苏、浙江、安徽、福建、江西、山东、河南、湖北、湖南、广东、广西、海南、重庆、四川、贵州、云南、西藏（东南部）、陕西、台湾、香港、澳门等地。

传播途径

人为传播或自然扩散。

管理措施

综合治理（Integrated Pest Management，IPM）或综合控制（Integrated Pest Control，IPC），但不包括植物卫生措施（Phytosanitary Measures）。

外来入侵植物

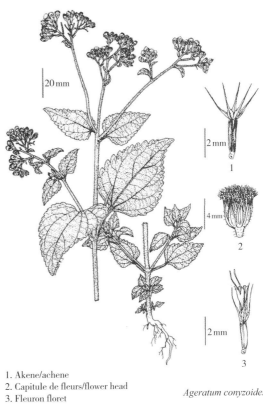

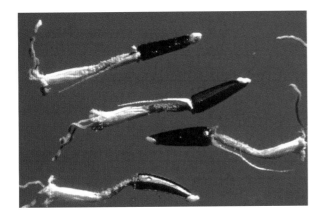

1. Akene/achene
2. Capitule de fleurs/flower head
3. Fleuron floret

Ageratum conyzoides

（图片选自www.plantwise.org，keys.lucidcentral.org，www.tisanes-indigenes.re，www.nzenzeflowerspauwels.be，www.hear.org，www.pioneercatchment.org.au，www.natureloveyou.sg，www.westafricanplants.senckenberg.de，stewartia.net，www.fpcn.net，plantjdx.com，www.ctahr.hawaii.edu，publish.plantnet-project.org）

藿香蓟（*Ageratum conyzoides*）

【 www.efloras.org 】*Ageratum conyzoides* Linnaeus, Sp. Pl. 2: 839. 1753.

Herbs, annual, 50-100 cm tall, sometimes less than 10 cm, with inconspicuous main root. Stems robust, ca. 4 cm in diam. at base, simple or branched from middle, stems and branches reddish, or green toward apex, white powdery puberulent or densely spreading long tomentose. Leaves often with axillary abortive buds; petiole 1-3 cm, densely white spreading villous; median leaves ovate, elliptic, or oblong, 3-8 cm × 2-5 cm; upper leaves gradually smaller, oblong, sometimes all leaves small, ca. 1 × 0.6 cm, both surfaces sparsely white puberulent and yellow gland-dotted, basally 3-veined or obscurely 5-veined, base obtuse or broadly cuneate, margin crenate-serrate, apex acute. Capitula small, 4-14, in dense terminal corymbs; peduncle 0.5-1.5 cm, powdery puberulent; involucre campanulate or hemispheric, ca. 5 mm in diam.; phyllaries 2-seriate, oblong or lanceolate-oblong, 3-4 mm, glabrous, margin lacerate; corollas 1.5-2.5 mm, glabrous or apically powdery puberulent; limb purplish, 5-lobed. Achenes black, 5-angled, 1.2-1.7 mm, sparsely white setuliferous; pappus scales 5 or awned, 1.5-3 mm. Fl. and fr. year-round. $2n = 20, 38, 40$.

【 www.efloras.org 】*Ageratum conyzoides* Linnaeus, Sp. Pl. 2: 839. 1753.

Tropical whiteweed

Ageratum latifolium Cavanilles

Annuals, perennials, or sub-shrubs, 20-150 cm（fibrous-rooted）.

Stems erect, sparsely to densely villous.

Leaf blades ovate to elliptic-oblong, 2-8 cm × 1-5 cm, margins toothed, abaxial faces sparsely pilose and gland-dotted.

Peduncles minutely puberulent and sparsely to densely pilose, eglandular.

Involucres 3-3.5 mm × 4-5 mm.

Phyllaries oblong-lanceolate（0.8-1.2 mm wide）, glabrous or sparsely pilose（margins often ciliate）, eglandular, tips abruptly tapering, subulate, 0.5-1 mm.

Corollas usually blue to lavender, sometimes white.

Cypselae sparsely strigoso-hispidulous;

pappi usually of scales 0.5-1.5（-3）mm, sometimes with tapering setae, rarely 0.

$2n = 20, 40$.

【 www.plantwise.org 】*Ageratum conyzoides*（billy goat weed）

A. conyzoides is an erect, branching, annual herb with shallow, fibrous roots. It may, depending upon environmental conditions, reach 50-1 500 mm tall at flowering. The stems, which may root where the bases touch the ground, are cylindrical, and become strong and woody with age; nodes and young parts of the stem are covered with short, white hairs. Leaves are opposite, 20-100 mm long, 5-50 mm wide, on hairy petioles 5-75 mm long, broadly ovate, with a rounded or narrowed acute base and an acute or obtuse or sometimes acuminate tip and toothed margins. Both leaf surfaces are sparsely hairy, rough with prominent veins and when crushed the leaves have a characteristic odour which is reminiscent of the male goat. The branched, terminal or axillary inflorescence bears 4-18 flower heads arranged in showy, flat-topped clusters. Individual flower heads are light blue, white or violet, are carried on 50-150 mm long peduncles and are 5 mm across, 4-6 mm long with 60-75 tubular flowers. The flower head is surrounded by two or three rows of oblong bracts which are green with pale or reddish-violet tops. The bracts are 3-5 mm high, outer ones 0.5-1.75 mm wide, sparsely hairy, evenly toothed in the upper part, with an abruptly acuminate, acute tip. Flowers are 1.5-3 mm long and scarcely protrude above the bracts. The fruit is a ribbed or angled, black achene, 1.25-2 mm long, roughly hairy, with a pappus of 5, rarely 6, rough bristles, white to cream coloured, 1.5-3 mm long with upward turning spines.

【keys.lucidcentral.org】*Ageratum conyzoides*（Billygoat Weed）

Ageratum conyzoides is an erect, branching, soft, slightly aromatic, annual herb with shallow, fibrous roots. It grows to approximately 1 m in height. The stems and leaves are covered with fine white hairs; the leaves are egg-shaped with broad end at base（ovate）up to 7.5 cm long. The flowers are purple, blue, pinkish or white, less than 6 mm across, with around 30 to 50 flowers and arranged in close terminal flower-heads. The fruits are small brown one-seeded achenes fruits.

薇甘菊

学　　名：***Mikania micrantha*** Kunth

异　　名：*Eupatorium denticulatum* Vahl，*Eupatorium orinocense*（Kunth）Gómez de la Maza，*Eupatorium orinocense*（Kunth）M.Gómez，*Eupatorium orinocense* var. *batataefolium*（DC.）M.Gómez，*Eupatorium orinocense* var. *tamoides*（DC.）M.Gómez，*Kleinia alata* G. Meyer，*Mikania alata*（G. Mey.）DC.，*Mikania cissampelina* DC.，*Mikania denticulata*（Vahl）Willd.，*Mikania glechomifolia* Sch. Bip. ex Baker，*Mikania micrantha*（Hieron.）B. L. Rob.，*Mikania micrantha* f. *hirsuta*（Hieron.）B. L. Rob.，*Mikania micrantha* var. *micrantha*，*Mikania orinocensis* Kunth，*Mikania scandens* var. *cynanchifolia* Hook. & Arn. ex Baker，*Mikania scandens* var. *sagittifolia* Hassl.，*Mikania scandens* var. *subcymosa*（L.）Wild.，*Mikania scandens* var. *umbellifera*（Gardner）Baker，*Mikania sinuata* Rusby，*Mikania subcrenata* Hook. & Arn.，*Mikania subcymosa* Gardner，*Mikania umbellifera* Gardner，*Mikania variabilis* Meyen & Walp.，*Willoughbya cissampelina*（DC.）Kuntze，*Willoughbya micrantha*（Kunth）Rusby，*Willoughbya scandens* var. *orinocensis*（Kunth）Kuntze，*Willoughbya variabilis*（Meyen & Walp.）Kuntze

别　　名：小花蔓泽兰、小花假泽兰

英文名：bitter vine，American rope，Chinese creeper，climbing hempweed，Mikania vine，mile-a-minute weed

| 形态特征 |

薇甘菊为多年生草本或木质藤本植物。

植物茎细长，匍匐或攀缘，多分枝，被短柔毛或近无毛。植物茎幼时为绿色，近圆柱形；老茎淡褐色，具多条肋纹。植物茎中部叶呈三角状卵形至卵形，长4.0~13.0cm，宽2.0~9.0cm。茎基部叶呈心形，偶近戟形；先端渐尖，边缘具数个粗齿或浅波状圆锯齿，两面无毛。叶基出3~7脉。叶具叶柄，叶柄长2.0~8.0cm；上部的叶渐小，叶柄短。

头状花序多数。在枝端常排成复伞房花序状，花序渐纤细；在顶部的头状花序花先开放，依次向下逐渐开放。头状花序长4.5~6.0mm，含小花4朵，全为结实的两性花。总苞片4枚，狭长椭圆形，顶端渐尖，部分急尖，绿色，长2~4.5mm。总苞基部有一线状椭圆形的小苞叶（外苞片），长1~2mm。花冠白色，脊状，长3~3.5（4）mm，檐部钟状，5齿裂。花带有香气。

瘦果长1.5~2.0mm，黑色，被毛，具5棱，被腺体。冠毛有32~38（40）条刺毛组成，白色，长2~3.5（4）mm。

花期在11~12月。

种子在12月底成熟。

| 生物危害 |

薇甘菊为我国检疫性有害生物。已被列入世界上最严重的100种外来入侵物种之一。薇甘菊是多年生草本植物或木质藤本植物，在其适生地攀援缠绕于乔木或灌木植物上，重压于其冠层顶部，阻碍被附着植物的光合作用，继而导致其死亡。在我国，它主要危害天然次生林、人工林，尤其是对当地6~8m以下的几乎所有树种，特别是对一些郁闭度小的林木危害最为严重。有严重危害的乔木树种包括红树、血桐、紫薇、山牡荆、小叶榕；有严重为害的灌木树种包括马缨丹、酸藤果、白花酸藤果、梅叶冬青、盐肤木、叶下珠、红背桂等。为害较重的乔木树种有龙眼、人心果、刺柏、苦楝、番石榴、朴树、荔枝、九里香、铁冬青、黄樟、樟树、乌桕；危害较重的灌木植物有桃金娘、四季柑、华山矾、地桃花、狗芽花等。

薇甘菊生长迅速，通过攀缘缠绕并覆盖被附着植物，对森林和农田土地造成巨大影响。其茎节随时可以生根并繁殖，快速覆盖生境；且能够产出大量种子，快速入侵。

通过竞争或他感作用，抑制自然植被和作物生长。

在东南亚地区，薇甘菊严重威胁当地木本植物，油棕、椰子、可可、茶叶、橡胶、柚木等均遭受其危害。由于薇甘菊常常攀缘至10m高的树冠或灌木丛的上层，因此，清除它时，常常会伤及其附着作物。

| 地理分布 |

薇甘菊原产于中美洲。现分布于印度、孟加拉国、斯里兰卡、泰国、菲律宾、马来西亚、印度尼西亚、巴布亚新几内亚和太平洋诸岛屿、毛里求斯、澳大利亚、中南美洲各国和美国南部地区。

大约在1919年，薇甘菊在我国香港发现。1984年在深圳发现。2008年以来，已广泛分布在广东、香港、澳门等珠江三角洲和广西地区。

| 传播途径 |

薇甘菊种子细小而轻，且基部有冠毛，易借风力、水流、动物（如昆虫）以及人类活动而远距离传播。也可随携带有种子、藤茎的载体及交通工具而进行传播。

| 管理措施 |

综合治理（Integrated Pest Management，IPM）或综合控制（Integrated Pest Control，IPC），包括植物卫生措施（Phytosanitary Measures）。

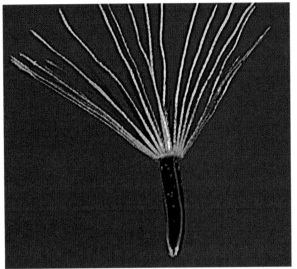

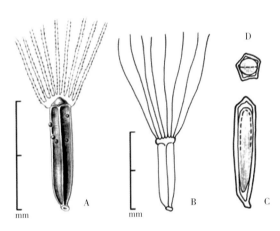

(图片选自weeds.brisbane.qld.gov.au, en.wikipedia.org, baike.sogou.com, www.freshfromflorida.com, www.hear.org, img.over-blog-kiwi.com, cnas-re.uog.edu, idtools.org, keyserver.lucidcentral.org, alberts.ac.in)

薇甘菊（*Mikania micrantha*）

【www.efloras.org】*Mikania micrantha* Kunth in Humboldt et al., Nov. Gen. Sp. 4, ed. f° : 105. 1818.

Vines, slender, branched. Stems yellowish or brownish, usually terete, slightly striate, glabrate to sparsely puberulent. Leaves opposite; petiole 1-6 cm; blade ovate, 3-13 × ca. 10 cm, both surfaces glabrate with numerous glandular spots, base cordate to deeply so, margin entire to coarsely dentate, apex shortly acuminate. Synflorescence a corymbose panicle, capitula clustered on subcymose branches; phyllaries oblong, ca. 3.5 mm, glabrous to puberulent, apex shortly acuminate; corollas white, 2.5-3 mm, tube narrow, limb broadly campanulate, inside papillate. Achenes 1.5-2 mm, 4-ribbed, with many scattered glands; pappus setae dirty white, ca. 3 mm. Fl. and fr. year-round. $2n = 36, 72$.

【www.cabi.org】*Mikania micrantha*（bitter vine）

Perennial，Seed propagated，Vegetatively propagated Vine / climber plant

M. micrantha is a vigorous, fast-growing, perennial, creeping or twining plant with numerous cordate lea Herbaceous ves and numerous large, loose heads of white or cream-coloured flowers that produce many seeds. This plant can climb and smother Hevea brasiliensis（rubber）trees as tall as 25 m.

Much-branched, perennial, scrambling, twining, slender-stemmed vine; stems herbaceous to semi-woody, branched, ribbed, sparsely pubescent or glabrous; leaves simple, opposite, glabrous, thin, broadly ovate, shallowly or coarsely toothed, triangular or ovate, tip acuminate, blade 4-13 cm long, 2-9 cm wide, 3-7 nerved; at the junction of the petioles with the nodes, unusual nodal appendages, membranous, up to 5 mm long; petioles tendriliform, 2-9 cm long; inflorescence a corymbose panicle with subcymose branches, 3-6 cm long by 3-10 cm wide; flowers small, white or cream-coloured, actinomorphic, 4.5-6 mm long, in leaf axils or on terminal shoots; florets white or greenish, fragrant; corolla mostly white, tubular, 2.5-4 mm long; involucral bracts 4, oblong to obovate, 2-4 mm long, acute, green, with one additional smaller bract 1-2 mm long; pappus（calyx）of 32-38 barbellate, capillary bristles, 2-3 mm long; stamens attached by their anthers, these exserted, with a triangular-ovate apical appendage as long as broad or longer and rounded or rarely emarginate or subsagittate at base; ovary inferior, the style base glabrous; fruit an achene that is somewhat flattened, elliptic, 4-ribbed with short, white hairs along the ribs, with a tuft of white pappus at the summit, glandular, 1.2-1.8 mm long, dark grey to black（Parham, 1958; Parham, 1962; Adams et al., 1972; Nair, 1988; Holm et al., 1991）.

【www.iucngisd.org】*Mikania micrantha*

A branched, slender-stemmed perennial vine. The leaves are arranged in opposite\r\npairs along the stems and are heart-shaped or triangular with an acute tip and a broad base. Leaves\r\nmay be 4-13cm long. The flowers, each 3-5mm long, are arranged in dense terminal or axillary\r\ncorymbs. Individual florets are white to greenish-white. The seed is black, linear-oblong, five-angled\r\nand about 2mm long. Each seed has a terminal pappus of white bristles that facilitates dispersal by\r\nwind or on the hair of animals（Pacific Island Ecosystems at Risk）.

【cnas-re.uog.edu】*Mikania micrantha*

Form: twining, scrambling/climbing herbaceous vine, fast growing

Stem: glabrous, ribbed, branched, pubescent, slightly four angled, reddish（often）; nodes swollen;

travels large distances along ground

Leaves: opposite, simple, heart-shaped, cordate or deltoid; surfaces glabrous; margins wavy

Inflorescence: axillary or terminal corymbs, tended by ribbed bracts; flowers in November and December on Guam

Flower: ray florets: absent, disc florets: perfect, 4 per head, corolla tubular, 5 lobed, white; style: branched, long, excerted

Fruit: achene, black, 5 angled, linear, pappus of many white bristles, resinous

Seed: disseminated by wind

【近似种】蔓菊

学　名：*Mikania scandens*（L.）Willd.

异　名：*Eupatorium scandens* L.，*Mikania angulosa* Raf.，*Mikania batatifolia* DC，*Mikania dioscoreaefolia* DC.，*Mikania floribunda* Bojer ex DC.，*Mikania scandens* var. *pubescens*（Nutt.）Torr. & A. Gray，*Willoughbya heterophylla* Small，*Willoughbya scandens*（L.）Kuntze，*Willoughbya scandens* var. *normalis* Kuntze

英文名：climbing hemp weed（USA），louse-plaster

【www.asianplant.net】*Mikania scandens*（L.）Willd., Sp. Pl., ed. 4 [Willdenow] 3（3）: 1743（1803）

Climbing herbaceous vine up to 3 m tall with opposite, simple leaves that have distinct petioles. The leaves are triangular to triangular-ovate in shape（3-15 cm long, 2-11 cm wide）with an attenuate apex and a cordate to hastate leaf base. The inflorescence is a loose corymb and the flowers are borne in clusters of 4. The corolla is pinkish to purplish or sometimes white.

【www.cabi.org】*Mikania scandens*

Broadleaved，Herbaceous，Perennial，Seed propagated，Vegetatively propagated Vine / climber

Herbaceous, perennial vine; branching stem obscurely 6-angled to terate, ranging from glabrous to densely pilose; 8-15 cm internodes. Petioles glabrous or puberulent, 20-50 mm. Triangular to triangular-ovate leaf blades, 3-15 cm × 2-11 cm with cordate to hastate bases; margins subentire to undulate, crenate, or dentate, apices acuminate（tips often caudate）, faces puberulent. Produces dense corymbiform flowers, with small heads 5-7 mm long. Corollas generally pinkish to purplish, occasionally white, 3-5.4 mm, dotted sparsely with glands, lobes triangular to deltate. Cypselae dark brown to blackish, 1.8-2.2 mm, also dotted with glands; pappi of 30-37 white or pinkish to purplish bristles 4-4.5 mm. Fruits are oblong 1-1.5-2.5 mm long, brownish black, five angled resinous achenes; Chromosome number $2n=38$（Holm et al., 1991; Flora of North America Editorial Committee, 2013）.

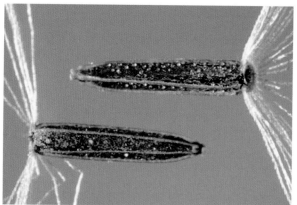
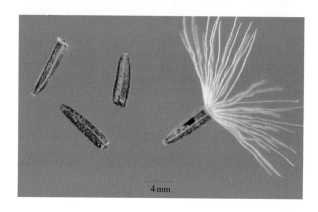

（图片选自keys.lucidcentral.org、www.cabi.org、farm9.staticflickr.com、susanleachsnyder.com、www.jeffpippen.com、www.asianplant.net、delawarewildflowers.org、luirig.altervista.org、idtools.org）

蔓菊（*Mikania scandens*）

空心莲子草

学　　名：***Alternanthera philoxeroides*** (Mart.) Griseb.
异　　名：*Achyranthes philoxeroides* (Mart.) Standley，*Bucholzia philoxeroides* Mart.，*Telanthera philoxeroides* (Mart.) Moq.
别　　名：水蕹菜、革命草、水花生、抗战草、野花生、空心苋、过江龙、湖羊草、水马兰头、东洋草、洋马兰、甲藤草、水冬瓜、水杨梅、花生藤草、通通草
英文名：alligator weed，pig weed

| 形态特征 |

空心莲子草为多年生宿根性草本植物。

植物茎基部匍匐，上部上升，管状，不明显4棱，长55～120cm，具分枝。幼茎及叶腋有白色或锈色柔毛；茎老时无毛，仅在两侧纵沟内有保留。

植物叶片矩圆形、矩圆状倒卵形或倒卵状披针形，长2.5～5cm，宽7～20mm。叶顶端急尖或圆钝，具短尖；基部渐狭，全缘，两面无毛或上面有贴生毛及缘毛，下面有颗粒状突起。叶柄长3～10mm，无毛或微有柔毛。

花密生，形成具总花梗的头状花序，单生在叶腋，球形，直径8～15mm。苞片及小苞片白色，顶端渐尖，具1脉；苞片卵形，长2～2.5mm。小苞片披针形，长2mm。花被片矩圆形，长5～6mm，白色，光亮，无毛，顶端急尖，背部侧扁。雄蕊花丝长2.5～3mm，基部连合成杯状；退化的雄蕊矩圆状条形，和雄蕊近等长，顶端裂成窄条。

子房倒卵形，具短柄。背面侧扁，顶端圆形。

果实未见。

花期5～10月。

| 生物危害 |

空心莲子草可排挤其他植物，使群落物种单一化；覆盖水面，影响鱼类生长；危害农田作物，造成作物减产；入侵湿地，破坏生态景观；促进蚊蝇滋生，危害人类健康。

| 地理分布 |

空心莲子草原产于巴西。1892年在上海附近岛屿发现。20世纪50年代，我国作猪饲料引种推广栽培于北京、江苏、浙江、江西、湖南、福建，后逸为野生。为全球性入侵物种。

| 传播途径 |

空心莲子草的匍匐茎及根可随人为和自然因素进行传播扩散。

| 管理措施 |

综合治理（Integrated Pest Management，IPM）或综合控制（Integrated Pest Control，IPC），但不包括植物卫生措施（Phytosanitary Measures）。

(图片选自farm7.staticflickr.com、img0.ph.126.net、www.nzpcn.org.nz、keyserver.lucidcentral.org、www.naturamediterraneo.com、flowers.la.coocan.jp、hyg.ycit.cn、www.asergeev.com、www.geo.arizona.edu、plants.ifas.ufl.edu)

空心莲子草（*Alternanthera philoxeroides*）

【www.efloras.org】*Alternanthera philoxeroides*（C. Martius）Grisebach, Abh. Königl. Ges. Wiss. Göttingen. 24: 36. 1879.

Bucholzia philoxeroides C. Martius, Nov. Actorum Acad. Caes. Leop.-Carol. Nat. Cur. 13（1）：107. 1825; *Achyranthes philoxeroides*（C. Martius）Standley; *Telanthera philoxeroides*（C. Martius）Moquin-Tandon.

Herbs perennial. Stem ascending from a creeping base, 55-120 cm, branched; young stem and leaf axil white hairy; old ones glabrous. Petiole 3-10 mm, glabrous or slightly hairy; leaf blade oblong, oblong-obovate, or ovate-lanceolate, 2.5-5 cm × 0.7-2 cm, glabrous or ciliate, adaxially muricate, base attenuate, margin entire, apex acute or obtuse, with a mucro. Heads with a peduncle, solitary at leaf axil, globose, 0.8-1.5 cm in diam. Bracts and bracteoles white, 1-veined, apex acuminate; bracts ovate, 2-2.5 mm; bracteoles lanceolate, ca. 2 mm. Tepals white, shiny, oblong, 5-6 mm, glabrous, apex acute. Filaments 2.5-3 mm, connate into a cup at base; pseudostaminodes oblong-linear, ca. as long as stamens. Ovary obovoid, compressed, with short stalk. Fruit not known. Fl. May-Oct. $2n = 100$.

【www.efloras.org】*Alternanthera philoxeroides*（Martius）Grisebach, Abh. Königl. Ges. Wiss. Göttingen. 24: 36. 1879.

Alligatorweed

Bucholzia philoxeroides Martius, Beitr. Amarantac., 107. 1825;

Achyranthes philoxeroides（Martius）Standley

Herbs, perennial, aquatic to semiterrestrial, stoloniferous, to 50 dm.

Stems prostrate, forming mats, often fistulose, glabrous.

Leaves sessile; blade narrowly elliptic, elliptic, or oblanceolate, 3.5-7.1 cm × 0.5-2 cm, herbaceous, apex acute to obtuse, glabrous.

Inflorescences terminal and axillary, pedunculate; heads white, globose, 1.4-1.7 cm diam.; bracts not keeled, less than 1/2 as long as tepals.

Flowers: tepals monomorphic, white, lanceolate or oblong, 6 mm, apex acute, glabrous; stamens 5; pseudostaminodes ligulate.

Utricles not seen.

Seeds not seen.

【www.cabi.org】*Alternanthera philoxeroides*（alligator weed）

Aquatic，Biennial，Broadleaved，Herbaceous，Perennial，Succulent，Vegetatively propagated Vine / climber plant

Decumbent or ascending glabrate aquatic perennials, the simple or branched, often fistulose stems to 100 cm long. Leaves glabrous or glabrate, lanceolate to narrowly obovate, apically rounded to acute, basally cuneate, rarely denticulate, 2-10 cm long, 0.5-2 cm broad; petioles 1-3 mm long. Inflorescences of terminal and occasionally axillary white glomes, 10-18 mm long, 10-18 mm broad, the usually unbranched peduncles 1-5 cm long. Flowers perfect, bracts and bracteoles subequal, ovate, acuminate, 1-2 mm long; sepals 5, subequal, oblong, apically acute and occasionally denticulate, neither indurate nor ribbed, 5-6 mm long, 1.5-2.5 mm broad; stamens 5, united below into a tube, the pseudostaminodia lacerate and exceeding the anthers; ovary reniform, the style about twice as long as the globose capitate stigma. Fruit an indehiscent reniform utricle 1 mm long, 1-1.5 mm broad (Flora of Panama, 2016).

【keyserver.lucidcentral.org】*Alternanthera philoxeroides*

Distinguishing Features

A semi-aquatic, aquatic, or terrestrial herbaceous plant that produces roots at its stem joints.

These stems are often hollow when growing in water, and form dense mats of vegetation out over the water surface.

Its oppositely arranged leaves are almost stalkless and elongated in shape (2-14 cm long and 1-4 cm wide). Its flowers are borne in dense globular clusters (1-2 cm across) on stalks 2-9 cm long in the forks of the upper leaves.

These small flowers have five white 'petals' that acquire a papery appearance as the fruit mature.

Stems and Leaves

The stems of this weed often grow as runners along the ground (i.e. stolons) or creeping below the ground surface (i.e. rhizomes). They may also spread out over the surface of water bodies and tend to form dense mats of vegetation (up to 1 m thick). These aquatic stems usually become hollow as they mature, which aids in floatation. The production of roots (i.e. adventitious roots) from the joints (i.e. nodes) of these stems is quite common. Stems can be up to 10 m long and mats of vegetation can be formed up to 15 m out over the water surface. Younger stems are light green to reddish in colour, hairless (i.e glabrous), and have slightly swollen joints (i.e. nodes). The dark green leaves are borne in pairs along the stems and usually do not have any leaf stalks (i.e. they are sessile or sub-sessile). They are elongated in shape (i.e. narrowly elliptic to lanceolate) with entire margins and pointed tips (i.e. acute apices). These leaves (2-14 cm long and 1-4 cm wide) are also hairless (i.e. glabrous) and have a somewhat waxy appearance.

Flowers and Fruit

The whitish flowers are borne in dense globular clusters (1-2 cm across) at the top of stalks (i.e. peduncles) 2-9 cm long. These flower clusters are usually produced in the forks (i.e. axils) of the upper leaves. Each flower has five small white 'petals' (i.e. perianth segments or tepals) and five yellow stamens. The 'petals' (5-7 mm long) tend to develop a papery appearance and may turn straw-coloured as they mature. Flowering occurs from late spring through to early autumn.

The small fruiting 'capsules' (i.e. utricles) are brownish in colour, bladder-like in appearance, and contain a single seed. These seeds are smooth in texture and oval (i.e. elliptic) in shape, but are rarely produced in Australia.

【en.wikipedia.org】*Alternanthera philoxeroides*

Though alligator weed tends to differ in appearance, the plant can often be identified by its fleshy stems and white flowers. These horizontal stems, which form dense mats on the surface of lakes and ponds, can grow up to 10 meters in length. Leaves are simple, elliptic, and have smooth margins. The plant flowers from December to April and usually grows around 13 mm in diameter and tend to be papery and ball-shaped. The weed's intricate root system does not have to establish in sediment or soil: many plants let their roots hang free in water to absorb nutrients. As an adaptive response to its environment, plants in water tend to have more hollow stems than its counterparts on land.

【www.eppo.int】*Alternanthera philoxeroides*（Mart.）Griseb.

Plant type Emergent aquatic perennial herb, amphibious or terrestrial.

Description *Alternanthera philoxeroides* is an emergent stoloniferous perennial herb. The leaves are dark green, elliptic, glabrous and opposite）, 3.5-7.1 cm long and 0.5-2 cm wide（Flora of North America Editorial Committee, 1993+）. Mature aquatic plants have hollow stems up to 10 m long that form thick interwoven mats throughout the water body and emerge up to 20 cm out of the water when the plant lowers. Inflorescences are white, terminal and axillary, 1.4-1.7 cm in diameter, on a short stalk（Flora of North America Editorial Committee, 1993+）. In the native range the species is known to set seed（Vogt, 1973）. In much of the invasive range seed production is not observed（Van Oosterhout, 2007）. However, the species has been recorded to set seed in China. Liu-qing et al.（2007）cite Zhang et al.（2004）（in China）and detail that *A. philoxeroides* showed a 6.5% seed set in Zhengzhou City, Henan Province. Contamination of bonsai plant soil sourced from China and detected in the Netherlands（van Valkenburg, pers. comm., 2015）indicates that viable seed is produced in China. In North America, *A. philoxeroides* flowers from early spring into the summer months, whereas in Australia, the species flowers around mid-summer（Flora of North America Editorial Committee, 1993+; Queensland Government, 2015）. *A. philoxeroides* can be confused with a number of semi-aquatic species within the EPPO region; in particular the closely related congeners including *Alternanthera caracasana* Kunth., *Alternanthera* nodiflora R.Br. and *Alternanthera sessilis*（L.）R.Br. ex DC.

豚 草

学　名：***Ambrosia artemisiifolia*** L.
异　名：*Ambrosia artemisiaefolia* L.，*Ambrosia chilensis* Hook. & Arn.，*Ambrosia elata* Salisb.，*Ambrosia elatior* L.，*Ambrosia glandulosa* Scheele，*Ambrosia monophylla* （Walter）Rydb.，*Ambrosia paniculata* Michx.，*Ambrosia peruviana* Cabrera 1941 not Willd. 1805 nor DC.，*Iva monophylla* Walter
别　名：豕草、普通豚草、艾叶破布草、美洲艾
英文名：common ragweed，annual ragweed，bitterweed，blackweed，carrot weed，hayfever weed，hayweed，hogweed，low ragweed，Roman wormwood，short ragweed，small ragweed，stammerwort，wild tansy

| 形态特征 |

豚草为一年生草本植物。

植物茎直立，上部有圆锥状分枝，有棱，被疏生密糙毛。植株高20～150cm。

植株下部叶对生，具短叶柄，二次羽状分裂，裂片狭小，长圆形至倒披针形，全缘，有明显的中脉，上面深绿色，被细短伏毛或近无毛，背面灰绿色，被密短糙毛；上部叶互生，羽状分裂，无柄。

头状花序，单性，雌雄同株。

雌性头状花序无花序梗。在头花序下面或在下部叶腋单生，或2～3个密集成团伞状，有1个无被能育的雌花。总苞闭合，具结合的总苞片，倒卵形或卵状长圆形，长4～5mm，宽约2mm，顶端有围裹花柱的圆锥状嘴部，在顶部以下有尖刺4～6个，稍被糙毛。花柱2深裂，丝状，伸出总苞的嘴部。

雄性头状花序呈半球形或卵形，径4～5mm。具短梗，下垂，在枝端密集成总状花序。总苞呈宽半球形或碟形。总苞片全部结合，无肋，边缘具波状圆齿，稍被糙伏毛。花托具刚毛状托片。每个头状花序有10～15个不育的小花。花冠淡黄色，长2mm，有短管部，上部钟状，有宽裂片。花药呈卵圆形。花柱不分裂，顶端膨大呈画笔状。

瘦果倒卵形，无毛，藏于坚硬的总苞中。

花期8～9月。果期9～10月。

| 地理分布 |

豚草原产于北美洲。现广布于世界大部分地区。亚洲：缅甸、马来西亚、越南、印度、巴基斯坦、菲律宾、日本、中国。非洲：埃及、毛里求斯；欧洲：法国、德国、意大利、瑞士、奥地利、瑞典、匈牙利、原苏联。美洲：加拿大、美国、墨西哥、古巴、牙买加、危地马拉、阿根廷、玻利

维亚、巴拉圭、秘鲁、智利、巴西。大洋洲：澳大利亚。

1935年，豚草在我国杭州发现。现分布于东北、华北、华中和华东等地约19个省、直辖市，包括辽宁、吉林、黑龙江、河北、山东、江苏、浙江、江西、安徽、湖南、湖北、北京、上海等地。以沈阳、铁岭、丹东、南京、南昌、武汉等市发生严重，形成以沈阳—铁岭—丹东、南京—武汉—南昌为核心的发生、传播和蔓延中心。

| 生物危害 |

豚草为我国检疫性有害生物。豚草生命力极强，侵入裸地后一年即可成为优势种，可以遮盖和压抑土生植物，造成原有生态系统的破坏。豚草能混杂所有旱地作物，特别是玉米、大豆、向日葵、大麻、洋麻等中耕作物和禾谷类作物，能导致作物大面积草荒，以致于绝收。

豚草花粉中含有水溶性蛋白，与人接触后可迅速释放，引起过敏性变态反应，它是秋季花粉过敏症的主要致病原，易导致"枯草热症"，患者轻的会引起咳嗽、哮喘症状，病情严重时，可引起肺气肿，而且感染以后会年年复发，且一年比一年加重。豚草植株和花粉还可使一些人患过敏性皮炎，引起全身起"风疱"。在美国，每年因豚草患病者达1 460万人；在加拿大，患者有80万人；在前苏联克拉斯诺尔达地区，在豚草花期，约有1/7的人因豚草花粉患病而无法劳动。有关专家告诫，在中国，因患花粉过敏的人虽然仅有10%多一点，但由于人口基数太大，如果不趁豚草泛滥之初尽快清除，将来我国的患病人多绝不亚于欧美，需要引起足够重视。

| 传播途径 |

豚草果实常能混杂于作物中，特别是混杂在大麻、洋麻、玉米、大豆、中耕作物和禾谷类作物种子中，并通过调运进行远距离传播。

| 管理措施 |

综合治理（Integrated Pest Management，IPM）或综合控制（Integrated Pest Control，IPC），包括植物卫生措施（Phytosanitary Measures）。

外来入侵植物

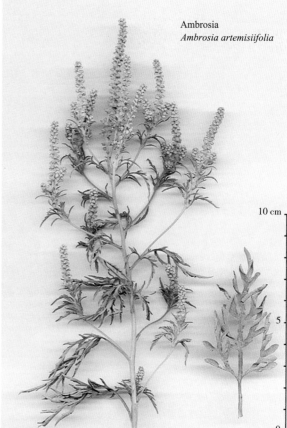

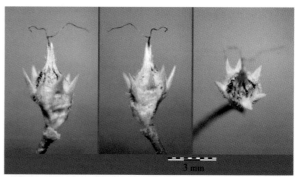

（图片选自www.ewrs.org，pestid.msu.edu，www.fnanaturesearch.org，baike.baidu.com，en.wikipedia.org，commons.wikimedia.org，www.naturalmedicinefacts.info）

豚草（*Ambrosia artemisiifolia*）

【www.efloras.org】*Ambrosia artemisiifolia* Linnaeus, Sp. Pl. 2: 988. 1753.

Ambrosia artemisiifolia var. *elatior*（Linnaeus）Descourtilz; *A. elatior* Linnaeus.

Annuals, 20-150 cm. Stems erect. Leaves opposite and alternate; petiole 2.5-3.5（-6）cm; blade deltate to lanceolate or elliptic, 2.5-5.5（-9）cm × 2-3（-5）cm, pinnately 1- or 2-lobed, abaxially sparsely pilosulose to strigillose, adaxially strigillose, both surfaces gland-dotted, base cuneate, ultimate margin entire or toothed. Female capitula clustered, proximal to male; floret 1. Male capitula: peduncles 0.5-1.5 mm; involucres shallowly cup-shaped（usually without black veins）, 2-3 mm in diam., glabrous or hispid to pilosulose; florets 12-20. Burs ± globose to pyriform, 2-3 mm, ± pilosulose, spines or tubercles 3-5, near middle or apex, conical to acerose, 0.1-0.5 mm, tips straight. Fl. Jul-Oct, fr. Sep-Oct. $2n = 34, 36$.

【www.efloras.org】*Ambrosia artemisiifolia* Linnaeus, Sp. Pl. 2: 988. 1753.

Petite herbe à poux

Ambrosia artemisiifolia var. *elatior*（Linnaeus）Descourtilz; *A. artemisiifolia* var. *paniculata*（Michaux）Blankinship; *A. elatior* Linnaeus; *A. glandulosa* Scheele; *A. monophylla*（Walter）Rydberg

Annuals, 10-60（-150+）cm.

Stems erect.

Leaves opposite（proximal）and alternate; petioles 25-35（-60+）mm; blades deltate to lanceolate or elliptic, 25-55（-90+）mm × 20-30（-50+）mm, 1-2-pinnately lobed, bases cuneate, ultimate margins entire or toothed, abaxial faces sparsely pilosulous to strigillose, adaxial faces strigillose, both gland-dotted.

Pistillate heads clustered, proximal to staminates; florets 1.

Staminate heads: peduncles 0.5-1.5 mm; involucres shallowly cup-shaped（usually without black nerves）, 2-3+ mm diam., glabrous or hispid to pilosulous; florets 12-20+.

Burs: bodies ± globose to pyriform, 2-3 mm, ± pilosulous, spines or tubercles 3-5+, near middles or distal, ± conic to acerose, 0.1-0.5+ mm, tips straight.

$2n = 34, 36$.

【www.cabi.org】*Ambrosia artemisiifolia*（common ragweed）

Annual，Broadleaved，Herbaceous，Seed propagated plant

Annual herb（therophyte），（10-）20-60（-150）cm tall. Stems erect. Leaves opposite（proximal）and alternate, with blades lanceolate or elliptic [（20-）25-55（-90）mm × 20-30（-50）mm], 1-2-pinnately lobed, sparsely pubescent abaxially, glandular-dotted on both faces, petioled [petiole 25-35（-60）mm long]. Flowers arranged in capitula, the male capitula（5-20 flowers per capitulum, the involucre being cup-shaped, glabrous to pubescent）forming a terminal spike-like inflorescence, the female capitula proximal to the male ones. Fruit globose to pyriform, 2-3 mm long, more or less pubescent.

【en.wikipedia.org】*Ambrosia artemisiifolia*

Ambrosia artemisiifolia is an annual plant that emerges in late spring. It propagates mainly by rhizomes, but also by seed. It is much-branched, and grows up to 7 decimetres（2.3 feet）in height. The pinnately divided soft and hairy leaves are 3-12 centimetres（1.2-4.7 inches）long. Its bloom period is July to October in North America. Its pollen is wind-dispersed, and can be a strong allergen to people with hay fever. It produces 2-4 mm obconic green to brown fruit. It sets seed in later summer or autumn. Since the seeds persist into winter and are numerous and rich in oil, they are relished by songbirds and upland game birds.

【keyserver.lucidcentral.org】*Ambrosia artemisiifolia* L.

Distinguishing Features

An upright herbaceous plant (growing up to 2 m tall) that forms a basal rosette of leaves during the early stages of growth.

Its rounded stems bear deeply divided leaves that are fern-like in appearance.

Separate male and female flower-heads are formed on the same plant.

The drooping male flower-heads are borne in elongated spike-like clusters (up to 20 cm long) at the tips of the branches.

The inconspicuous female flower-heads are borne in the upper leaf forks.

Stems and Leaves

This plant forms a basal rosette of leaves during the early stages of growth. The much-branched, upright (i.e. erect), stems are rounded in cross-section (i.e. cylindrical) and reddish or brownish-green in colour. These stems vary from being almost hairless (i.e. sub-glabrous) to roughly hairy (i.e. hirsute).

The leaves are oppositely arranged at the base of the plant, but are alternately arranged further up the stems. The leaf blades (1-16 cm long and 1-7 cm wide) are deeply divided (i.e. pinnatifid to bi-pinnatifid) and fern-like in appearance. They are borne on leaf stalks (i.e. petioles) usually about 1-3 cm long (occasionally up to 10 cm long). The uppermost leaves are usually much reduced in size and less divided than the lower leaves. All leaves are usually covered in hairs (i.e. pubescent), particularly on their undersides, and these hairs may be long and spreading or short and soft.

Flowers and Fruit

Separate male and female (i.e. unisexual) flower-heads are formed on different parts of the same plant (i.e. this species is monoecious). The male (i.e. staminate) flower-heads outnumber the female (i.e. pistillate) flower-heads and droop from branching spike-like flower clusters (up to 20 cm long) that are borne at the tips of the stems. These male flower-heads are small, hemispherical in shape, and either cream, yellowish or pale green in colour. The female flower-heads are less conspicuous and consist of a single tiny flower (i.e. floret). The base of these flower-heads (i.e. the involucre) is saucer shaped with 5-7 small bristle-like spines (each 3-5 mm long). These female flower-heads are held upright and borne singly in the forks (i.e. axils) of the uppermost leaves (i.e. below the male flower-heads). Flowering occurs mostly during summer, autumn and early winter. The fruit is a small brown or blackish achene (2-5 mm long) that is top-shaped (i.e. turbinate) and contains a single seed.

These fruit become woody as they mature and have a pointed beak (1-2 mm long) and a ring of four to eight small blunt spines (each less than 1 mm long).

Similar Species

Annual ragweed (*Ambrosia artemisiifolia*) is very similar to the other ragweeds (*Ambrosia* spp.) present in Australia, including burr ragweed (*Ambrosia confertiflora*), perennial ragweed (*Ambrosia psilostachya*) and lacy ragweed (*Ambrosia tenuifolia*). It is also very similar to parthenium weed (*Parthenium hysterophorus*) when in the vegetative stage of growth. These species can be distinguished by the following differences:

Annual ragweed (*Ambrosia artemisiifolia*) is a large short-lived (i.e. annual) herbaceous plant (growing up to 2 m tall) with rounded stems and leaves that are usually twice-divided (i.e. bipinnatifid). The single-sex (i.e. unisexual) greenish or yellowish male flower-heads are borne in elongated spikes. Its hairless (i.e. glabrous) fruit (2-5 mm long) are borne in small clusters and have a single row of 4-8 short blunt spines.

Burr ragweed (*Ambrosia confertiflora*) is a large long-lived (i.e. perennial) herbaceous plant (growing up to 2 m tall) with rounded stems and leaves that are usually twice-divided (i.e. bipinnatifid). The single-sex (i.e. unisexual) greenish or yellowish male flower-heads are borne in elongated spikes. Its abundant fruit (about 4 mm long) are borne in large clusters and are covered with short hooked spines.

Perennial ragweed (*Ambrosia psilostachya*) is a relatively large long-lived (i.e. perennial) herbaceous plant (growing up to 2 m tall) with rounded stems and leaves that are only once-divided (i.e. pinnatifid). The single-sex (i.e. unisexual) greenish or yellowish male flower-heads are borne in elongated spikes. Its small hairy (i.e. pubescent) fruit (about 2 mm long) are borne in small clusters and have five short blunt spines. It also produces a large network of creeping underground stems.

Lacy ragweed (*Ambrosia tenuifolia*) is a relatively small long-lived (i.e. perennial) herbaceous plant (growing up to 75 cm tall) with rounded stems and leaves that are usually twice-divided (i.e. bipinnatifid). These leaves are covered in long whitish hairs and are very finely divided, thereby giving them a greyish and lacy appearance. The single-sex (i.e. unisexual) greenish or yellowish male flower-heads are borne in elongated spikes. Its small fruit (about 2 mm long) are borne singly or in small clusters and have a few very short teeth.

Parthenium weed (*Parthenium hysterophorus*) is a large short-lived (i.e. annual) herbaceous plant (growing up to 2 m tall) with ribbed stems and leaves that are usually twice-divided (i.e. bipinnatifid). Masses of small, white, flower-heads are borne at the tips of the branches and each of these flower-heads usually gives rise to five small 'seeds'.

Annual ragweed (*Ambrosia artemisiifolia*) may also be confused with some of the wormwoods (*Artemisia* spp.). However, these species have bisexual flowers in rounded flower-heads and they do not produce burr-like fruit.

三裂叶豚草

学　　名：*Ambrosia trifida* L.
异　　名：*Ambrosia aptera* DC.，*Ambrosia integrifolia* Muhl. ex Willd.
别　　名：大破布草、豚草、三裂豚草
英文名：giant ragweed，blood ragweed，buffalo-weed，crownweed，great ragweed，horseweed

| 形态特征 |

三裂叶豚草为一年生粗壮草本植物。

植物茎直立，有分枝，被短糙毛，有时近无毛。植株高50～120cm，有时可达170cm甚至更高。

植物叶对生，有时互生。下部叶3～5裂，上部叶3裂或有时不裂，裂片卵状披针形或披针形，顶端急尖或渐尖，边缘有锐锯齿；叶基出3脉，粗糙，上面深绿色，背面灰绿色，两面被短糙伏毛。叶具叶柄，叶柄长2～3.5cm，被短糙毛，基部膨大，边缘有窄翅，被长缘毛。

雌性头状花序在雄性头状花序下面上部的叶状苞叶的腋部，聚作团伞状，具一个无被，肋，每肋顶端有瘤或尖刺，无毛；花柱2深裂，丝状，上伸出总苞的嘴部之外。

雄性头状花序多数，圆形，径约5mm。有长2～3mm的细花序梗，下垂，在枝端密集成总状花序。总苞浅碟形，绿色，总苞片结合，外面有3肋，边缘有圆齿，被疏短糙毛。花托无托片，具白色长柔毛，每个头状花序有20～25不育的小花。小花黄色，长1～2mm，花冠钟形，上端5裂，外面有5紫色条纹。花药离生，卵圆形。花柱不分裂，顶端膨大呈画笔状。

瘦果倒卵形，无毛，藏于坚硬的总苞中。

花期8月。果期9～10月。

| 生物危害 |

三裂叶豚草为我国检疫性有害生物。是一种恶性农田杂草，可危害小麦、大麦、大豆及各种园艺作物。

三裂叶豚草也是致敏性植物。其花粉能引起人的过敏性哮喘、鼻炎和皮炎等症，致使患者眼耳鼻发痒，阵发性喷嚏，流泪，流大量清水样鼻涕，咳嗽，憋气，哮喘等，部分病人会并发肺气肿、肺心病，甚至导致死亡；有的病人则表现为皮炎、荨麻疹、湿疹等症状。

| 地理分布 |

三裂叶豚草原产于北美。20世纪30年代三裂叶豚草在我国辽宁铁岭地区发现，首先在辽宁省蔓延，随后向河北、北京地区扩散。现分布于吉林、辽宁、河北、北京、天津等地。

传播途径

三裂叶豚草的种子易混杂在玉米、大豆、小麦等粮食及包装材料中传播，也可随水流扩散。

管理措施

综合治理（Integrated Pest Management，IPM）或综合控制（Integrated Pest Control，IPC），包括植物卫生措施（Phytosanitary Measures）。

外来入侵 植物

中国外来入侵物种图鉴

TREFLIKSAMBRQSIA, AMBROSIA TRIFIDAL

（图片选自bugwoodcloud.org，s3.amazonaws.com，meltonwiggins.com，www.thismia.com，hasbrouck.asu.edu，upload.wikimedia.org，www.ewrs.org，www.pfaf.org，www.sbs.utexas.edu，flora.nhm-wien.ac.at，en.wikipedia.org，www.backyardnature.net，plants.usda.gov，gobotany.newenglandwild.org，www.mapaq.gouv.qc.ca，www.plantarium.ru）

三裂叶豚草（*Ambrosia trifida*）

外来入侵植物

【www.efloras.org】*Ambrosia trifida* Linnaeus, Sp. Pl. 2: 987. 1753.

Annuals, 30-150（-400+）cm.

Stems erect.

Leaves mostly opposite; petioles 10-30（-70+）mm; blades rounded-deltate to ovate or elliptic, 40-150（-250+）mm × 30-70（-200+）mm, usually some blades palmately 3（-5）-lobed, bases truncate to cuneate（sometimes decurrent onto petioles）, margins usually toothed, rarely entire, abaxial and adaxial faces ± scabrellous and gland-dotted.

Pistillate heads clustered, proximal to staminates; florets 1.

Staminate heads: peduncles 1-3+ mm; involucres ± saucer-shaped, 2-4 mm diam., scabrellous（often with 1-3 black nerves）; florets 3-25+.

Burs: bodies ± pyramidal, 3-5（-7+）mm, glabrous or glabrate, spines 4-5, ± distal, ± acerose, 0.5-1 mm, tips straight（bases ± decurrent as ribs）.

$2n$ = 24, 48.

【www.efloras.org】*Ambrosia trifida* Linnaeus, Sp. Pl. 2: 987. 1753.

Giant ragweed, grande herbe à poux

Ambrosia aptera de Candolle; *A. trifida* var. *integrifolia*（Muhlenberg ex Willdenow）Torrey & A. Gray; *A. trifida* var. *texana* Scheele

Annuals, 30-150（-400+）cm.

Stems erect.

Leaves mostly opposite; petioles 10-30（-70+）mm; blades rounded-deltate to ovate or elliptic, 40-150（-250+）mm × 30-70（-200+）mm, usually some blades palmately 3（-5）-lobed, bases truncate to cuneate（sometimes decurrent onto petioles）, margins usually toothed, rarely entire, abaxial and adaxial faces ± scabrellous and gland-dotted.

Pistillate heads clustered, proximal to staminates; florets 1.

Staminate heads: peduncles 1-3+ mm; involucres ± saucer-shaped, 2-4 mm diam., scabrellous（often with 1-3 black nerves）; florets 3-25+.

Burs: bodies ± pyramidal, 3-5（-7+）mm, glabrous or glabrate, spines 4-5, ± distal, ± acerose, 0.5-1 mm, tips straight（bases ± decurrent as ribs）.

$2n$ = 24, 48.

【www.cabi.org】*Ambrosia trifida*（giant ragweed）

Annual，Broadleaved，Herbaceous，Seed propagated plant

A. trifida is an annual herb（therophyte）,（30-）150-400 cm tall. Stems erect, branched or not. Leaves opposite, with blades rounded, deltate or elliptic [40-150（-250）mm ×（10-）30-70（-200）mm], palmately 2-5-lobed, lobes with margins toothed, sparsely pubescent and glandular-dotted on both faces, petioled [petiole（10-）25-30（-70）mm long]. Flowers arranged in capitula, the male capitula forming a terminal leafless spike-like or raceme-like inflorescences at the end of the branches, the female capitula clustered at the base of the male inflorescences（pistillate flowers are tubular and without pappi）; anthers yellow; ovary inferior with one ovule. Fruit（achene）pyramidal,（5-）6-12 mm long, brown to gray, glaborus or slightly pubescent, with 4-5 usually distal spines.

【en.wikipedia.org】*Ambrosia trifida*

This is an annual herb usually growing up to 2 m (6 feet 7 inches) tall, but known to reach over 6 m (20 feet) in rich, moist soils. The tough stems have woody bases and are branching or unbranched. Most leaves are oppositely arranged. The blades are variable in shape, sometimes palmate with five lobes, and often with toothed edges. The largest can be over 25 cm (9.8 inches) long by 20 cm (7.9 inches) wide. They are borne on petioles several centimeters long. They are glandular and rough in texture. The species is monoecious, with plants bearing inflorescences containing both pistillate and staminate flowers. The former are clustered at the base of the spike and the latter grow at the end. The fruit is a bur a few millimeters long tipped with several tiny spines.

【gobotany.newenglandwild.org】*Ambrosia trifida* L.

毒 麦

学　名：***Lolium temulentu*** L.

异　名：*Lolium triticoides* Janka，*Lolium temulentum* var. *speciosum*，*Lolium temulentum* subsp. *speciosum*，*Lolium temulentum* var. *semiglabrum*，*Lolium temulentum* var. *muticum*，*Lolium temulentum* var. *macrochaeton*，*Lolium temulentum* var. *leptochaeton*，*Lolium temulentum* subsp. *cuneatum*，*Lolium temulentum* f. *arvense*，*Lolium temulentum* var. *arvense*，*Lolium temulentum* subsp. *arvense*，*Lolium speciosum* Steven ex M. Bieb.，*Lolium robustum* Rchb.，*Lolium remotum* f. *asperum*，*Lolium pseudolinicola* Gennari，*Lolium maximum* Willd.，*Lolium maximum* Guss.，*Lolium lucidum* Dumort.，*Lolium longiglume* St.-Lag.，*Lolium linicola* A. Braun，*Lolium infelix* Rouville，*Lolium gussonei* Nyman，*Lolium gracile* Hegetschw.，*Lolium gracile* Dumort.，*Lolium giganteum* Roem. & Schult.，*Lolium decipiens* Dumort.，*Lolium cuneatum* Nevski，*Lolium berteronianum* Steud.，*Lolium asperum* Roth ex Kunth，*Lolium arvense* With.，*Lolium annuum* Lam.，*Lolium album* Steud.，*Lolium aegyptiacum* Bellardi ex Rouville，*Craepalia temulenta*（L.）Schrank，*Bromus temulentus* Bernh.

别　名：黑麦子、小尾巴麦子、闹心麦

英文名：darnel，poison ryegrass，Darnel ryegrass

| 形态特征 |

毒麦为一年生或越年生草本植物。

植物秆疏丛生，直立。植株高20~120cm。形似小麦，但须根较稀，光滑坚硬，不易倒伏。成株秆无毛，3~4节，一般比小麦矮10~15cm。

植物叶片长6~40cm，宽3~13mm，质地较薄，无毛或微粗糙。叶鞘疏松，大部分长于节间。叶舌长约2.7mm，膜质截平。叶耳狭窄。

花为穗状花序，长5~40cm，宽1~1.5cm，有12~14个小穗；穗轴节间长5~7mm；小穗有小花2~6朵，小穗轴节间长1~1.5mm，光滑无毛。颖质地较硬，有5~9脉，具狭膜质边缘，颖长8~10mm，宽1.5~2mm；外稃质地薄，基盘较小，有5脉，顶端膜质透明，第1外稃长约6mm，有长达0.7~1.5cm的芒，自近外稃顶稍下方伸出；内稃约等长于外稃，脊上具有微小纤毛。小穗的第一颖均退化（但顶生小穗除外）。

颖果长椭圆形，长4~6mm，宽约2mm；颖果腹面凹陷成一宽沟，并与内稃嵌合，不易剥落。

种子褐黄色到棕色，坚硬，无光泽，腹腔沟较宽。种子上皮细胞宽大，排列整齐，且紧贴胚乳组织。糊粉层细胞大而明显，多达4层，有较多的糊粉细胞网状深入层。

毒麦幼苗基部紫红色，胚芽鞘长1.5~1.8cm。第1叶线形，长6.5~9.5cm，宽2~3mm，先端渐尖，光滑无毛。毒麦分蘖力较强，一般生有4~9个分蘖，平均每株分蘖5.47个，比小麦多1.34个；毒麦繁殖能力强，单株结籽数14~100粒，平均62.6粒，而小麦仅28.13粒，其繁殖能力是小麦的2.23倍。染色体2n=14（Hubbard 1954，Jen Kins 1985）。

| 生物危害 |

毒麦为我国检疫性有害生物。其颖果内种皮与淀粉层之间寄生座盘菌（*Stromatinia temulenata*）的菌丝，产生毒麦碱（$C_7H_{12}N_2O$）。人、畜误食后都能中毒。中毒轻者引起头晕、昏迷、呕吐、痉挛等症；重者则会使中枢神经系统麻痹而导致死亡。面粉中若含有4%毒麦会使人畜食物中毒，尤其未成熟的毒麦或在多雨季节收获时收获物中混入的毒麦毒力最大。此外，毒麦引起的中毒，还可造成人视力障碍。

毒麦生于麦田中，影响小麦产量和品质。毒麦的混生株率与小麦产量损失呈正相关。毒麦混生株率为0.1%时，小麦产量损失0.64%~2.94%；混生株率为5%，产量损失达19.12%~26.12%，减产幅度相当明显。

| 地理分布 |

毒麦原产于欧洲地中海地区。1954年，我国在从保加利亚进口的小麦中发现毒麦。目前除西藏外，我国大陆各省均有发现。为全球性入侵物种。

| 传播途径 |

毒麦种子混入谷物中，随谷物调运而传播扩散。

| 管理措施 |

综合治理（Integrated Pest Management，IPM）或综合控制（Integrated Pest Control，IPC），包括植物卫生措施（Phytosanitary Measures）。

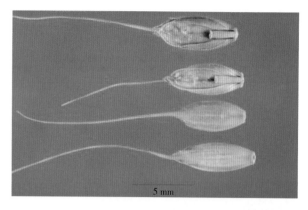

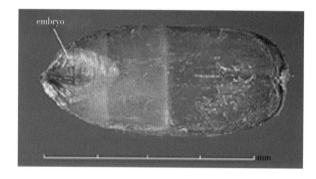

（图片选自www2.dijon.inra.fr，static.jardipedia.com，keys.lucidcentral.org，baike.baidu.com，www.speciesrecoverytrust.org.uk，www.blumeninschwaben.de，gobotany.newenglandwild.org，luirig.altervista.org，idtools.org，wilde-planten.nl）

毒麦（*Lolium temulentu*）

【www.cabi.org】 *Lolium temulentum* (darnel)

Annual Grass / sedge, Seed propagated plant

L. temulentum is an annual plant with a fibrous root system. Culms tufted, solitary, ascending, erect, 60-90 cm tall, glabrous and smooth, or rough at the top.

Leaves are lanceolated, simple with a shiny surface, leaf blades narrowly linear, not contracted at the base, acute at apex, with smooth or scabrid margins, somewhat rough above, glabrous and smooth beneath, 10-30 cm long, 3-10 mm wide, young leaves with involute margins, ligule 1-2 mm long.

Inflorescence is a terminal spike, rigidly erect, 12-30 cm long with 6-30 spikelets, dorsally placed in shallow excavations along a non-articulate rachis with a zigzag shape. Spikelets 12-30 mm in length, usually with 4-10 flowers. Outer glume of the lateral spikelets usually 2.5 cm in length, as long as or longer than the entire spikelets, 7-9-nerved, thinly coriaceous with narrow membranous margins. Flowering glumes are shorter and broader, oblong, usually obtuse with an awn as long as or longer than the glume itself. Some specimens may have awnless glumes or in rare instances the whole spikelet is without awns (Bentham, 1967). The lemmas are up to 8 mm long, obtuse with awns 6-12 mm long. Palea two-keeled. Seeds elliptic-oblong in shape, grooved.

【www.plantwise.org】 darnel (*Lolium temulentum*)

L. temulentum is an annual plant with a fibrous root system. Culms tufted, solitary, ascending, erect, 60-90 cm tall, glabrous and smooth, or rough at the top.

Leaves are lanceolated, simple with a shiny surface, leaf blades narrowly linear, not contracted at the base, acute at apex, with smooth or scabrid margins, somewhat rough above, glabrous and smooth beneath, 10-30 cm long, 3-10 mm wide, young leaves with involute margins, ligule 1-2 mm long.

Inflorescence is a terminal spike, rigidly erect, 12-30 cm long with 6-30 spikelets, dorsally placed in shallow excavations along a non-articulate rachis with a zigzag shape. Spikelets 12-30 mm in length, usually with 4-10 flowers. Outer glume of the lateral spikelets usually 2.5 cm in length, as long as or longer than the entire spikelets, 7-9-nerved, thinly coriaceous with narrow membranous margins. Flowering glumes are shorter and broader, oblong, usually obtuse with an awn as long as or longer than the glume itself. Some specimens may have awnless glumes or in rare instances the whole spikelet is without awns (Bentham, 1967). The lemmas are up to 8 mm long, obtuse with awns 6-12 mm long. Palea two-keeled. Seeds elliptic-oblong in shape, grooved.

【www.efloras.org】 *Lolium temulentum* Linnaeus, Sp. Pl. 1: 83. 1753.

Annual. Culms tufted, erect or decumbent, slender to moderately robust, 20-120 cm tall, 3-5-noded. Leaf blades flat, thin, 10-25 cm × 4-10 mm, smooth or scabridulous on abaxial surface, margins scabrid, young blades rolled; auricles present or absent; ligule 0.5-2.5 mm, obtuse to truncate. Raceme stiff, straight, 10-30 cm; rachis thick, smooth or scabridulous, spikelets about their own length apart. Spikelets turgid, 0.8-2.5 cm, florets 4-10, rachilla internodes 1-1.5 mm, smooth, glabrous; glume linear-oblong, rigid, as long as spikelet, often exceeding florets, 5-9-veined, margins narrowly membranous, apex obtuse; lemmas elliptic to ovate, turgid at maturity, 5.2-8.5 mm, apex obtuse; awn usually present, stiff, scabrid; palea ciliolate along keels. Caryopsis very plump, length 2-3 times width, 4-7 mm. Fl. and fr. May–Aug. $2n = 14$.

【research.vet.upenn.edu】*Lolium temulentum*（Darnel）

Darnel is an annual grass with stems: solitary or a few clumped together, 4-8 dm tall; blades: glabrous beneath, scabrous above, 3-9 mm wide; spike: 1-2 dm; spikelets: placed edgewise to the rachis, 5-8 flowered; glume: firm, straight, 5-7 nerved, equalling or surpassing the uppermost lemma, 12-22 mm; lemmas: obtuse, awned, or awnless.

【chestofbooks.com】Grass Family（Gramineae）- Darnel（*Lolium Temulentum* L.）Plate IV

Darnel is an annual grass. It has smooth simple stems from two to four feet high. The leaf blades are four to ten inches long and about one-quarter inch wide, rough on the upper surface and smooth oh the lower. The flower, spike is four to twelve inches long, with four to eight flowers to each spikelet, which fits tightly into a slight curve on either side of the stalk. The seed is about the size of a small grain of wheat, it is rounded at each end, with a shallow groove on the inner surface, and is closely covered by two scales, the outer one usually possessing a short awn. The kernel itself is greenish, tinged with brown or purple. It is in bloom from June to August. Darnel is very closely related to English and Italian rye grasses, but may be readily distinguished from both in having no leafy shoots from the base, and consequently it does not grow in tufts or bunches.

【keys.lucidcentral.org】*Lolium temulentum*（Darnel Ryegrass）

Lolium temulentum is a weedy annual grass. The plant stem can grow up to 1 meter tall. The flower is a large panicle（"ears"）. The ears are light and upright and mature ones are black in colour.

Similar species

The similarity between wheat and *Lolium temulentum* is so great that in some regions *L. temulentum* is referred to as true wheat（Triticum species）. *L. temulentum* and wheat look alike until the ear appears and are distinguished as follows:

The ears on the real wheat are so heavy that it makes the entire plant droop downward, but *L. temulentum*, whose ears are light, stands up straight.

Wheat ripens to a brown colour, whereas *L. temulentum* turns black.

When *L. temulentum* matures, the spikelets turn edge ways to the rachis whereas the wheat spikelets remain as they grew previously.

互花米草

学　　名：***Spartina alterniflora*** Loisel.

异　　名：*Spartina maritima* var. *alterniflora*（Loisel.）St.-Yves，*Spartina stricta* var. *alterniflora*（Loisel.）A. Gray，*Spartina glabra* var. *alterniflora*（Loisel.）Merr.，*Trachynotia alterniflora*（Loisel.）DC.

别　　名：平滑网茅、盐滩大米草

英文名：smooth cordgrass，Atlantic cordgrass，saltmarsh cordgrass，salt-water cordgrass

| 形态特征 |

互花米草为多年生草本植物。

互花米草植物地下部分通常由短而细的须根和长而粗的地下茎（根状茎）组成。根系发达，常密布于地下30cm深的土层内，有时可深达50~100cm。

植株茎秆坚韧、直立，高达0.5~3m，直径在1cm以上。茎节具叶鞘，叶腋有腋芽。

植物叶互生，呈长披针形，长可达90cm，宽1.5~2cm。具盐腺，根吸收的盐分大都由盐腺排出体外，因而叶表面往往有白色粉状的盐霜出现。

花为圆锥花序，花序长20~45cm，具10~20个穗形总状花序，有16~24个小穗，小穗侧扁，长约1cm。两性花。子房平滑，两柱头很长，呈白色羽毛状。雄蕊3个，花药成熟时纵向开裂，花粉黄色。

颖果长0.8~1.5cm。胚呈浅绿色或蜡黄色。

种子通常8~12月成熟。

| 生物危害 |

互花米草威胁本土海岸生态系统，致使大片盐沼植物消失，影响滩涂养殖。

| 地理分布 |

互花米草原产于美国东南部海岸，在美国西部和欧洲海岸归化。1979年作为经济作物引入我国上海（崇明岛）、浙江、福建、广东、香港。现在已经扩散至东部沿海盐沼（北至天津塘沽，南至广西山口保护区）的广大地区，危及当地海岸生态系统，致使大片红树林消失。

| 传播途径 |

互花米草的种子可随风浪传播，根可无性繁殖。种子和无性繁殖材料均可人为传播或自然扩散。

| 管理措施 |

综合治理（Integrated Pest Management，IPM）或综合控制（Integrated Pest Control，IPC），但不包括植物卫生措施（Phytosanitary Measures）。

中国外来入侵物种图鉴

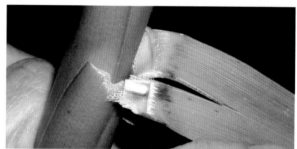

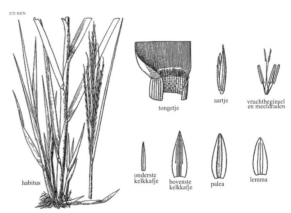

（图片选自en.wikipedia.org，newfs.s3.amazonaws.com，streamwebs.org，plants.ifas.ufl.edu，baike.baidu.com，www.soortenbank.nl）

互花米草（*Spartina alterniflora*）

【www.efloras.org】*Spartina alterniflora* Loiseleur, Fl. Gall. 719. 1807.

Spartina glabra Muhlenberg ex Elliott var. *alterniflora*（Loiseleur）Merrill; *S. maritima*（Curtis）Fernald var. *alterni-flora*（Loiseleur）St.-Yves; *S. stricta* Roth var. *alterniflora*（Loiseleur）A. Gray; *Trachynotia alterniflora*（Loiseleur）Candolle.

Perennial with soft fleshy rhizomes. Culms stout, forming large clumps, erect,（0.5-）1-2（-3）m tall, ca. 1 cm in diam. Leaf sheaths mostly longer than internodes, smooth; leaf blades linear-lanceolate, flat, 10-90 cm × 1-2 cm, smooth or margins minutely scabrous, tapering to long hard involute apex; ligule ca. 1 mm. Racemes racemosely arranged,（5-）10-20, 5-20 cm, slender, erect or slightly spreading; spikelets scarcely overlapping; rachis smooth, terminating in a bristle up to 3 cm. Spikelets ca. 10 mm, glabrous or nearly so; lower glume linear, 1/2-2/3 as long as spikelet, acute; upper glume ovate-lanceolate, as long as spikelet, glabrous or with very short hairs on keel, subacute; lemma lanceolate-oblong to narrowly ovate, glabrous; palea slightly longer than lemma. Anthers 5-6 mm. $2n = 62$.

【www.cabi.org】*Spartina alterniflora*（smooth cordgrass）

Grass / sedge，Vegetatively propagated plant

S. alterniflora is a rhizomatous perennial grass that grows initially in round, genetically similar, clumps ranging between 0.5-3m in height, eventually forming extensive monoculture meadows. The stems are hollow and hairless. The leaf blades are 3 to 25 mm wide. The leaves lack auricles and have ligules（1-2 mm）that consist of a fringe of hairs. The flowers（classified yellow, although visually seem white）are inconspicuous and are borne in greatly congested spikes, 2-5 cm long（Hitchcock et al., 1969）. Along its introduced east coast range *S. alterniflora* flowers between late August and September. The plant is deciduous; its stems die back at the end of each growing season（Ebasco Environmental, 1992; Daehler and Strong, 1994）.

Within its native range of the Atlantic and Gulf coastlines of USA, *S. alterniflora* exhibits two growth forms, at different salt marsh zones. A tall form occurs along creek banks and drainage channels. Landward of the tall form, an intermediate form occurs, which grades into a stunted form at the salt marsh interior（Smart, 1982）.

A detailed description of *S. alterniflora* is provided by the Grass Manual on the Web（http://herbarium.usu.edu/）.

Plants rhizomatous; rhizomes elongate, flaccid, white, scales inflated, not or only slightly imbricate. Culms to 250 cm tall,（0.3）5-15（20）mm thick, erect, solitary or in small clumps, succulent, glabrous, having an unpleasant, sulphurous odor when fresh. Sheaths mostly glabrous, throat glabrous or minutely pilose, lower sheaths often wrinkled; ligules 1-2 mm; blades to 60 cm long, 3-25 mm wide, lower blades shorter than those above, usually flat basally, becoming involute distally, abaxial surfaces glabrous, adaxial surfaces glabrous or sparsely pilose, margins usually smooth, sometimes slightly scabrous, apices attenuate. Panicles 10-40 cm, with 3-25 branches, often partially enclosed in the uppermost sheath; branches 5-15 cm, loosely appressed, not twisted, more or less equally subremote to moderately imbricate throughout the panicle, axes often prolonged beyond the distal spikelets, with 10-30 spikelets. Spikelets 8-14 mm, straight, usually divergent, more or less equally imbricate on all the branches. Glumes straight, sides usually glabrous, sometimes pilose near the base or appressed pubescent, hairs to 0.3 mm; lower glumes 4-10 mm, acute; upper glumes 8-14 mm, keels glabrous, lateral veins not present, apices acuminate to obtuse, occasionally apiculate; lemmas glabrous or sparsely pilose, apices usually acuminate; paleas slightly exceeding the lemmas, thin, papery, apices obtuse or rounded; anthers 3-6 mm. $2n = 62$.

【www.iucngisd.org】*Spartina alterniflora*

S. alterniflora is an erect, perennial salt tolerant grass that characteristically grows in dense stands. The inflorescence is a flowering panicle made of many spikes and it is 10-40cm long with dense colourless flowers, which are closely appressed and overlapping. S. alterniflora blooms from July through November (The Invasive Spartina Project, 2003). Leaf blades which are grey-green in colour can be 20-55cm long and and be up to 5cm in width. The stems range in height from 60-250cm and are upto 2cm wide at the base (Brian Silliman., pers. comm., 2005).

【近似种】大米草

学　　名：*Spartina anglica* Hubb.

异　　名：*Spartina townsendii* H. Groves & J. Groves（misapplied）

别　　名：普通大米草、英国大米草

英文名：Common Cord-grass，common cordgrass，cord-grass，English cord-grass，English cordgrass，rice grass，ricegrass，salt marsh-grass

| 形态特征 |

大米草为多年生草本植物。

植物秆直立，分蘖多而密聚成丛，10～120cm，径3～5mm，无毛。植株高达10～120cm。

植物叶片线形，先端渐尖，基部圆形，两面无毛，长约20cm，宽8～10mm，中脉在上面不显著。叶鞘大多长于节间，无毛，基部叶鞘常撕裂成纤维状而宿存。叶舌长约1mm，具长约1.5mm的白色纤毛。

穗状花序长7～11cm，劲直而靠近主轴，先端常延伸成芒刺状，穗轴具3棱，无毛，2～6枚总状着生于主轴上；小穗单生，长卵状披针形，疏生短柔毛，长14～18mm，无柄，成熟时整个脱落；第一颖草质，先端长渐尖，长6～7mm，具1脉；第二颖先端略钝，长14～16mm，具1～3脉；外稃草质，长约10mm，具1脉，脊上微粗糙；内稃膜质，长约11mm，具2脉；花药黄色，长约5mm，柱头白色羽毛状；子房无毛。

颖果圆柱形，长约10mm，光滑无毛，胚长达颖果的1/3。

染色体$2n$=120，122，124，126，127（Goodman et al., 1969）。

花果期8～10月。

| 生物危害 |

引进大米草，本是为了沿海抵御风浪、保滩护岸、促淤造陆、改良土壤。但因其抗逆性与繁殖力极强，生长密集，导致滩涂生态失衡，诱发赤潮，使得海洋生物窒息而亡，已对沿海滩涂多种海洋生物构成严重威胁。

| 地理分布 |

原产于欧洲。现在欧洲、北美和澳洲均有分布。我国于1963年从丹麦、荷兰和英国引进，在江苏海滨试种。现已经在全国沿海各省栽种。

| 传播途径 |

种子和植物根茎可人为或自然传播扩散。

| 管理措施 |

综合治理（Integrated Pest Management，IPM）或综合控制（Integrated Pest Control，IPC），但不包括植物卫生措施（Phytosanitary Measures）。

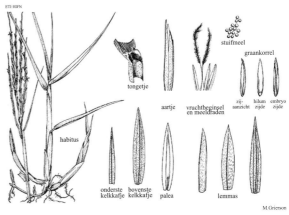

大米草（*Spartina anglica*）

（图片选自canope.ac-besancon.fr，bugwoodcloud.org，baike.baidu.com，www.soortenbank.nl，gallery.nen.gov.uk，www.diversitasnaturae.be）

【www.efloras.org】*Spartina anglica* C. E. Hubbard, Bot. J. Linn. Soc. 76: 364. 1978.

Spartina townsendii H. Groves & J. Groves var. *anglica* (C. E. Hubbard) Lambinon & Maquet.

Perennial with soft fleshy rhizomes, deeply rooted. Culms forming large clumps, erect, 10-50 (-120) cm tall, 3-35 mm in diam. Leaf sheaths mostly longer than internodes, smooth; leaf blades linear, flat or inrolled upward, 10-45 cm × 0.7-1.5 cm, smooth, apex fine, hard, upper blades usually patent; ligule 2-3 mm. Racemes racemosely arranged, 2-6 (-12), 7-23 cm, stiff, erect or slightly spreading; spikelets closely overlapping; rachis terminating in a hard bristle up to 5 cm. Spikelets 12-21 mm, pubescent; lower glume 2/3-4/5 as long as spikelet, acute; upper glume lanceolate-oblong, as long as spikelet, acute; lemma lanceolate-oblong, ca. 1 cm, keel scaberulous, pubescent, entirely or in upper half; palea slightly longer than lemma. Anthers 7-13 mm. $2n = 124$.

Spartina anglica is an extremely vigorous species, which arose in England at the end of the 19th century by the natural hybridization of *S. alterniflora* and *S. maritima* (Curtis) Fernald, followed by a doubling of chromosomes in the resulting sterile hybrid to form a fertile amphidiploid. It was introduced from England to China in 1963 and was planted in coastal areas. At first it spread rapidly, occurring in all coastal provinces by 1985. In recent years it has died back, leaving only small residual colonies. The reasons for the dieback are not fully understood.

【keyserver.lucidcentral.org】*Spartina anglica* C.E. Hubb.

Distinguishing Features

A long-lived grass spreading by fleshy creeping underground stems and growing 30-130 cm tall.

Its upright flowering stems are relatively stout (about 5mm thick) and usually hollow.

Its green or greyish-green leaves (10-45 cm long) are long and narrow with flat or inrolled margins.

Its seed-heads (12-40 cm long) have several contracted branches and end in a bristle up to 5 cm long.

Its elongated flower spikelets (14-21 mm long) are borne in two rows along one side of the seed-head branches.

Stems and Leaves

The upright flowering stems (i.e. erect culms) are relatively stout (about 5mm thick) and hairless (i.e. glabrous). They are round, smooth and usually hollow.

The green or greyish-green leaves consist of a leaf sheath, which partially encloses the stem, and a spreading leaf blade. These leaves are clustered at the base of the plant and alternately arranged along the stems. The overlapping leaf sheaths and the elongated (i.e. linear) leaf blades are hairless (i.e. glabrous). These leaf blades (10-45 cm long and 6-15 mm wide) are flat or inrolled upwards, with entire margins and pointed tips (i.e. acute apices). Where the leaf sheath meets the leaf blade there is a dense line of hairs (i.e. ciliate ligule) 2-3 mm long.

Flowers and Fruit

The seed-heads (12-40 cm long) have several (2-12) contracted branches and are borne at the tips of the flowering stems (i.e. they are terminal panicles). The main stem of the seed-head (i.e. rachis) is somewhat three-angled and ends in a bristle up to 5 cm long. Each of the upright (i.e. erect) or slightly spreading branches is up to 25 cm long. Numerous closely-overlapping, stalkless (i.e. sessile), flower spikelets are borne in two rows along one side of the seed-head branches. The narrowly oblong flower spikelets (13-21 mm long and 2.5-3 mm wide) are flattened and tightly held (i.e. appressed) to the branches. Each of these flower spikelets consists of a pair of bracts (i.e. glumes) and one, or rarely two, tiny flowers (i.e. florets). Each of the florets has an inner and outer floral bract (i.e. palea and lemma), three large stamens with anthers 8-13 mm long, and an ovary topped with a feathery two-branched stigma.

These flower spikelets fall from the seed-heads intact when mature. They contain the 'seed' (i.e. grain or caryopsis) enclosed in the other flower parts (i.e. lemma, palea and glumes).

【www.iucngisd.org】 *Spartina anglica*

A deep-rooting perennial, 30-130 cm high, spreading by soft stout fleshy rhizomes, forming large clumps and extensive meadows. Culms erect, stout, many-noded, smooth. Leaves green or greyish-green; sheaths overlapping, rounded on the back, smooth; ligules densely silkily ciliate, with hairs 2-3 mm long; blades with a fine hard point, 10-45 cm long, 6-15 mm wide, flat or inrolled upwards, firm, closely flat-ribbed above, smooth, the upper widely spreading. Panicles erect, finally contracted and dense, 12-40 cm long, of 2-12 spikes, overtopping the leaves. Spikes erect or slightly spreading, stiff, up to 25cm long; axis 3-angled, smooth, terminating in a bristle up to 5cm long. Spikelets closely overlapping, in two rows on one side of and appressed to the axis, narrowly oblong, flattened, 14-21 mm long, mostly 2.5-3 mm wide, 1- rarely 2- flowered, falling entire at maturity, loosely to closely pubescent. Glumes keeled, pointed; lower two-thirds to four-fifths the lenght of the upper, 1-nerved; upper as long as the spikelet, lanceololate-oblong, tough except for the membranous margins, 3-6 nerved. Lemma shorter than the upper glume, lanceolate-oblong, 1-3 nerved, with broad membranous margins, shortly hairy. Palea a little longer than lemma, 2-nerved. Anthers 8-13 mm long. Grain with a long green embryo, enclosed between the lemma, palea, and glumes. Ch. no. $2n = 122\text{-}124$ (Hubbard, C.E. 1968, Grasses, Penguin Books Ltd, England).

凤眼莲

学　　名：***Eichhornia crassipes***（Mart.）Solms

异　　名：*Eichhornia cordifolia* Gandoger，*Eichhornia crassicaulis* Schlect.，*Eichhornia speciosa* Kunth，*Heteranthera formosa* Miq.，*Piaropus crassipes*（Mart.）Raf.，*Piaropus mesomelas* Raf.，*Pontederia crassicaulis* Schlect.，*Pontederia crassipes* Mart.，*Pontederia elongata* Balf.

别　　名：水葫芦、水浮莲、水葫芦苗、布袋莲、浮水莲花

英文名：water hyacinth，floating water hyacinth，lilac devil，Nile lily，pickerelweed，water orchid，water violet

| 形态特征 |

凤眼莲为多年生浮水草本植物。

植物须根发达，棕黑色，长达30cm。

植物茎极短，具长匍匐枝，匍匐枝淡绿色或带紫色，与母株分离后可长成新植物体。植株高30~60cm。

植物叶在基部丛生，莲座状排列，一般5~10片。叶片圆形、宽卵形或宽菱形，长4.5~14.5cm，宽5~14cm，顶端钝圆或微尖，基部宽楔形或在幼时为浅心形，全缘，具弧形脉，表面深绿色，光亮，质地厚实，两边微向上卷，顶部略向下翻卷。叶柄长短不等，中部膨大呈囊状或纺锤形，内有许多多边形柱状细胞形成的气室，维管束散布其间，黄绿色至绿色，光滑；叶柄基部有鞘状苞片，长8~11cm，黄绿色，薄而呈半透明。

花葶从叶柄基部的鞘状苞片腋内伸出，长34~46cm，多棱。

穗状花序长17~20cm，通常具9~12朵花。花被裂片6枚，花瓣状，卵形、长圆形或倒卵形，紫蓝色。花冠略两侧对称，直径4~6cm，上方1枚裂片较大，长约3.5cm，宽约2.4cm，三色即四周淡紫红色，中间蓝色，在蓝色的中央有1黄色圆斑，其余各片长约3cm，宽1.5~1.8cm，下方1枚裂片较狭，宽1.2~1.5cm。花被片基部合生成筒，外面近基部有腺毛。雄蕊6枚，贴生于花被筒上，3长3短，长的从花被筒喉部伸出，长1.6~2cm，短的生于近喉部，长3~5mm。花丝上有腺毛，长约0.5mm，3（2~4）细胞，顶端膨大。花药箭形，基着，蓝灰色，2室，纵裂。花粉粒长卵圆形，黄色。子房上位，长梨形，长6mm，3室，中轴胎座，胚珠多数。花柱1，长约2cm，伸出花被筒的部分有腺毛。柱头上密生腺毛。

蒴果卵形。花期7~10月，果期8~11月。

| 生物危害 |

凤眼莲爆发及腐烂后，会大量消耗水体中的溶解氧，从而抑制浮游生物生长，造成水生动物如

鱼类活动、繁殖空间减少，甚至会导致鱼类等大量死亡；还能为蚊蝇提供滋生条件，导致血吸虫和脑炎流行；污染水源，影响水质。

凤眼莲大量逸生后，常常生长在海拔200～1 500m的水塘、沟渠及稻田中，堵塞河道，破坏水生生态系统，威胁本地生物多样性。

| 地理分布 |

凤眼莲原产于巴西东北部。1901年我国台湾作花卉从日本引入。20世纪50年代作为猪饲料在我国大陆推广栽培。目前已经在辽宁南部、华北、华东、华中和华南的19个省（自治区、直辖市）有栽培，在长江流域及其以南地区逸生为杂草。凤眼莲已经在全球亚热带地区广泛生长甚至爆发成灾。

| 传播途径 |

凤眼莲兼有性和无性两种繁殖方式，因此可以通过种子，也可以通过匍匐枝与母枝分离的方式进行无性繁殖而人为传播或自然扩散。

| 管理措施 |

综合治理（Integrated Pest Management，IPM）或综合控制（Integrated Pest Control，IPC），但不包括植物卫生措施（Phytosanitary Measures）。

（图片选自www.baumschule-horstmann.de，oformi-akvarium.ru，www.plantright.org，www.qjure.com，upload.wikimedia.org，gobotany.newenglandwild.org，idtools.org）

凤眼莲（*Eichhornia crassipes*）

【www.efloras.org】*Eichhornia crassipes* （Martius） Solms in A. de Candolle & C. de Candolle, Monogr. Phan. 4: 527. 1883.

Pontederia crassipes Martius, Nov. Gen. Sp. Pl. 1: 9. 1823; *Eichhornia speciosa* Kunth; *Heteranthera formosa* Miquel.

Herbs floating, 0.3-2 m. Roots many, long, fibrous. Stems very short; stolons greenish or purplish, long, apically producing new plants. Leaves radical, rosulate; petiole yellowish green to greenish, 10-40 cm, spongy, usually very much swollen at or below middle; leaf blade orbicular, broadly ovate, or rhomboidal, 4.5-14.5 cm × 5-14 cm, leathery, glabrous, densely veined, base shallowly cordate, rounded, or broadly cuneate. Inflorescences bracteate, spirally 7-15-flowered; peduncle 35-45 cm. Perianth 6-parted, segments purplish blue, petaloid, ovate to elliptic, upper one larger with yellow blotch at center adaxially, others subequal but lower one narrower. Stamens 6, 3 long and 3 short; filaments curved, glandular hairy. Pistil heterostylic; stigma glandular hairy. Capsule ovoid. Fl. Jul-Oct, fr. Aug-Nov.

【www.efloras.org】*Eichhornia crassipes* （Martius） Solms in A. L. P. P. de Candolle and C. de Candolle, Monogr. Phan. 4: 527. 1883.

Pontederia crassipes Martius, Nov. Gen. Sp. Pl. 1: 9, plate 4. 1823; *Eichhornia speciosa* Kunth; *Piaropus crassipes* （Martius） Rafinesque

Plants perennial, typically free-floating. Vegetative stems condensed, except when branching. Flowering stems erect, bending over after flowering, to 25 cm, distal internode less than 4 cm. Sessile leaves in basal rosette. Petiolate leaves floating or emersed; stipule 2.5-14 cm, apex truncate; petiole at least slightly swollen, 3.5-33 cm; blade ovate to round, 2.5-11 cm × 3.5-9.5 cm. Spikes 4-15-flowered; spathes obovate, 4-11 cm; peduncle 5-12.5 cm, glabrous. Flowers opening individually within 2 hours after sunrise, wilting by night; perianth blue or mauve-blue, limb lobes obovate, 16-37 mm, margins entire, central distal lobe with dark blotch in center and yellow spot within blotch; proximal stamens 20-35 mm, distal 14-19 mm; anthers 1.7-2.1 mm; style 3-lobed. Seeds 11-14-winged, 1.1-2.1 cm × 0.6-0.9 mm. $2n = 32$.

【www.cabi.org】*Eichhornia crassipes* （water hyacinth）

Aquatic，Herbaceous，Perennial，Seed propagated，Vegetatively propagated plant

The initial leaves of seedling E. crassipes are elongated and strap-like, but soon develop the familiar spathulate form and, under suitable unshaded conditions, swollen petioles which ensure that, once dislodged, the seedlings will float from the mud into open water. The plant is very variable in size, seedlings having leaves that are only a few centimetres across or high, whereas mature plants with good nutrient supply may reach 1 m in height. Plants in an uncrowded situation tend to have short, spreading petioles with pronounced swelling, while in a dense stand they are taller, more erect and with little or no swelling of the petioles.

The plant system consists of individual shoots/crowns each with up to ten expanded leaves arranged spirally （3/8 phyllotaxy） and separated by very short internodes. As individual shoots develop, the older leaves die off leaving a stub of leafless dead shoot projecting downwards. This may eventually cause the whole shoot to sink and die.

Leaves consist of petiole （often swollen, 2-5 cm thick） and blade （roughly round, ovoid or kidney-shaped, up to 15 cm across）. The base of the petiole and any subsequent leaf is enclosed in a stipule up to 6 cm long.

Roots develop at the base of each leaf and form a dense mass: usually 20-60 cm long, though they can extend to 300 cm. The ratio of root to shoot depends on the nutrient conditions, and in low nutrient conditions they may account for over 60% of the total plant weight. They are white when formed in total darkness but often purplish under field conditions, especially in conditions of low nutrients.

Periodically, axillary buds develop as stolons, growing horizontally for 10-50 cm before establishing daughter plants. Extremely large populations of inter-connected shoots can develop very rapidly, though the connecting stolons eventually die.

The inflorescence is a spike which develops from the apical meristem, but tends to appear lateral owing to the immediate development of an axillary bud as a 'renewal' or 'continuation' shoot. Each spike, up to 50 cm high, is subtended at the base by two bracts and has 8-15 sessile flowers (rarely 4-35). Each flower has a perianth tube 1.5 cm long, expanding into six mauve or purple lobes up to 4 cm long. The main lobe has a bright-yellow, diamond-shaped patch surrounded by deeper purple. Once the inflorescence is fully emerged from the leaf sheath, flowers all open together, starting at night, completing the process in the morning and withering by the next night when the peduncle starts to bend down. Each capsule may contain up to 450 small seeds, each about 1×3 mm.

The flowers are tristylous. They have six stamens and one style, arranged in three possible configurations (floral trimorphism) - with short style (and medium and long stamens), medium style (short and long stamens) or long style (short and medium stamens). The medium style form is genetically dominant and is by far the commonest form in almost all infested areas. The short-styled form is only known from South America, whereas the long-styled form is found commonly in South America, more rarely in South-East Asia and very rarely in Africa. Only in Sri Lanka is the long-styled the commonest form. Some other tristylous species show incompatibility between the different forms but E. crassipes does not. Hence pollination (mainly by wind) can result in good seed set, though in some populations there may be a higher degree of self-incompatibility.

假高粱

学　名：***Sorghum halepense***（L.）Pers.

异　名：*Andropogon arundinaceus* Scop.，*Andropogon halepensis*（L.）Brot.，*Andropogon halepensis*（L.）Brot. var. *anatherus* Piper，*Andropogon halepensis*（L.）Brot. var. *genuinus* Stapf ex Hook. f.，*Andropogon halepensis*（L.）Brot. var. *muticus*（Hack.）Asch & Graebn.，*Andropogon halepensis*（L.）Brot. var. *typicus* Asch & Graebn.，*Andropogon sorghum*（L.）Brot. ssp. *halepensis*（L.）Hack，*Andropogon sorghum*（L.）Brot. subvar. *genuinus* Hack，*Andropogon sorghum*（L.）Brot. subvar. *leiostachys* Hack，*Andropogon sorghum*（L.）Brot. subvar. *muticus* Hack，*Blumenbachia halepensis*（L.）Koeler，*Holcus halepensis* L.，*Milium halepense*（L.）Cav.，*Sorghum almum* Parodi，*Sorghum almum* Parodi var. *typicum* Parodi，*Sorghum controversum*，*Sorghum halepense*（L.）Pers. var. *muticum*（Hack.）Grossh.，*Sorghum miliaceum*（Roxb.）Snowden，*Sorghum saccharatum*（L.）Moench var. *halepense*（L.）Kuntze

别　名：石茅、石茅高粱、宿根高粱、阿拉伯高粱、琼生草

英文名：Johnson grass，Aleppo grass，Arabian millet，Egyptian millet，evergreen millet，false guinea，Morocco millet，Syrian grass

|形态特征|

假高粱为多年生草本植物。

植物有地下横走根状茎。

植物秆直立，高0.5~3m，直径约5mm。

植物叶片阔线状披针形，长25~80cm，宽1~4cm。基部有白色绢状疏柔毛，中脉白色而厚。叶舌长约1.8mm，具缘毛。

圆锥花序长20~50cm，淡紫色至紫黑色。分枝轮生，基部有白色柔毛；分枝上生出小枝，小枝顶端着生总状花序。穗轴具关节，较纤细，具纤毛。小穗柄在总状花序轴基部有明显折断。小穗成对，一具柄，一无柄。有柄小穗较狭，长约4mm，颖片草质，无芒；无柄小穗椭圆形，长3.5~4mm。二颖片革质，近等长；第一颖的顶端具3齿，第二颖的上部1/3处具脊，第二颖基部带有一枝小穗轴节段和一枚有柄小穗的小穗柄，二者均具纤毛。每小穗有1小花，第一外稃膜质透明，被纤毛，第二外稃长约为颖片的1/3，顶端微2裂，主脉由齿间伸出呈小尖头或芒。

果实带颖片，椭圆形，长约1.4mm，暗紫色（未成熟的呈麦秆黄色或带紫色），光亮，被柔毛。去颖颖果倒卵形至椭圆形，长2.6~3.2mm，宽1.5~2mm，棕褐色，顶端圆，具2枚宿存花柱。

染色体数$n=20$（Kalia，1978）；$2n=40$（R. P. Celarier，1958）。

| 生物危害 |

假高粱为我国检疫性有害生物。是谷类作物、棉花、苜蓿、甘蔗、麻类等30多种作物田里的主要杂草。它不仅使作物降低产量，还是高粱属作物的许多重要害虫和病害的寄主。

其花粉可与高粱属作物杂交，给农业生产带来很大危害，被普遍认为是世界农作物最危险的恶性杂草之一。

假高粱的地下茎分节、分枝，具有相当强的繁殖力。即使将它切成小段，甚至只有一节，仍能在适生条件下生长，形成新植株，具有很强的适应性，是一种极其难于防治的恶性杂草。

此外，其根产生的分泌物，腐烂的叶片、地下茎、根等，均能抑制作物种子萌发和幼苗生长。

假高粱的嫩芽聚积有氰化物，牲畜误食后会引起中毒。

| 地理分布 |

假高粱原产于地中海地区。现为世界性分布。欧洲：希腊、前南斯拉夫、意大利、保加利亚、西班牙、葡萄牙、法国、瑞士、罗马尼亚、波兰、俄罗斯。亚洲：土耳其、以色列、阿拉伯半岛、黎巴嫩、约旦、伊拉克、伊朗、印度、巴基斯坦、阿富汗、泰国、缅甸、斯里兰卡、印度尼西亚、菲律宾。非洲：摩洛哥、坦桑尼亚、莫桑比克、南非。美洲：古巴、牙买加、危地马拉、洪都拉斯、尼加拉瓜、波多黎各、萨尔瓦多、多米尼亚、委内瑞拉、哥伦比亚、秘鲁、巴西、玻利维亚、巴拉圭、智利、阿根廷、墨西哥、美国、加拿大。大洋洲及太平洋岛屿：澳大利亚、新西兰、巴布亚新几内亚、斐济、美拉尼西亚、密克罗尼西亚、夏威夷。

20世纪初从日本引入我国台湾南部栽培，同一时期在香港和广东北部发现。假高粱常混在进口粮油作物种子中扩散，现分布于我国山东、贵州、福建、吉林、河北、广西、北京、甘肃、安徽、江苏等局部地区。

| 传播途径 |

假高粱种子可混杂在粮谷等农作物种子中进行远距离传播，也可随水流传播。同时，假高粱的根茎可以在地下扩散蔓延，还能随货物携带进行传播。

| 管理措施 |

综合治理（Integrated Pest Management，IPM）或综合控制（Integrated Pest Control，IPC），包括植物卫生措施（Phytosanitary Measures）。

中国外来入侵 物种图鉴

(78)

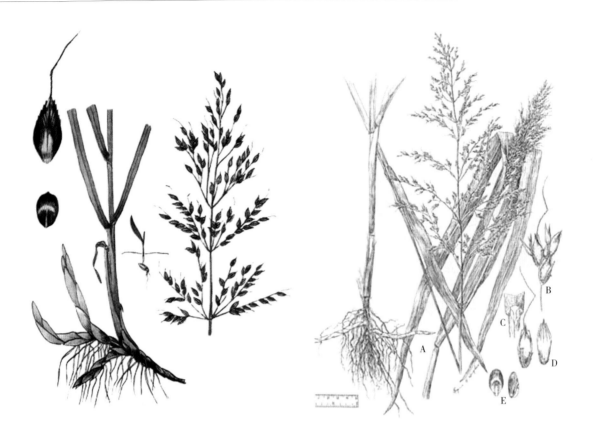

（图片选自www.cropscience.bayer.com，www.weedimages.org，www.barbechoquimico.com，www.unavarra.es，www.missouriplants.com，upload.wikimedia.org，flowers.la.coocan.jp，www.pesticide.ro，bugwoodcloud.org，baike.baidu.com，calphotos.berkeley.edu，biopix.com，www.aphotoflora.com，idtools.org，www.agroatlas.ru，plants.ces.ncsu.edu）

假高粱（*Sorghum halepense*）

【www.efloras.org】*Sorghum halepense*（Linnaeus）Persoon, Syn. Pl. 1: 101. 1805.

Holcus halepensis Linnaeus, Sp. Pl. 2: 1047. 1753; *Andro-pogon halepensis*（Linnaeus）Brotero; *A. sorghum*（Linnaeus）Brotero subsp. *halepensis*（Linnaeus）Hackel.

Perennial with vigorous spreading rhizomes. Culms 0.5-1.5 m tall, 4-6 mm in diam.; nodes puberulous. Leaf sheaths glabrous; leaf blades linear or linear-lanceolate, 25-80 cm × 1-4 cm, glabrous; ligule 0.5-1 mm, glabrous. Panicle lanceolate to pyramidal in outline, 20-40 cm, soft white hairs in basal axil; primary branches solitary or whorled, spreading, lower part bare, upper part branched, the secondary branches tipped by racemes; racemes fragile, composed of 2-5 spikelet pairs. Sessile spikelet elliptic, 4-5 mm; callus obtuse, bearded; lower glume subleathery, often pale yellow or yellowish brown at maturity, shortly pubescent or glabrescent, 5-7-veined, veins distinct in upper part, apex 3-denticulate; upper lemma acute and mucronate or 2-lobed and awned; awn 1-1.6 cm. Pedicelled spikelet staminate, narrowly lanceolate, 4.5-7 mm, often violet-purple. Fl. and fr. summer-autumn. $2n = 40$.

【www.cabi.org】*Sorghum halepense*（Johnson grass）

Grass / sedge，Herbaceous，Perennial，Seed propagated，Vegetatively propagated plant

S. halepense is a perennial grass with extensively creeping, fleshy rhizomes which are covered with brown scale-like sheaths, are up to 1 cm in diameter, 2 m in length, and often root from the nodes. The fibrous root system branches freely to depths of 1.2 m. The leaf blades, 20-60 cm long, 1.0-3.3 cm wide, have prominent midribs, are many nerved and hairless with projections on the lower surface and margins. The ribbed, hairless leaf sheaths have open overlapping margins and a membranous ligule with a hairy fringe, 2-5 mm long. Flowering stems are unbranched, 0.5-3.0 m tall, 0.5-2.0 cm in diameter, often with basal adventitious prop roots, nodes sometimes with fine hairs. The inflorescence is a pale green to purplish, hairy, pyramidal, many branched panicle, 15-50 cm long. The primary branches are up to 25 cm long, usually without spikelets for 2-5 cm from the base. The spikelets are usually in pairs but towards the top of the inflorescence they occur in threes, one spikelet of each pair or triplet is sessile and perfect with stamens and a stigma, the others are stalked and sterile or only carry stamens. The fertile spikelets are ovoid, hairy, 4.5-5.5 mm long; awns if present are 1-2 cm long and abruptly bent. The stalked spikelets are narrower, 5-7 mm long. The grain remains enclosed by glumes 4-6.6 mm long, 2-2.6 mm wide, the glumes are reddish brown to shiny black, glossy and finely lined on the surface.

【www.cropscience.bayer.com】*Sorghum halepense*

S. halepense is a perennial grass with extensively creeping, fleshy rhizomes which are covered with brown scale-like sheaths, up to 1 cm（0.39 inch）in diameter, 2 m（6.56 feet）in length, and often root from the nodes. The fibrous root system branches freely to depths of 1.2 m（3.93 feet）.

Characteristic Features

Ribbed leaf sheath, conspicuous midrib, large, purplish panicle and extensive rhizome system.

Stems Flowering stems are unbranched, 0.5-3.0（-4.0）m（1.64-9.84-13.12 feet）tall, 0.5-2.0 cm（0.19-0.78 inch）in diameter, often with basal adventitious prop roots, nodes sometimes with fine hairs.

Leaves Leaf blades, 20-60 cm（7.87-23.62 inches）long, 1.0-3.3 cm（0.39-1.56 inches）wide, prominent midribs and white, whitish midvein, many nerved and hairless.Membranous ligule with hairy fringe, 2-5 mm（0.078-0.19 inch）long. No distinct auricles. Ribbed, hairless leaf sheaths with overlapping margins.

Flowers Inflorescence pale green to purplish, hairy, pyramidal, many branched panicle, 15-50 cm (5.9-19.68 inches) long.Primary branches up to 25 cm (9.84 inches) long, usually without spikelets for 2-5 cm (0.78-1.96 inches) from the base.Spikelets usually in pairs but towards the top of the inflorescence in threes, one spikelet of each pair or triplet is sessile and perfect with stamens and stigma, others stalked and sterile or only carry stamens. Fertile spikelets are ovoid, hairy, 4.5-5.5 mm (0.17-0.21 inch) long. Awns if present are 1-2 cm (0.39-0.78 inch) long, twisted and abruptly bent. Glumes reddish brown to shiny black, glossy and finely lined.

Fruit The grain remains enclosed by glumes 4-6.6 mm (0.16-0.25 inch) long, 2-2.6 mm (0.078-0.10 inch) wide. The glumes are reddish brown to shiny black, glossy and finely lined on the surface.

【www.iucngisd.org】 *Sorghum halepense*

Perennial grass with strong rhizomes; rhizomes fleshy, to 1 cm in diameter, to 2 m long, often rooting from the nodes; culms erect, to 1.5 m tall; nodes with short pubescence; sheaths glabrous; ligule ciliate-membranous, 2 mm long; blades elongate, usually 1-1.5 cm wide, the midrib prominent; panicles 15-25 cm long, branches ascending; spikelets 5 mm long, acute; first glume hard; fertile lemma awned or awnless, awn if present 1 cm long or less\" (Stone, 1970). Grain remains enclosed by glumes, 4.0-6.6 mm long, 2.0-2.6 mm wide, oblong-ovate, glumes reddish brown to shiny black. The plant has both diploid and tetraploid races, with a chromosome number of either $2n = 20$ or 40 (Stone, 1970; Warwick and Black, 1983).

【gobotany.newenglandwild.org】 *Sorghum halepense* (L.) Pers.

Characteristics

Habitat   Terrestrial

Leaf blade   Width 8-40 mm

Inflorescence branches   The flowers are attached to branches rather than to the main axis of the inflorescence

Spikelet length   3.6-6.5 mm

Glume relative length   Both glumes are as long or longer than all of the florets

Awn on glume   The glume has no awn

One or more florets   There is one floret per spikelet

Lemma awn length   Up to 13 mm

Leaf sheath hair type   There are no hairs on the surface of the leaf sheath

Leaf ligule length   2-6 mm

Anther length   1.9-2.7 mm

【近似种】黑高粱

学　　名：*Sorghum almum* Parodi

异　　名：*Sorghum almum* var. *almum* Parodi，*Sorghum almum* var. *parvispiculum* Parodi，*Sorghum x almum* Parodi

别　　名：哥伦布草、五年高粱、苏丹黑高粱

英文名：Almum grass，almum sorghum，columbus grass（Australia），five-year sorghum，sorgo negro，Sudan negro（Argentina）

| 形态特征 |

黑高粱为多年生草本植物。具短根状茎。

植物秆高80～150cm，甚至高达4.5m。不分枝或有时自下部节上分枝。

植物叶鞘无毛或基部有柔毛；叶舌硬膜质，顶端近平截，无毛；叶片线形至线状披针形，长25～50cm，宽2～5cm，中部最宽，向两端渐狭窄；中脉白色，边缘软骨质，具微细的小锯齿。

花序为圆锥花序，长约40cm，分枝较细而斜升，单生或数枚在主轴上轮生，或在近一侧着生；基部腋间具灰白色柔毛。每一总状花序具2～5节，下部裸露部分长1.5～4cm。小穗柄在总状花序轴顶部折断。无柄小穗椭圆形或长椭圆形，长4.5～6mm，上端具柔毛或近于无毛，宽1.5～2mm，成熟后黑色或黄绿色，不脱落或迟缓脱落，基盘钝，具短柔毛。颖纸质或薄革质，第一颖具5～7脉，上部明显，除背部的毛较疏或无毛外，其余部分的毛较多；第二颖上部具脊；第二外稃披针形，顶端钝或微尖，膜质，具2脉，第二内稃顶端多少2浅裂。芒细弱，长约1.5cm，自第二外稃裂缝间伸出；鳞被2枚，宽倒卵形，顶端微凹，无毛。雄蕊3枚。有柄小穗雄性，较无柄小穗狭窄，颖的质地较薄。子房椭圆形。

染色体数2n=40（Parodi，1943）。

花果期为夏、秋季。

| 地理分布 |

黑高粱原产于阿根廷。现分布于南美、中美和北美洲等地区，包括美国、阿根廷等。南非、澳大利亚也有分布。我国和印度少数园圃有引种栽培作牧草的记录。

| 生物危害 |

黑高粱为我国检疫性有害生物。它与假高粱的生长习性基本相似。其宿根多年生，根茎发达，结实多。以种子和地下根茎进行繁殖，生活力极强，适应性广，具有很强的竞争能力，是一种危害大、防治难的恶性杂草。

| 传播途径 |

黑高粱种子可混杂在粮食中进行远距离传播，也可随水流等自然传播。

| 管理措施 |

综合治理（Integrated Pest Management，IPM）或综合控制（Integrated Pest Control，IPC），包括植物卫生措施（Phytosanitary Measures）。

中国外来入侵物种图鉴

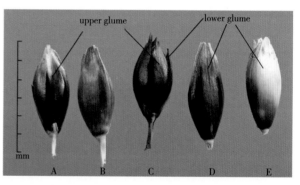

Sorgum×almum Parodi-1, base of plant;2, culm-and leaf-part; inflorescence;4, sessile spikelet;5, pedicelled spikelet;6, caryopsis

（图片选自erec.ifas.ufl.edu，bugwoodcloud.org，www.feedipedia.org，flowers.la.coocan.jp，agron-www.agron.iastate.edu，www.invasive.org，cropgenebank.sgrp.cgiar.org，www.tropicalforages.info，calphotos.berkeley.edu，idtools.org）

黑高粱（*Sorghum almum*）

【idtools.org】*Sorghum almum* Parodi

Spikelets heteromorphic, occurring in pairs of 1 sessile fertile and 1 sterile or staminate spikelet or a triplet of 1 sessile fertile and 2 sterile or staminate spikelets in the terminal spikelet unit. Disseminule usually consists of a fertile spikelet, rachis segment from beneath a sessile spikelet, and pedicel or pedicellate sterile/staminate spikelet.

Fertile spikelet narrowly ovate to elliptic, dorsally compressed, 5-6.5 mm long, 1.5-2.5 mm wide, shiny, light reddish brown to dark brown or black, consisting of one fertile floret and one basal sterile lemma. Spikelet callus pubescent.

Lower glume coriaceous, length of spikelet, its margins enclosing upper glume, 2-keeled near apex, with hairs 0.5-0.8 mm long. Upper glume coriaceous, surface pubescent.

Sterile lemma hyaline.

Fertile lemma hyaline, apex entire or dentate, awnless or awn twisted, early deciduous.

Fertile palea absent or minute.

Sterile/staminate spikelet on pubescent pedicel, well-developed, dorsally compressed, glumes chartaceous, 4-6 mm long.

Caryopsis brown to reddish-brown, 3-3.8 mm long, elliptic to obovate, dorsally compressed, embryo 0.5 length of caryopsis, hilum ± round.

【wiki.bugwood.org】*Sorghum almum*

Appearance *Sorghum almum* is a cross between *S. bicolor* and *S. propinquum*. It is a perennial rhizomatous grass with stout stems up to 14 feet (4.3 m) tall. *S. almum* is native to South America.

Foliage The waxy leaf blades of S. almum are flat with a sandpapery feel. The blades are 0.5-1.5 inches (1.3-3.8 cm) wide by 18-32 inches (45.7-81.3 cm) long. Foliage occasionally has long hairs at the base of the upper leaf surface. The ligule is membranous and fringed along the top.

Flowers The inflorescence of *S. almum* is a panicle 6-24 inches (15.2-61 cm) long, with lax, spreading branches.

Fruit Seeds are 0.25-0.4 inch (0.6-1 cm) long and smooth. *S. almum* reproduces by seed and rhizomes.

马缨丹

学　名：***Lantana camara*** L.

异　名：*Camara vulgaris* Benth.，*Lantana aculeata* L，*Lantana antidotalis* Thonning（1827），*Lantana armata* Schauer，*Lantana camara* var. *aculeata*，*Lantana crocea* Jacq.，*Lantana glandulosissima* Hayek，*Lantana mexicana* Turner，*Lantana mixta* Medik.，*Lantana moritziana* Otto & A. Dietr.，*Lantana sanguinea* Medik.，*Lantana scabrida* Ait.，*Lantana scabrida* Sol.，*Lantana spinosa* L. ex Le Cointe，*Lantana tiliifolia* Cham.，*Lantana undulata* Raf.，*Lantana urticifolia* Mill.，*Lantana* x *aculeata* f. *crocea*（Jacq.）Voss

别　名：五色梅、山大丹、如意草、五彩花、五雷丹、五色绣球、变色草、大红绣球、臭草、七姐妹、七变花

英文名：lantana，arch man，common lantana，large leaf lantana，pink-flowered lantana，prickly lantana，red sage，red-flowered sage，shrub verbena，tickberry，white sage，wild sage，yellow sage

| 形态特征 |

马缨丹，又名五色梅，为马鞭草科直立或半蔓性常绿灌木植物。

植物茎枝呈四方形，有短柔毛，通常有短而倒钩状刺。植株高2～5m，有时枝条生长呈藤状。

植物叶为单叶对生，卵形或卵状长圆形，先端渐尖，基部圆形，两面粗糙有毛。

头状花序较为密集，花序腋生于枝梢上部，每个花序20多朵花，花冠筒细长，顶端多五裂，状似梅花。花冠筒状，雄蕊4枚。花冠颜色多变，黄色、橙黄色、粉红色、深红色。

果为圆球形浆果，熟时紫黑色。

植物全株被短毛。

植株有强烈气味。

| 生物危害 |

马缨丹在原产地美洲繁殖速度快、分布广，侵占了大面积的牧场、耕地和森林。它还会排挤本地植物，减少生物多样性，破坏生态平衡。被列入世界100种最严重入侵物种之一的恶性杂草。

马缨丹的叶及未成熟果实均有毒性，人畜误食会引起中毒。

| 地理分布 |

马缨丹原产于美洲热带。现已经分布于全球近50个国家。马缨丹于明末由西班牙人引入我国台湾，由于花比较美丽而被广泛栽培引种，后逃逸。我国现分布于台湾、福建、广东、海南、香港、

广西、云南、四川南部等热带及亚热带地区。是南方牧场、林场、茶园和橘园的恶性竞争者。

| 传播途径 |

马缨丹除栽培引种外，亦或因鸟类、猴类和羊群摄食果实后空投或排粪而迅速传播扩散。

| 管理措施 |

综合治理（Integrated Pest Management，IPM）或综合控制（Integrated Pest Control，IPC），但不包括植物卫生措施（Phytosanitary Measures）。

（图片选自www.darwinfoundation.org，keyserver.lucidcentral.org，upload.wikimedia.org，plants.ces.ncsu.edu，en.hortipedia.com，www.wm-sec.com，chalk.richmond.edu，www.carolinanature.com，en.wikipedia.org，www.vilmorin-tree-seeds.com，etc.usf.edu）

马缨丹（*Lantana camara*）

【 www.efloras.org 】*Lantana camara* Linnaeus, Sp. Pl. 2: 627. 1753.

Shrubs with long weak branches, armed with stout recurved prickles, pubescent. Petiole 1-2 cm, pubescent; leaf blade ovate to oblong, 3-8.5 cm × 1.5-5 cm, papery, wrinkled, very rough, with short stiff hairs, aromatic when crushed, base rounded to subcordate, margin crenate; lateral veins 5 pairs, very prominent, elevated. Capitula terminal, 1.5-2.5 cm across. Flowers yellow or orange, often turning deep red soon after opening. Ovary glabrous. Drupes deep purple, globose, ca. 4 mm in diam. $2n = 44$.

【 www.cabi.org 】*Lantana camara*（lantana）

Perennial, Seed propagated Shrub, Vine / climber Woody plant

L. camara is a medium-sized perennial aromatic shrub, 2-5 m tall, with quadrangular stems, sometimes having prickles. The posture may be sub-erect, scrambling, or occasionally clambering (ascending into shrubs or low trees, clinging to points of contact by means of prickles, branches, and leaves). Frequently, multiple stems arise from ground level. The leaves are generally oval or broadly lance-shaped, 2-12 cm in length, and 2-6 cm broad, having a rough surface and a yellow-green to green colour. The flat-topped inflorescence may be yellow, orange, white, pale violet, pink, or red. Flowers are small, multicoloured, in stalked, dense, flat-topped clusters to 4 cm across. Fruit is a round, fleshy, 2 seeded drupe, about 5 mm wide, green turning purple then blue-black (similar in appearance to a blackberry).

【 plants.ces.ncsu.edu 】*Lantana camara*

Small, perennial shrub with spiny, square stems; leaves simple, opposite or whorled, toothed, fragrant when crushed; flowers in flat-topped clusters on a long stalk, each flower small, tubular, 4-parted, white, pink, or yellow, changing to orange or red; fruit fleshy, green becoming bluish black. Lantana will perennialize in the warm coastal areas of N.C. The cultivar 'Miss Huff' will survive the winter in the Piedmont. Lantana is very attractive to butterflies and flowers from summer until frost.

【 www.iucngisd.org 】*Lantana camara*

Lantana camara is a low erect or subscandent, vigorous shrub with stout recurved prickles and a strong odour of black currents; it grows to 1.2-2.4 metres (or even more); its root system is very strong, and it gives out a new flush of shoots even after repeated cuttings; Leaf ovate or ovate-oblong, acute or subacute, crenate-serrate, rugose above, scabrid on both sides; Flower small, usually orange, sometimes varying from white to red in various shades and having a yellow throat, in axillary heads, almost throughout the year; Fruit small, greenish-blue black, blackish, drupaceous, shining, with two nutlets, almost throughout the year, dispersed by birds. Seeds germinates very easily (Sastri and Kavathekar, 1990).

【 keyserver.lucidcentral.org 】*Lantana camara* L.

Distinguishing Features

A rough-textured and usually prickly shrub with oppositely arranged leaves.

Its dense flower clusters consist of numerous small tubular flowers (9-14 mm long and 4-10 mm across).

These flower clusters are borne on stalks originating in the leaf forks.

The flowers can be a wide variety of colours (i.e. white, yellow, orange, red, pink or multi-coloured).

Its mature fruit (5-8 mm across) are glossy in appearance and black, purplish-black or bluish-black in colour.

大 藻

学　名：*Pistia stratiotes* L.

异　名：*Apiospermun obcordatum*（Schleid.）Klotzsch，*Limnonesis commutate*（Schleid.）Klotzsch，*Limnonesis friedrichsthaliana* Klotzsch，*Pistia aegyptiaca* Schleid，*Pistia aethiopica* Fenzl ex Klotszch，*Pistia africana* C. Presl，*Pistia amazonica* C. Presl，*Pistia asiatica* Lour.，*Pistia brasiliensis* Klotzsch，*Pistia commutata* Schleid，*Pistia crispate* Blume，*Pistia cumingii* Klotszch，*Pistia gardneri* Klotszch，*Pistia horkeliana* Miq.，*Pistia leprieuri* Blume，*Pistia linguiformis* Blume，*Pistia minor* Blume，*Pistia natalensis* Klotzsch，*Pistia obcordata* Schleid，*Pistia occidentalis* Blume，*Pistia schleideniana* Klotzsch，*Pistia spathulata* Michx.，*Pistia stratiotes* var *cuneata* Engl.，*Pistia stratiotes* var *obcordata*（Schleid.）Engl.，*Pistia stratiotes* var *spathulata*（Michx.）Engl.，*Pistia texensis* Klotzsch，*Pistia turpini* Blume，*Pistia turpinii* K. Koch，*Pistia weigeltiana* C. Presl

别　名：水荷莲、大萍、水莲、肥猪草、水芙蓉

英文名：Water lettuce，floating aroid，Nile cabbage，pistia，shell-flower，tropical duckweed，water bonnet，water cabbage，water fern，water lily

| 形态特征 |

大藻为多年生浮水草本植物。

植物根须发达，呈羽状，垂悬于水中。

植物主茎短缩而叶簇生于其上呈莲座状，从叶腋间向四周分出匍匐茎，茎顶端发出新植株，有白色成束的须根。

植物叶簇生。叶片倒卵状楔形，长2～8cm，顶端钝圆而呈微波状，两面都有白色细毛。

花序生叶腋间，有短的总花梗，佛焰苞长约1.2cm，白色，背面生毛。

果为浆果，内含种子10～15粒。

种子椭圆形，黄褐色。

花期为6～7月。

| 生物危害 |

大藻被列入世界最严重的100种入侵物种名录中。它雌雄同株，繁殖迅速；强大的根系，能够大量消耗水体中的氧气，致使鱼类及沉水生植物死亡甚至灭绝，从而影响水产养殖业，危害水生生态系统。

| 地理分布 |

大藻原产于巴西热带和亚热带的小溪或淡水湖中。现在在南亚、东南亚、南美及非洲都有分布。

大藻于明末引入中国。目前在我国珠江三角洲到长江流域，在湖南、湖北、四川、福建、江苏、浙江、安徽等省均有分布。

| 传播途径 |

大藻在自然条件下，主要依靠无性繁殖。而成熟的种子也可以进行繁殖。因此，大藻可以通过种子和无性繁殖材料进行人为传播或自然扩散。

| 管理措施 |

综合治理（Integrated Pest Management，IPM）或综合控制（Integrated Pest Control，IPC），但不包括植物卫生措施（Phytosanitary Measures）。

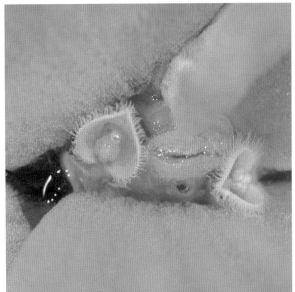

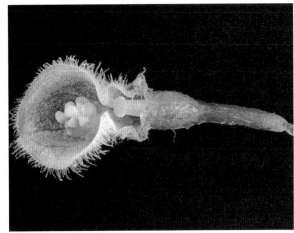

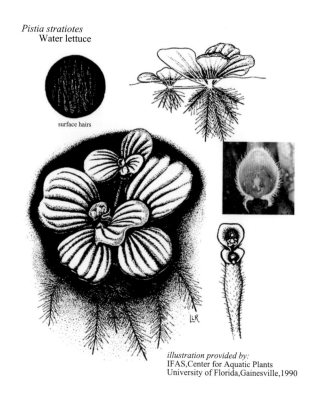

（图片选自aquaworldbg.com，rybicky.net，i83.servimg.com，www.visoflora.com，www.mafengwo.cn，www.kalake.ee，plants.ifas.ufl.edu，www7a.biglobe.ne.jp，www.floresefolhagens.com.br，www.ebay.co.uk，www.q-bank.eu，idtools.org，upload.wikimedia.org）

大藻（*Pistia stratiotes*）

【www.efloras.org】*Pistia stratiotes* Linnaeus, Sp. Pl. 2: 963. 1753.

Water-lettuce

Pistia spathulata Michaux

Roots to 50 cm, with short branches. Leaves light green to grayish green, 2-15 (-20) cm, spongy, pubescence dense, white; major veins 5-13 (-15), nearly parallel, abaxially prominent. Inflorescences: spathe white to pale green, convolute basally, slightly constricted above middle, spreading apically, pwhite to pale green, pubescent outside, glabrous inside; spadix adnate to spathe more than 1/2 its length, shorter than spathe; axis naked at base of staminate part and sometimes extending beyond staminate flowers. Flowers: staminate flowers (2-) 6-8, in single whorl around central stalk, stamens 2, connate; pistillate flower solitary; ovariesy 1-locular, 4-5 mm; ovules 4-15 (-20), orthotropous; styles ca. 3 mm; stigmas obtuse, with small hairs. ; staminate flowers (2-) 6-8, in single whorl around central stalk; stamens 2, connate. Fruits with thin pericarp. Seeds light brown, cylindric, 2-1 mm. $2n = 28$ (India, Borneo).

【www.efloras.org】*Pistia stratiotes* Linnaeus, Sp. Pl. 2: 963. 1753.

Apiospermum obcordatum Klotzsch, nom. illeg. superfl.; *Pistia obcordata* Schleiden, nom. illeg. superfl.

Morphological characters are the same as those of the genus. Plants aquatic, floating. Roots many, pendulous in water, feathery. Leaves in rosettes, 1.3-10 cm × 1.5-6 cm. Spathe white, 5-12 mm. Fl. May-Nov.

【www.cabi.org】*Pistia stratiotes* (water lettuce)

P. stratiotes is a free-floating, stoloniferous plant with sessile leaves in rosettes. Leaves pale-green, up to 20 cm long and 10 cm wide, mostly spathulate to broadly obovate with a rounded to truncate apex, with 7-15 prominent veins radiating fanwise from the base; both surfaces, in particular the lower surface, covered by a dense mat of white woolly hairs (Cook et al., 1974; Aston, 1977; Holm et al., 1977; Sainty and Jacobs, 1981). Inflorescence axillary, solitary, ascending; spathe 1.3-1.5 cm long, convolute and adnate to the spadix below, spreading above, whitish; spadix with a single pistillate flower at base, and with 2-8 staminate flowers above, shorter than the spathe. Flowers unisexual, the perianth wanting; stamens 2; ovary 1-locular, with numerous ovules, the style slender, the stigma penicillate. Fruit thin-walled, many-seeded. Seeds cylindrical, rugulose. (Acevedo-Rodríguez and Nicolson, 2005)

The morphology of Pistia varies largely owing to the influence of environmental factors. In a survey of two populations in ponds of distinct hydrochemical characteristics, two biotypes were identified that propagate true. The biotypes were distinct regarding biomass, productivity allocation, pH of the cell saps, chlorophyll, nucleic acids, total free amino acid content of the leaves and total nitrogen, crude protein and phosphorus in whole plants (Rao and Reddy, 1984). The leaves rise into the air, but under conditions less favourable for optimal growth they may lie flat on the water.

【en.wikipedia.org】*Pistia stratiotes*

It is a perennial monocotyledon with thick, soft leaves that form a rosette. It floats on the surface of the water, its roots hanging submersed beneath floating leaves. The leaves can be up to 14 cm long and have no stem. They are light green, with parallel veins, wavy margins and are covered in short hairs which form basket-like structures which trap air bubbles, increasing the plant's buoyancy. The flowers are dioecious, and are hidden in the middle of the plant amongst the leaves. Small green berries form after successful fertilization. The plant can also undergo asexual reproduction. Mother and daughter plants are connected by a short stolon, forming dense mats.

【nas.er.usgs.gov】*Pistia stratiotes* L.

Stem/Roots: *Pistia stratiotes* is a free-floating, herbaceous monocot with a rosette of gray-green leaves, resembling a head of lettuce (thus the common name), occurring as a single plant or connected to others by stolons (Dressler et al., 1987; Langeland and Burks 1998).

Roots: numerous and feathery.

Leaves: Leaves are ovate to obovate, up to 15 cm in length, without a leaf stalk, spongy near the leaf base, densely pubescent, with deeply furrowed parallel veins and wavy leaf margins (Godfrey and Wooten 1981; Dressler et al., 1987; Langeland and Burks 1998).

Flowers: Flowers inconspicuous, perfect, clustered in leaf axils with a single female flower and multiple male flowers (Langeland and Burks 1998).

Fruit/Seeds: Produces abundant seeds with high percentage of seed viability (Dray and Center 1989a, 1989b). Look-a-likes: none.

Size: Rosette generally 6 to 30 cm in diameter (Godfrey and Wooten 1981).

加拿大一枝黄花

学　　名：***Solidago canadensis*** L.

异　　名：*Solidago altissima* var. *gilvocanescens*（Rydb.）Semple，*Solidago dumetorum* Lunell，*Solidago elongata* Nutt.，*Solidago gilvocanescens*（Rydb.）Smyth，*Solidago lepida* DC.，*Solidago pruinosa* Greene

别　　名：黄莺、麒麟草

英文名：Canadian goldenrod，lechuguilla，vara de oro del Canadá，vara de San Jose

| 形态特征 |

加拿大一枝黄花为多年生草本植物。

植物有长的地下根状茎。

植物茎直立，植株高0.25～2.5m。

植物秆粗壮，中下部直径可达2cm，下部一般无分枝，常成紫红色。

植物叶披针形或线状披针形，长5～12cm，互生，顶渐尖，基部楔形，近无柄，大多呈三出脉，边缘具锯齿。

头状花序很小，长4～6mm，在花序分枝上单面着生，多数弯曲的花序分枝与单面着生的头状花序，形成开展的圆锥状花序。总苞片线状披针形，长3～4mm。边缘舌状花很短。蝎尾状圆锥花序，长10～50cm，具向外伸展的分支。

花期9～10月。果期10～11月。

| 生物危害 |

加拿大一枝黄花具有极强的竞争优势，主要表现在：第一，繁殖能力强，无性有性繁殖相结合；第二，传播能力强，远近结合；第三，生长期长，在其他秋季杂草枯萎或停止生长的时候，加拿大一枝黄花依然茂盛，花黄叶绿，而且地下根茎继续横走，不断蚕食其他杂草的领地，而此时其他杂草已无力与之竞争。这三个特点使得它对所到之处的本土物种造成严重威胁，并可能最终形成加拿大一枝黄花单一生长区。

另外，由于加拿大一枝黄花的根部能分泌生物抑制物质。这种物质可以抑制糖槭幼苗生长，也抑制其他草本植物发芽。从而破坏本地生态平衡，威胁生物多样性。据上海植物专家统计，加拿大一枝黄花20世纪80年代逃逸后成为恶性杂草，近几十年来，已导致30多种乡土植物物种消亡，对我国社会经济、自然生态系统和生物多样性构成了巨大威胁。

| 地理分布 |

原产于北美。1935年，作为观赏植物引入上海、南京等地。20世纪80年代逸生为恶性杂草。现分布于浙江、上海、安徽、湖北、湖南郴州、江苏、江西等省市。为全球性入侵物种。

| 传播途径 |

加拿大一枝黄花可以借根状茎和种子两种方式进行繁殖。因此，可以通过根状茎和种子进行人为传播或自然扩散。

| 管理措施 |

综合治理（Integrated Pest Management，IPM）或综合控制（Integrated Pest Control，IPC），但不包括植物卫生措施（Phytosanitary Measures）。

（图片选自www.nwvisualplantid.com，pestid.msu.edu，www.sevenoaksnativenursery.com，upload.wikimedia.org，luirig.altervista.org，image.baidu.com，familist.ro，ontariowildflowers.com，www.plantarium.ru，cn.bing.com，www.about-garden.com，commons.wikimedia.org，plants-of-styria.uni-graz.at，actaplantarum.org，wikis.evergreen.edu）

加拿大一枝黄花（*Solidago canadensis*）

【www.efloras.org】 *Cenchrus echinatus* Linnaeus, Sp. Pl. 2: 1050. 1753.

Herbs, perennial; rhizomes creeping, branched. Stems to 150 cm tall, erect, simple, shortly and softly downy above. Leaves numerous, lanceolate or linear-lanceolate, 5-12 cm, abaxially downy, veins sparsely hairy, adaxially shortly pilose, tapering at both ends, margin of basal sometimes entire, of lower and upper cauline sharply serrate, longitudinal veins 3（triplinerved）, of which 2 lateral veins protrude weakly, apex acuminate. Capitula in paniculiform synflorescences, branches（racemes）curved downward, capitula attached on upper side of branch. Involucre 2.5-3 mm; phyllaries linear-lanceolate, slightly obtuse. Florets golden yellow; ray florets hardly longer than involucre. Pappus inner（longest）bristles not obviously clavate. Fl. Aug-Sep.

【www.efloras.org】 *Solidago canadensis* Linnaeus, Sp. Pl. 2: 878. 1753.

Aster canadensis（Linnaeus）Kuntze

Plants 30-150（-200）cm; rhizomes short to long creeping.

Stems 1-20+, erect, glabrate proximally or sparsely strigoso-villous, becoming more densely so distal to mid stem.

Leaves: basal 0; proximal to mid cauline usually withering by flowering, tapering to sessile bases, blades narrowly ovate-lanceolate, 50-190 mm × 5-30 mm, margins sharply serrate, 3-nerved, apices acuminate, abaxial faces glabrous or more commonly hairy along main nerves, adaxial glabrous or slightly scabrous; mid to distal similar, 30-50（-120）mm × 8-12 mm, largest near mid stem, reduced distally, margins usually serrate or serrulate（teeth 3-8）, sometimes entire proximal to arrays.

Heads（70-）150-1 300+, secund, in secund pyramidal-paniculiform arrays（obscurely so and club-shaped thyrsiform in small plants or shoots with small arrays）, branches divergent and recurved, branches and peduncles hairy.

Peduncles 3-3.4 mm, bracteoles 0-3, linear-triangular.

Involucres narrowly campanulate, 1.7-2.5（-3）mm.

Phyllaries in 3-4 series, strongly unequal, acute to obtuse; outer lanceolate, inner linear-lanceolate.

Ray florets（5-）8-14（-18）; laminae 0.5-1.5 mm × 0.15-0.3（-0.5）mm.

Disc florets（2-）3-6（-8）; corollas 2.2-2.8（-3）mm, lobes 0.4-0.8（-1）mm.

Cypselae（narrowly obconic）1-1.5 mm（ribbed）, sparsely strigose;

pappi 1.8-2.2 mm.

【www.cabi.org】 *Solidago canadensis*（Canadian goldenrod）

Herbaceous，Perennial，Seed propagated，Vegetatively propagated plant

S. canadensis is a 25-250 cm（mean 150 cm）tall, erect rhizomatous perennial with annual aboveground shoots and persistent belowground rhizomes. One to several rhizomes emerge near the base of the dying shoots in autumn, thus leading to a branched rhizome system rooted mainly at the old and current shoot bases. Each rhizome has the potential to produce a single aerial stem arising from the apex of the rhizome in the following spring. Roots arise from the shoot base and reach a minimum depth of 20 cm. Stems are branched only in the inflorescence, glabrous at the base, weakly to densely pubescent at least in the upper half and often reddish. Plants of var. *scabra* have nodding shoot tips during growth. Leaves are triple-nerved, pubescent beneath, lanceolate, often acuminate, with margins mostly serrate, occasionally entire. Inflorescences form broad pyramidal panicles with recurving branches and a central axis. Bracts of the involucre are linear, obtuse or somewhat acute. Ray florets are lemon yellow, female and fertile, disc florets bisexual and fertile. The corolla is 2.4-2.8 mm long. Achenes are pubescent, 0.9-1.2 mm long, with a pappus of 2.0-2.5 mm.

【 www.missouribotanicalgarden.org 】 *Solidago canadensis*

Central stems are clad with numerous, narrow, alternate, lance-shaped, sharply-toothed, stalkless to short-stalked green leaves (to 6 feet long and 1 feet wide) which are hairless above but hairy beneath and tapered at each end. Central stems are hairless near the base but soft hairy above the middle. Central stems rise to 4-5 inches (less frequently to 7 inches) tall and are topped in late summer to fall (August to October) with large horizontally branched terminal pyramidal panicles containing one-sided recurving branches filled with masses of tiny yellow flowers (each to 1/8 feet).

蒺藜草

学　名：*Cenchrus echinatus* Linn.

异　名：*Cenchrus brevisetus* Fourn.，*Cenchrus pungens* HBK，*Cenchrus quinquevalvis* Ham. ex Wall.，*Cenchrus viridis* Spreng.

别　名：野巴夫草

英文名：southern sandbur，bur grass，hedgehog grass，mossman rivergrass，piquant cousin，sandbur grass，sandspur

| 形态特征 |

蒺藜草为一年生禾草植物。

植物秆压扁，一侧具深沟，基部曲膝状或横卧地面；近地面节上生根，下部各节常具分枝。植株高15～90cm。

植物叶鞘松弛，叶舌短小，叶片线形，质地柔软，长10～30cm，宽4～8mm，上面粗糙，无毛或疏被长柔毛。

总状花序直立，长4～8cm，刺苞球形，直径5～7mm，其裂片于中部以下连合，背部被细毛，边缘被白色纤毛，顶端具倒向糙毛，基部具一圈小刺毛，裂片直立或反曲，但彼此不相连接；每一刺苞具小穗2～4个，无柄，披针形，长5～6mm。第一颖薄膜质，卵状披针形，具一脉；第二颖卵状披针形，具3～5脉。第一小花的外稃具5脉，与小穗近等长，内稃狭长，长与外秆近等；第二小花的外稃质地较厚，具5脉，内稃稍短，成熟时质地变硬。花药长约1mm，柱头帚刷状。

颖果椭圆状扁球形，背腹压扁。

| 生物危害 |

蒺藜草为我国检疫性有害生物，可为害花生、甘薯等多种作物，是田地和果园中的一种危害严重的杂草。入侵后降低生物多样性；还可成为热带牧场中的有害杂草。

| 地理分布 |

蒺藜草原产于美洲热带和亚热带地区。1934年在我国台湾兰屿采到该入侵物种标本。现已经分布于我国福建、台湾、广东、香港、广西和云南南部等地。

| 传播途径 |

蒺藜草的刺苞具多数微小的倒刺，可附着在衣服、动物皮毛和货物上传播。种子常在刺苞内萌发。

| 管理措施 |

综合治理（Integrated Pest Management，IPM）或综合控制（Integrated Pest Control，IPC），包括植物卫生措施（Phytosanitary Measures）。

外来入侵 植物

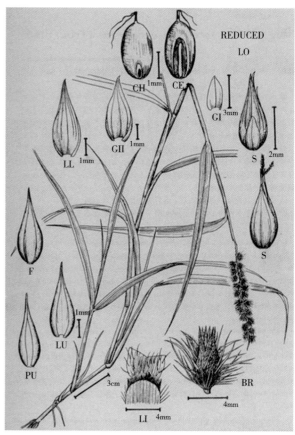

（图片选自www.fmcagricola.com.br，upload.wikimedia.org，www.conabio.gob.mx，keyserver.lucidcentral.org，gardenbreizh.org，www.backyardnature.net，www.agrokurier.de，flowers.la.coocan.jp，luirig.altervista.org，flora.huh.harvard.edu）

蒺藜草（*Cenchrus echinatus*）

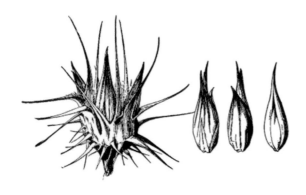

【 www.efloras.org 】 *Cenchrus echinatus* Linnaeus, Sp. Pl. 2: 1050. 1753.

Annual. Culms geniculate, usually rooting at basal nodes, 15-90 cm tall. Leaf sheaths keeled, usually imbricate at base; leaf blades linear or linear-lanceolate, 5-20 (-40) cm × 0.4-1 cm, glabrous to pubescent; ligule ca. 1 mm. Inflorescence 3-10 × ca. 1 cm, burrs contiguous, rachis scabrous. Burrs globose, 0.4-1 cm, truncate, stipe pubescent, all spines and bristles retrorsely barbed; inner spines connate for 1/3-1/2 their length forming a globose cupule, the flattened free tips triangular, erect or bent inward, cupule and tips pubescent, outer spines in 2 divergent whorls, a median whorl of stout rigid spines equaling the inner teeth, and an outermost whorl of relatively few short, slender bristles. Spikelets 2-4 in burr, 4.5-7 mm; lower glume 1/2 spikelet length; upper glume 2/3-3/4 spikelet length. Fl. and fr. summer. $2n = 34, 68$.

【 www.cabi.org 】 *Cenchrus echinatus* (southern sandbur)

Erect, 30-90 cm high, forming loose tufts, lower parts of the culm sometimes prostrate, rooting at the lower nodes. The stems are usually flattened and dark green. The leaves are flat, smooth to hairy, 5-30 cm long, 3-11 mm wide, with hairs at the mouth of sheath, the sheath compressed with moderately stiff hairs on the margin of the upper part. The youngest leaf is rolled. The ligule is replaced by a ring of hairs (0.7-1.7 mm long) ; auricles are absent.

Inflorescence forms a dense cylindrical spike, 3-10 cm long, 1-2 cm wide, the rachis is strongly undulate and rough, spikelets enclosed in spinous burs; distance between individual burs is 2-3 cm. Each bur contains 2-4 spikelets, 5-7 mm long without pedicels. The burs are compressed at the base, globular, clustered, 5-10 mm long, 3.5-6 mm wide, irregular in length and thickness, the inner ones larger than the outer. The tips of the spines turn purple with increasing maturity.

【 www.cropscience.bayer 】 *Cenchrus echinatus*

Annual grass, erect, 30-90 cm (11.81-35.43 inches) high, forming loose tufts, lower parts of the culm sometimes prostrate, rooting at the lower nodes (no stolones) .

Characteristic Features Broad red leaves. The burs are reddish and broader than those of field sandbur (*C. pauciflorus*) . It grows more prostrate to the ground and also roots at the node and forms a mat.

Young Plant Leaves are folded in a bud. Ligule is a fringe of hairs. The blade is rough with long hairs at the base near the ligule and collar.

Stems Usually flattened and dark green.

Leaves The leave blades are flat, smooth to hairy, 5-30 cm (1.96-11.81 inches) long, 3-11 mm (0.12- 0.43 inch) wide. The youngest leaf is rolled. The ligule is replaced by a ring of hairs (0.7-1.7 mm, 0.027-0.067 inch) long. Auricles are absent. With hairs at the mouth of sheath, the sheath compressed with moderately stiff hairs on the margin of the upper part.

Propagation Organs Flowers The inflorescence forms a dense cylindrical spike, 3-10 cm (1.18-3.93 inches) long, 1-2 cm (0.39-0.78 inch) wide. The rachis is strongly undulate and rough. Spikelets enclosed in spinous burs. The distance between individual burs is 2-3 cm (0.78-1.18 inch) . Each bur contains 2-4 spikelets, 5-7 mm (0.19-0.27 inch) long without pedicels. The burs are compressed at the base, globular, clustered, 5-10 mm (0.19-0.39 inch) long, 3.5-6 mm (0.14-0.23 inch wide, irregular in length and thickness, the inner ones larger than the outer. The tips of the spines turn purple with increasing maturity.

Fruit Burlike fruit, each spike has 5-20 burs.

长刺蒺藜草

学　　名：**_Cenchrus spinifex_** Cavanilles

异　　名：_Cenchrus carolinianus_ Walter，_Cenchrus echinatus_（Linnaeus）_forma longispinus_（Hackel），_Cenchrus incertus_ M. A. Curtis，_Cenchrus longispinus_（Hackel）Fernald，_Cenchrus microcephalus_ Nash ex Hitchcock & Chase，_Cenchrus pauciflorus_ Bentham，_Cenchrus pauciflorus_ Bentham, var. _longispinus_（Hackel）Jansen & Wachter，_Cenchrus strictus_ Chapman，_Cenchrus tribuloides_ L.（misapplied）

别　　名：草蒺藜

英 文 名：coastal sandbur，field sandbur

| 形态特征 |

长刺蒺藜草为一年生草本植物。

植株丛生，具须根。株高20～90cm。

植物茎压扁，节上生根，下部各节分枝。植物秆扁圆形，中空，有时外倾呈匍匐状，常自基部分枝。

植物叶片长4～27cm，宽1.5～5（7.5）mm，上面粗糙，下面无毛。叶舌长0.6～1.8mm。叶鞘扁平，除鞘口缘毛外，其余无毛。

穗形总状花序，长1.5～8（10）cm。小穗2～3（4）枚簇生成束，其外围由不孕小枝愈合形成刺苞。刺苞近球形，长8.3～11.9mm，宽3.5～6mm。刺苞具刺45～75枚。外轮刺多数，常为刚毛状，有时反折，比内轮刺短；内轮刺10～20枚，钻形，长3.5～7mm，基部宽0.5～0.9（1.4）mm。刺苞及刺的下部具柔毛。小穗卵形，无柄，长（4）5.8～7.8mm，宽2.5～2.8mm。第一颖长0.8～3mm，第二颖长4～6mm，具3～5脉。第一小花常雄性，外稃长4～6.5mm，具3～7脉，花药长1.5～2mm；第二小花外稃质硬，背面平坦，顶端尖，长4～7（7.6）mm，具5脉，花药长0.7～1mm。

颖果卵形，长2～3.8mm，宽1.5～2.6mm，黄褐色或黑褐色，包藏于刺苞内。

花期为5月。果期为6月。

| 生物危害 |

长刺蒺藜草为我国检疫性有害生物。是农田恶性杂草。危害玉米、旱稻、番薯、花生、大豆、甘蔗、棉花、苜蓿、菠萝、咖啡、可可以及葡萄等作物。

刺苞入侵草原牧场，常扎伤人畜。

| 地理分布 |

长刺蒺藜草原产于美洲。现广布于东半球。20世纪70年代分别在辽宁和北京发现。现分布于北京、山东、河北、辽宁、吉林、内蒙古等地。

| 传播途径 |

长刺蒺藜草带刺的果实可随其他作物种子等自然扩散或人为传播。

| 管理措施 |

综合治理（Integrated Pest Management，IPM）或综合控制（Integrated Pest Control，IPC），包括植物卫生措施（Phytosanitary Measures）。

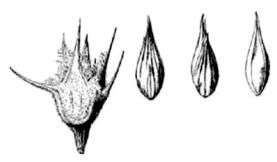

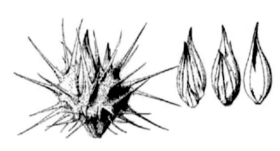

（图片选自www.weedinfo.ca，hasbrouck.asu.edu，luirig.altervista.org，www.discoverlife.org，www.ptrpest.com，www.backyardnature.ne，src.sfasu.edu，okeechobee.ifas.ufl.edu，www.wnmu.edu，delawarewildflowers.org，cn.bing.com，idtools.org，plants.usda.gov，newfs.s3.amazonaws.com）

长刺蒺藜草（*Cenchrus spinifex*）

【www.kew.org】*Cenchrus spinifex*

INFLORESCENCE Inflorescence a panicle.

Panicle Spiciform; linear; 2-8.5 cm long; 0.8-2 cm wide. Primary panicle branches accrescent to a central axis; with sessile scars on axis. Panicle axis angular; bearing deciduous spikelet clusters.

Spikelets Subtended by an involucre. Fertile spikelets sessile; 2-4 in the cluster. Involucre composed of bristles; connate into a cup below (cleft on 2 sides) ; with 2-7 mm connate; ovate, or globose; 5.5-10.2 mm long; base obconical; base glabrous, or pubescent. Involucral bristles deciduous with the fertile spikelets; emerging irregularly from body of a burr; 8-40 in principal whorl; with longest bristle scarcely emergent; 2-5 mm long; flattened; rigid; retrorsely scaberulous; pubescent; spinose.

Fertile Spikelets Spikelets comprising 1 basal sterile florets; 1 fertile florets; without rhachilla extension. Spikelets ovate; dorsally compressed; acuminate; 3.5-5.8 mm long; falling entire; deciduous with accessory branch structures.

Glumes Glumes shorter than spikelet; thinner than fertile lemma. Lower glume ovate; 1-1.3 mm long; 0.33-0.5 length of spikelet; membranous; without keels; 1 -veined. Lower glume lateral veins absent. Lower glume apex obtuse, or acute. Upper glume ovate; 2.8-5 mm long; 0.75-0.9 length of spikelet; membranous; without keels; 5-7 -veined. Upper glume apex obtuse, or acute.

Florets Basal sterile florets barren; with palea. Lemma of lower sterile floret ovate; 1 length of spikelet; membranous; 5-7 -veined; acute. Fertile lemma ovate; 3.4-5.8 mm long; coriaceous; much thinner on margins; without keel; 3 -veined. Lemma margins flat. Lemma apex acute. Palea coriaceous.

Flower Anthers 3; 0.5-2 mm long.

Fruit Caryopsis with adherent pericarp; ovoid; dorsally compressed; 1.3-3 mm long.

【en.wikipedia.org】*Cenchrus spinifex*

Cenchrus spinifex, is a perennial grass that grows from 5 to 30 inches (13 to 76 cm) high in sandy or gravelly terrain. It is found throughout the southern United States southward into Mexico and the Caribbean, and also grows in the Philippines and South Africa. It is a noxious weed in Europe where it was introduced.

The grass produces a bur, a type of grain fruit, consisting of eight to forty sharp, barbed spines that lodge in clothes, exposed feet, and fur.

银胶菊

学　　名：*Parthenium hysterophorus* L.
异　　名：*Argyrochaeta bipinnatifida* Cav., *Parthenium hysterophorus* var. *lyratum* A.Gray, *Parthenium lobatum* Buckl., *Villanova binnatifida* Ortega
别　　名：银色橡胶菊、西南银胶菊、野益母艾、野益母岩
英文名：parthenium weed, barley flower, bastard feverfew, broomweed, congress grass, congress weed, dog flea weed, mugwort, Santa Maria feverfew, whiteheads, wormwood

| 形态特征 |

银胶菊为一年生草本植物。

植物茎直立，基部径约5mm，多分枝，具条纹，被短柔毛，节间长2.5~5cm。植株高0.3~1.2m。

植物下部和中部叶二回羽状深裂，全形卵形或椭圆形。连叶柄长10~19cm，宽6~11cm。羽片3~4对，卵形，长3.5~7cm，小羽片卵形或长圆形，常具齿，顶端略钝，上面被基部为疣状的疏糙毛，下面的毛较密而柔软；上部叶无柄，羽裂，裂片线状长圆形，全缘或具齿，或指状3裂，中裂片较大，通常长于侧裂片的3倍。

头状花序多数，径3~4mm。在茎枝顶端排成开展的伞房花序，花序柄长3~8mm，被粗毛。

总苞宽钟形或近半球形，径约5mm，长约3mm。总苞片2层，各5个，外层较硬，卵形，长2.2mm，顶端叶质，钝，背面被短柔毛，内层较薄，几近圆形，长宽近相等，顶端钝，下凹，边缘近膜质，透明，上部被短柔毛。

舌状花1层，5个，白色，长约1.3mm，舌片卵形或卵圆形，顶端2裂。管状花多数，长约2mm，檐部4浅裂，裂片短尖或短渐尖，具乳头状突起。雄蕊4个。

瘦果倒卵形，基部渐尖，干时黑色、长约2.5mm，被疏腺点。冠毛2，鳞片状，长圆形，长约0.5mm，顶端截平或有时具细齿。

花期4~10月。

该种是菊科银胶菊属植物中唯一产胶的植物。并因其叶具银灰色而得名。

| 生物危害 |

银胶菊属于恶性杂草，对其他植物有化感作用，还可引起人和家畜（牛）的过敏性皮炎。银胶菊外表的微细状体含有银胶素，人吸入过多可能造成肝脏及遗传病变，被视为有毒杂草。

此外银胶菊的花粉也有毒，会造成过敏、支气管炎。直接接触大量花粉会引起皮肤发炎、红肿。在澳洲、印度，都曾发现牛羊等牲畜，因接触大量银胶菊而中毒死亡的案例。

| 地理分布 |

银胶菊原产于美国德克萨斯州及墨西哥北部。1924年，该种植物在越南北部被报道。1926年，在云南采到此植物标本。我国现分布于云南、贵州、广西、广东、海南、香港和福建等地。为全球性入侵物种。

| 传播途径 |

银胶菊的种子和幼苗均可以通过人为或自然因素进行传播。

| 管理措施 |

综合治理（Integrated Pest Management，IPM）或综合控制（Integrated Pest Control，IPC），但不包括植物卫生措施（Phytosanitary Measures）。

中国外来入侵 物种图鉴

外来入侵植物

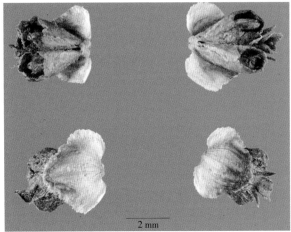

2 mm

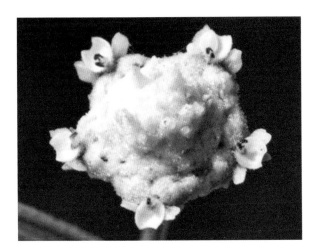

1–3.银胶菊 Parthenium hysterophorus L. 1.花枝；2.头花；3.果。
4–5.刺苞果 Acanthospermum australe (L.) Kuntze. 4.植株上部；5.果。（吴彰桦、王金凤绘）

（图片选自en.wikipedia.org，www.invasives.org.za，keys.lucidcentral.org，baike.baidu.com，pic.sogou.com，newfs.s3.amazonaws.com，delange.org，nathistoc.bio.uci.edu，www.redorbit.com，blog.growingwithscience.com，www.zhiwutong.com）

银胶菊（*Parthenium hysterophorus*）

【 www.efloras.org 】 *Parthenium hysterophorus* Linnaeus, Sp. Pl. 2: 988. 1753.

Herbs, annual, 30-120 cm tall. Leaf blade ovate to elliptic, 3-18 cm × 1-5（-9）cm, pinnately（1 or）2-lobed, ultimate lobes lanceolate to linear, 3-50 mm × 2-15 mm, both surfaces sparsely to densely scaberulose and gland-dotted. Synflorescences of open panicles. Capitula obscurely radiate; peduncles 1-8（-15+）mm; outer phyllaries 5（or 6）, elliptic-lanceolate, 2-4 mm, inner 5（or 6）ovate to orbicular, 2.5-4 mm. Female florets 5（or 6）; corolla limbs reniform or orbicular to oblong, 0.3-1 mm. Disk florets 12-30（-60）. Achenes obovoid, 1.5-2（-3.5）mm; pappuslike enations erect, deltate to ovate, 0.5-1 mm. Fl. Apr-Aug. $2n = 34$.

【 www.efloras.org 】 *Parthenium hysterophorus* Linnaeus, Sp. Pl. 2: 988. 1753.

Annuals,（10-）30-120+ cm.

Leaf blades ovate to elliptic, 30-180+ × 10-50（-90+）mm,（1-）2-pinnately lobed（ultimate lobes lanceolate to linear, 3-50 mm × 2-15 mm）, faces sparsely to densely scabrellous and gland-dotted（seldom with additional erect hairs 1-2 mm）.

Heads obscurely radiate, borne in open, paniculiform arrays.

Peduncles 1-8（-15+）mm.

Phyllaries: outer 5（-6）, lance-elliptic, 2-4 mm, inner 5（-6）ovate to ± orbiculate, 2.5-4 mm.

Pistillate florets 5（-6）; corolla laminae reniform or orbiculate to oblong, 0.3-1 mm.

Disc florets 12-30[-60].

Cypselae obovoid, 1.5-2（-3.5）mm; pappus-like enations erect, deltate to ovate, 0.5-1 mm（sometimes a third, subulate spur near apex adaxially）.

$2n = 34$.

【 www.cabi.org 】 *Parthenium hysterophorus*（parthenium weed）

Annual，Broadleaved，Herbaceous，Seed propagated plant

P. hysterophorus is an erect, much-branched with vigorous growth habit, aromatic, annual（or a short-lived perennial）, herbaceous plant with a deep taproot. The species reproduces by seed. In its neotropical range it grows to 30-90 cm in height（Lorenzi, 1982; Kissmann and Groth, 1992）, but up to 1.5 m, or even 2.5 m, in exotic situations（Haseler, 1976; Navie et al., 1996）. Shortly after germination the young plant forms a basal rosette of pale green, pubescent, strongly dissected, deeply lobed leaves, 8-20 cm in length and 4-8 cm in width. The rosette stage may persist for considerable periods during unfavourable conditions（such as water or cold stress）. As the stem elongates, smaller, narrower and less dissected leaves are produced alternately on the pubescent, rigid, angular, longitudinally-grooved stem, which becomes woody with age. Both leaves and stems are covered with short, soft trichomes, of which four types have been recognized and are considered to be of taxonomic importance within the genus（Kohli and Rani, 1994）.

Flower heads are both terminal and axillary, pedunculate and slightly hairy, being composed of many florets formed into small white capitula, 3-5 mm in diameter. Each head consists of five fertile ray florets（sometimes six, seven or eight）and about 40 male disc florets. The first capitulum forms in the terminal leaf axil, with subsequent capitula occurring progressively down the stem on lateral branches arising from the axils of the lower leaves. Thousands of inflorescences, forming in branched clusters, may be produced at the apex of the plant during the season. Seeds（achenes）are black, flattened, about 2 mm long, each with two thin, straw-coloured, spathulate appendages (sterile florets) at the apex which act as air sacs and aid dispersal.

【keys.lucidcentral.org】*Parthenium hysterophorus* (Parthenium Weed)

Parthenium hysterophorus is a much-branched, short-lived (annual), upright (erect) herbaceous plant that forms a basal rosette of leaves during the early stage of growth. It usually grows 0.5-1.5 m tall, but can occasionally reach up to 2 m or more in height.

Mature stems are greenish and longitudinally grooved, covered in small stiff hairs (hirsute), and become much branched at maturity.

Alternately arranged leaves are simple with stalks (petioles) up to 2 cm long and form a basal rosette during the early stages of growth. The lower leaves are relatively large (3-30 cm long and 2-12 cm wide) and are deeply divided (bi-pinnatifid or bi-pinnatisect). Leaves on the upper branches decrease in size and are also less divided than the lower leaves. The undersides of the leaves, and to a lesser degree their upper surfaces, are covered with short, stiff hairs that lie close to the surface (they are appressed pubescent).

Numerous small flower-heads (capitula) are arranged in clusters at the tips of the branches (in terminal panicles). Each flower-head (capitulum) in borne on a stalk (pedicel) 1-8 mm long. These flower-heads (4-5 mm across) are white or cream in colour and have five tiny 'petals' (ray florets) 0.3-1 mm long. They also have numerous (12-60) tiny white flowers (tubular florets) in the centre and are surrounded by two rows of small green bracts (an involucre). Colour changes to light brown when seeds are mature and about to shed. Flowering can occur at any time of the year, but is most common during the rainy seasons.

Five small 'seeds' (achenes) are usually produced in each flower-head (capitulum). These achenes (1.5-2.5 mm long) consist of a black seed topped with two or three small scales (a pappus) about 0.5-1 mm long, two straw-coloured papery structures (actually dead tubular florets), and a flat bract.

Similar Species Parthenium weed (*Parthenium hysterophorus*) can be confused with annual ragweed (*Ambrosia artemisiifolia*), perennial ragweed (*Ambrosia psilostachya*), burr ragweed (*Ambrosia confertiflora*) and lacy ragweed (*Ambrosia tenuifolia*) when in the vegetative stage of growth. However, parthenium weed (*Parthenium hysterophorus*) can be distinguished from all these species by its ribbed stems, and also by white flower-heads (i.e. capitula) when it is in flower.

The fleabanes (e.g. *Conyza bonariensis*, *Conyza canadensis*, and *Conyza sumatrensis*) are also reasonably similar, but do not have highly dissected leaves or ribbed stems, and their seeds are topped with a ring (i.e. pappus) of whitish hairs.

【gobotany.newenglandwild.org】*Parthenium hysterophorus* L.

【www.agric.wa.gov.au】Parthenium weed: declared pest

Appearance An erect aromatic annual herb 30-150 centimetres tall, with a deep tap root, reproducing by seed. Native to tropical North and South America. Parthenium weed is a Weed of National Significance (WoNS).

Stems: Erect, much-branched in the upper half, hairy, longitudinally grooved and becoming woody with age.

Leaves: Pale green, fern-like in appearance, alternate, shortly hairy with some of the hairs containing allergy-causing substances. Rosette leaves are deeply lobed, 8-20 centimetres long and four to five centimetres wide. Stem leaves are shorter and less divided.

Flowers: White florets, in compact daisy-like heads 4-10 millimetres diameter, which are grouped together in many-branched clusters arising from the stem nodes and all terminating about the same height. Each head contains about 45 florets.

Seed: Black, two millimetres long, flattened with two white spoon-shaped appendages at the apex.

Root: A deep tap root with many finely branched feeding roots.

An aggressive invader of disturbed land and perennial pastures and roadsides. Very prevalent in drier parts of Queensland and New South Wales, but absent from WA. Causes dermatitis and other allergic reactions to humans and cattle.

黄顶菊

学　　名：***Flaveria bidentis***（L.）Kuntze.

异　　名：*Ethulia bidentis* Linnaeus，*Flaveria bonariensis* DC.，*Flaveria capitata* Juss.，*Flaveria chilensis* Juss.，*Flaveria chilensis*（Molina）J. F. Gmel.，*Flaveria contrayerba*（Cav.）Pers.，*Milleria chiloensis* Ruiz & Pav. ex Juss.，*Milleria contrayerba* Cav.，*Vermifuga corumbosa* Ruiz & Pav.，*Vermifuga corymbosa* Ruiz & Pav.

别　　名：南美黄顶菊、野菊花

英文名：Coastalplain Yellowtops，Coastal Plain Flaveria

| 形态特征 |

黄顶菊为一年生草本植物。

植物茎直立，常带紫色，具有数条纵沟槽。茎带有短绒毛。植株高低差异很大，株高20～100cm，条件适合的地方株高可达180～250cm，最高可达到3m。

植物叶交互对生，长椭圆形，长6～18cm、宽2.5～4cm；叶边缘有稀疏而整齐的锯齿，基部生3条平行叶脉。

主茎及侧枝顶端上有密密麻麻的黄色花序。一朵花就是一个头状花序。其头状花序多数于主枝及分枝顶端密集成蝎尾状，它是由很多个只有米粒大小的花朵组成，每一朵花可以产生一粒瘦果，无冠毛。

花冠鲜艳，花鲜黄色，极为醒目。

黄顶菊的种子极其多。一粒果实中有一粒种子。种子为黑色，极小，每粒大小仅1～3.6mm。一株黄顶菊最多可结种子12万粒。

花果期为夏季至秋季。

| 生物危害 |

黄顶菊为我国检疫性有害生物。其根系发达，最高可以长到3m。在与周围植物争夺阳光和养分的竞争中，严重挤占其他植物的生存空间，影响其他植物生长。特别是黄顶菊的根产生的分泌物，能抑制其他植物生长，最终导致植物死亡，对绿地生态系统有极大的破坏性，使得许多生物物种灭绝。黄顶菊一旦入侵农田，将威胁整个农牧业生产及生态安全，因此又有"生态杀手"之称。

在生长过黄顶菊的土壤里种植小麦、大豆等作物，其发芽能力会变得非常低。

黄顶菊的花期很长，花粉量大，花期与大多数土著菊科植物重叠交叉。如果黄顶菊与其发生区域内的其他土著菊科植物天然杂交，就可能形成新的危害性更大的植物杂交品种。

| 地理分布 |

原产于南美洲巴西、阿根廷等国，后扩散到美洲中部、北美洲南部及西印度群岛。再后来，由于引种等原因而传播到埃及、南非、英国、法国、澳大利亚和日本等地。

2000年发现于天津南开大学校园。现分布于天津、河北等地。为全球性入侵物种。

| 传播途径 |

可自然扩散或人为传播。

| 管理措施 |

综合治理（Integrated Pest Management，IPM）或综合控制（Integrated Pest Control，IPC），包括植物卫生措施（Phytosanitary Measures）。

(图片选自asb.com.ar、sd.ifeng.com、www.he.xinhua.org、publish.plantnet-project.org、upload.wikimedia.org、baike.baidu.com、sr.yuanlin.com、www.yumpu.com)

黄顶菊(*Flaveria bidentis*)

【www.efloras.org】 *Flaveria bidentis* (Linnaeus) Kuntze, Revis. Gen. Pl. 3 (3): 148. 1898.
Coastal plain yellowtops

Ethulia bidentis Linnaeus, Mant. Pl. 1: 110. 1767

Annuals, to 100 cm (delicate or robust, sparsely villous). Stems erect. Leaves petiolate (proximal, petioles 3-15 mm) or sessile (distal); blades lanceolate-elliptic, 50-120 (-180) mm × 10-25 (-70) mm, bases (distal) connate, margins serrate or spinulose serrate. Heads 20-100+ in tight subglomerules in scorpioid, cymiform arrays. Calyculi of 1-2 linear bractlets 1-2 mm. Involucres oblong-angular, 5 mm. Phyllaries 3 (-4), oblong. Ray florets 0 or 1; laminae pale yellow, ovate-oblique, to 1 mm (not or barely surpassing phyllaries). Disc florets (2-) 3-8; corolla tubes ca. 0.8 mm, throats funnelform, 0.8 mm. Cypselae oblanceolate or subclavate, 2-2.5 mm (those of ray florets longer); pappi 0. $2n = 36$.

【www.efloras.org】 *Flaveria bidentis* (Linnaeus) Kuntze, Revis. Gen. Pl. 3 (3): 148. 1898.

Ethulia bidentis Linnaeus, Syst. Nat., ed. 12, 2: 536; Mant. Pl. 1: 110. 1767.

Annuals. Stems erect, to 100 cm tall, sparsely villous. Leaves petiolate (proximal, petioles 3-15 mm) or sessile (distal); blades lanceolate-elliptic, 50-120 (-180) mm × 10-25 (-70) mm, bases (distal) connate, margins serrate or spinulose serrate. Capitula 20-100+ in tight subglomerules in scorpioid cymes; calycular bracts 1 or 2, 1-2 mm; involucres oblong-angular, ca. 5 mm; phyllaries 3 (or 4), oblong. Ray florets 0 or 1; lamina pale yellow, obliquely ovate, to 1 mm (not or barely surpassing phyllaries). Disk florets (2 or) 3-8; corolla tubes ca. 0.8 mm, throats funnelform, ca. 0.8 mm. Achenes oblanceolate or subclavate, 2-2.5 mm (those of ray florets longer); pappus absent. Fl. Jul-Nov. $2n = 36$.

【en.hortipedia.com】 *Flaveria bidentis*

Growth The plants reach heights of 60 to 100 centimetres.

Leaves Flaveria bidentis has fern-green, simple leaves that are opposite. The leaves are lanceolate, serrulate and petiolate.

Flowers and Fruits Flaveria bidentis produces umbels of ligth-yellow many-stellate flowers. The plants produce achenes.

【plants.jstor.org】 *Flaveria bidentis*

Annuals, to 100 cm (delicate or robust, sparsely villous). Stems erect. Leaves petiolate (proximal, petioles 3-15 mm) or sessile (distal); blades lanceolate-elliptic, 50-120 (-180) mm × 10-25 (-70) mm, bases (distal) connate, margins serrate or spinulose serrate. Heads 20-100+ in tight subglomerules in scorpioid, cymiform arrays. Calyculi of 1-2 linear bractlets 1-2 mm. Involucres oblong-angular, 5 mm. Phyllaries 3 (-4), oblong. Ray florets 0 or 1; laminae pale yellow, ovate-oblique, to 1 mm (not or barely surpassing phyllaries). Disc florets (2-) 3-8; corolla tubes ca. 0.8 mm, throats funnelform, 0.8 mm. Cypselae oblanceolate or subclavate, 2-2.5 mm (those of ray florets longer); pappi 0. $2n = 36$.

【publish.plantnet-project.org】 *Flaveria bidentis* (L.) Kuntze

F. bidentis is an erect herb, grows up to 1 m height. It is single stem plant. The lower part of the stem is purple. Leaves simple, opposite and bright. The leaves are sessile or subsessile arranged. Flower are yellow, central flowers are tubular. Fruits are in achenes. Eeach bud of seeds in five gold, achenes are black a little flat, oblanceolate or nearly rod-like, without pappus.

Stem Single stem, erect, solid and groove shaped. It is yellowish and finely ribbed.

Leaves Leaves are simple, opposite and vary in size. The blade is elliptic to lanceolate shaped, it is joint around the stem. Leaves are bright green and leaf margin is meatly serrated. The apex is acuminate and base is gradually narrow three basal veins. The veins are parallel. the majority of leaves with 0.3 to 1.5 cm in length of the petiole, the petiole nearly connate.

Inflorescence Inflorescence are clustered, axillary and terminal. It is heliconia-like cyme.

Flowers Flowers are yellow, the central flowers are tubular, 4 to 6 mm long. The corolla is bright yellow, 2.3 mm and the tube is 0.8 mm. Flowers are 5 to 15 pieces.

Fruits Fruit is an achenes of about 2.5 mm long. Each bud seeds is in five gold. Achenes are black, little flat, oblanceolate or nearly rod-like and no pappus.

土荆芥

学　名：***Dysphania ambrosioides***（L.）Mosyakin & Clemants
异　名：*Ambrina ambrosioides*（L.）Spach，*Ambrina parvula* Phil.，*Ambrina spathulata* Moq.，*Atriplex ambrosioides*（L.）Crantz，*Blitum ambrosioides*（L.）Beck，*Botrys ambrosioides*（L.）Nieuwl.，*Chenopodium ambrosioides* L.，*Chenopodium integrifolium* Vorosch.，*Chenopodium spathulatum* Sieber ex Moq.，*Chenopodium suffruticosum* subsp. *remotum* Vorosch.，*Chenopodium suffruticosum* Willd.，*Orthosporum ambrosioides*（L.）Kostel.，*Orthosporum suffruticosum* Kostel.，*Teloxys ambrosioides*（L.）W. A. Weber，*Vulvaria ambrosioides*（L.）Bubani
别　名：臭草、杀虫芥、鸭脚草、红泽兰、天仙草、钩虫草、虱子草
英文名：Mexican tea，American wormseed，bluebush，Indian goosefoot，Jerusalem-tea，Mexican tea，Spanish-tea，wormseed

| 形态特征 |

土荆芥为一年生或多年生草本植物。

植物茎直立，多分枝，具条纹，近无毛。茎下部圆柱形，粗壮光滑；上部方柱形有纵沟，具毛茸。植株高30～100cm。

植物叶互生，披针形或狭披针形，上部叶渐小而近全缘，上面光滑无毛，下面有黄色腺点，沿脉稍被柔毛。下部叶较大，长达15cm，宽达5cm，顶端渐尖，基部渐狭成短柄，边缘有不整齐的钝齿；下部叶大多脱落，仅茎梢留有线状披针形的苞片。

花夏季开放，绿色。两性或部分雌性，组成腋生、分枝或不分枝的穗状花序。花被裂片5，少有3，结果时常闭合。雄蕊5枚，突出，花药长约0.5mm。子房球形，两端稍压扁。花柱不明显，柱头3或4裂，线形，伸出于花被外。

茎梢或枝梢常见残留簇生果穗，触之即脱落，淡绿色或黄绿色。剥除宿萼，内有棕黑色的细小果实1枚。

胞果扁球形，完全包藏于花被内。

种子肾形，直径约0.7mm，黑色或暗红色，光亮。

植物整个植株有强烈异臭气，味微苦、辛，尤以茎嫩、带果穗、色黄绿者为甚。

| 生物危害 |

土荆芥在长江流域经常是杂草群落的优势种或建群种，常常侵入并威胁种植在长江大堤上的草坪。

土荆芥含土荆芥油，油中主要成分为驱蛔素、对聚伞花素及其他萜类物质，有驱虫作用。但超

量内服中毒后，可刺激消化道黏膜，对呼吸系统先兴奋后麻痹，严重时对肾脏有损害，并毒害视神经和听神经，同时抑制血管运动中枢及心肌。

| 地理分布 |

土荆芥原产于中、南美洲。现广布于各热带地和温带地。1864年，在我国台湾省台北淡水采到标本。现分布于北京、山东、陕西、上海、浙江、江西、福建、台湾、广东、海南、香港、广西、湖南、湖北、重庆、贵州、云南等地。

| 传播途径 |

自然扩散或人为传播。

| 管理措施 |

综合治理（Integrated Pest Management，IPM）或综合控制（Integrated Pest Control，IPC），但不包括植物卫生措施（Phytosanitary Measures）。

（图片选自commons.wikimedia.org，gobotany.newenglandwild.org，en.wikipedia.org，www.bing.com，botanika.wendys.cz，plantgenera.org）

土荆芥（*Dysphania ambrosioides*）

【www.efloras.org】*Dysphania ambrosioides* (Linnaeus) Mosyakin & Clemants, Ukrayins' k. Bot. Zhurn. 59: 382. 2002.

Chenopodium ambrosioides Linnaeus, Sp. Pl. 1: 219. 1753; *Ambrina ambrosioides* (Linnaeus) Spach, nom. illeg.; *Atriplex ambrosioides* (Linnaeus) Crantz; *Blitum ambrosioides* (Linnaeus) G. Beck.

Herbs annual or perennial, 50-80 cm tall, with strong odor. Stem erect, much branched, striate, obtusely ribbed; branches usually slender, pubescent and articulated villous, sometimes subglabrous. Petiole short; leaf blade oblong-lanceolate to lanceolate, abaxially with scattered glands, slightly hairy around veins, adaxially glabrous, base attenuate, margin sparsely and irregularly coarsely serrate, apex acute or acuminate; lower leaves ca. 15 × 5 cm, upper ones gradually reduced and margin subentire. Flowers borne in upper leaf axils, usually 3-5 per glomerule, bisexual and female. Perianth segments (3 or) 5, usually nearly closed in fruit. Stamens 5; anthers ca. 0.5 mm. Style obscure; stigmas 3 (or 4), filiform, exserted from perianth. Utricle enclosed by perianth, depressed globose. Seed horizontal or oblique, black or dark red, sublustrous, ca. 0.7 mm in diam., glabrous, rim margin obtuse. Fl. and fr. over a lengthy period.

【www.efloras.org】Dysphania ambrosioides (Linnaeus) Mosyakin & Clemants, Ukrayins' k. Bot. Zhurn., n. s. 59: 382. 2002. Mexican-tea, wormseed

Chenopodium ambrosioides Linnaeus, Sp. Pl. 1: 219. 1753; *C. ambrosioides* var. *suffruticosum* (Willdenow) Ascherson & Graebner; *Teloxys ambrosioides* (Linnaeus) W. A. Weber

Plants annual.

Stems erect to ascending, much-branched, 3-10 (-15) dm, ± glandular-pubescent.

Leaves aromatic, distal leaves sessile; petiole to 18 mm; blade ovate to oblong-lanceolate or lanceolate, proximal ones mostly lanceolate, 2-8 (-12) cm × 0.5-4 (-5.5) cm, base cuneate, margins entire, dentate, or laciniate, apex obtuse to attenuate, copiously gland-dotted (rarely glabrous).

Inflorescences lateral spikes, 3-7 cm; glomerules globose, 1.5-2.3 mm diam.; bracts leaflike, lanceolate, oblanceolate, spatulate, or linear, 0.3-2.5 cm, apex obtuse, acute, or attenuate.

Flowers: perianth segments 4-5, connate for ca. 1/2 their length, distinct portion ovate, rounded abaxially, 0.7-1 mm, apex obtuse, glandular-pubescent, covering seed at maturity; stamens 4-5; stigmas 3.

Achenes ovoid; pericarp nonadherent, rugose to smooth.

Seeds horizontal or vertical, reddish brown, ovoid, 0.6-1 mm × 0.4-0.5 mm; seed coat rugose to smooth.

【www.cabi.org】*Dysphania ambrosioides* (Mexican tea)

Annual, Herbaceous, Seed propagated plant

Erect subshrub to 1 m tall, with strong, fetid smell, many-branched from a woody base; stem ribbed to cylindrical, more or less pubescent. Leaf blades 2-9 mm × 0.6-3.8 cm, chartaceous, lanceolate or oblanceolate, glabrous or nearly so, lower surface with abundant yellowish gland dots, the apex obtuse or acute, the base tapering into a more or less elongate (to 2 cm), winged petiole, the margins deeply lobed or serrate to entire on upper leaves. Flowers minute, greenish, in axillary glomerules or in spikes of glomerules, the spikes 1-2 cm long. Calyx greenish, ca. 1 mm long, the sepals oblong; stamens ca. 1 mm long; styles 3, whitish. Utricle whitish, ca. 1 mm long, covered with persistent sepals. Seeds 1 mm long, nearly lenticular, reddish brown (Acevedo-Rodríguez, 2005).

【gobotany.newenglandwild.org】*Dysphania ambrosioides* (L.) Mosyakin & Clemants

刺 苋

学　名：**Amaranthus spinosus** L.

异　名：*Amaranthus caracasanus* Kunth，*Amaranthus diacanthus* Raf.，*Galliaria spinosa*（L.）Nieuwel.，*Galliaria spitosa*（L.）Nieuwl.

别　名：苋菜、勒苋菜

英文名：spiny amaranth，calaloo，needle burr，pigweed，prickly calaloo，prickly callau，prickly caterpillar，spiny amaranthus，spiny calaloo，spiny pigweed，sticker weed，thorny pigweed，wild callau

| 形态特征 |

刺苋为多年生草本植物。

植物主根长圆锥形，有的具分枝，稍木质。

植物茎直立，圆柱形，多分枝，有纵条纹，棕红色或棕绿色。茎下部光滑，上部稍有毛。植株高0.3~1.0m。

植物叶互生。叶片卵状披针形或菱状卵形，长4~10cm，宽1~3cm，先端圆钝，基部楔形，全缘或微波状，中脉背面隆起，先端有细刺。叶柄与叶片等长或稍短，叶腋有坚刺1对。叶柄长1~8cm，无毛。

圆锥花序腋生及顶生，长3~25cm。花单性，雌花簇生于叶腋，呈球状。雄花集为顶生的直立或微垂的圆柱形穗状花序。花小，刺毛状苞片约与萼片等长或过之。苞片常变形成2锐刺，少数具1刺或无刺。花被片绿色，先端急尖，边缘透明。萼片5。雄蕊5。柱头3，有时2。

胞果长圆形，在中部以下为不规则横裂，包在宿存花被片内。

种子近球形，黑色带棕黑色。

花期5~9月。果期8~11月。

刺苋气微，味淡。

| 地理分布 |

刺苋原产于热带美洲。19世纪30年代，在我国澳门发现。1857年，在我国香港采到。现分布于陕西、河北、北京、山东、河南、安徽、江苏、浙江、江西、湖南、湖北、四川、重庆、云南、贵州、广西、广东、海南、香港、福建、台湾等地。为全球性入侵物种。

| 传播途径 |

自然扩散或人为传播。

外来入侵植物

| 管理措施 |

综合治理（Integrated Pest Management，IPM）或综合控制（Integrated Pest Control，IPC），但不包括植物卫生措施（Phytosanitary Measures）。

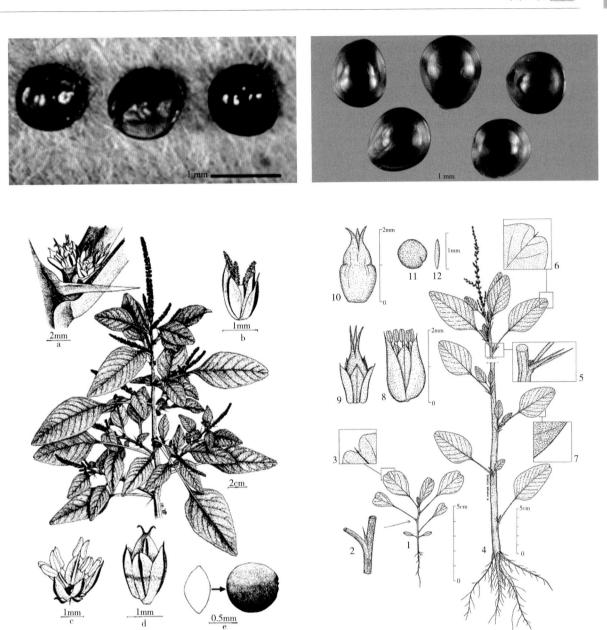

(图片选自cn.bing.com, cdn2.arkive.org, www.tropical-plants-flowers-and-decor.com, extension.umass.edu, www.utas.edu.au, image.baidu.com, www.cabi.org, www.redorbit.com, plants.usda.gov, malherbologie.cirad.fr)

刺苋(*Amaranthus spinosus*)

【www.efloras.org】*Amaranthus spinosus* Linnaeus, Sp. Pl. 2: 991. 1753.

Spiny amaranth, thorny amaranth

Plants Glabrous or sparsely pubescent in the distal younger parts of stems and branches.

Stems Erect or sometimes ascending proximally, much-branched and bushy, rarely nearly simple, 0.3-1 (-2) m; each node with paired, divergent spines (modified bracts) to 1.5 (-2.5) cm.

Leaves: Petiole ± equaling or longer than blade; blade rhombic-ovate, ovate, or ovate-lanceolate, 3-10 (-15) cm × 1.5-6 cm, base broadly cuneate, margins entire, plane or slightly undulate, apex acute or subobtuse to indistinctly emarginate, mucronulate.

Inflorescences Simple or compound terminal staminate spikes and axillary subglobose mostly pistillate clusters, erect or with reflexed or nodding tips, usually green to silvery green. Bracts of pistillate flowers lanceolate to ovate-lanceolate, shorter than tepals, apex attenuate.

Pistillate flowers: Tepals 5, obovate-lanceolate or spatulate-lanceolate, equal or subequal, 1.2-2 mm, apex mucronate or short-aristate; styles erect or spreading; stigmas 3.

Staminate flowers: Often terminal or in proximal glomerules; tepals 5, equal or subequal, 1.7-2.5 mm; stamens 5.

Utricles Ovoid to subglobose, 1.5-2.5 mm, membranaceous proximally, wrinkled and spongy or inflated distally, irregularly dehiscent or indehiscent.

Seeds Black, lenticular or subglobose-lenticular, 0.7-1 mm diam., smooth, shiny.

【www.efloras.org】*Amaranthus spinosus* Linnaeus, Sp. Pl. 2: 991. 1753.

Stem erect, green or somewhat tinged purple, 30-100 cm tall, terete or obtusely angulate, much branched, glabrous or slightly pubescent. Petiole 1-8 cm, glabrous, 2-armed at base; leaf blade ovate-rhombic or ovate-lanceolate, 3-12 cm × 1-6 cm, glabrous or slightly pubescent along veins when young, base cuneate, margin entire, apex obtuse, with a mucro. Complex thyrsoid structures terminal or axillary, 8-25 cm; terminal spike usually with all male flowers at or toward apex. Bracts becoming very sharply spiny in proximal part of spike. Tepals green, transparent at margin and with green or purple median band, apex acute, with a mucro; male flowers oblong, 2-2.5 mm; female flowers oblong-spatulate, ca. 1.5 mm. Filaments nearly as long as or slightly shorter than perianth. Stigmas 3 (or 2). Utricles included in perianth, oblong, 1-1.2 mm, circumscissile slightly below middle. Seeds brownish black, subglobose, ca. 1 mm in diam. Fl. and fr. Jul-Nov. $2n = 34, 68$.

【www.cabi.org】*Amaranthus spinosus* (spiny amaranth)

Annual, Herbaceous, Seed propagated plant

Plants glabrous or sparsely pubescent in the distal younger parts of stems and branches. Stems erect or sometimes ascending proximally, much-branched and bushy, rarely nearly simple, 0.3-1 (-2) m; each node with paired, divergent spines (modified bracts) to 1.5 (-2.5) cm. Leaves: petiole ± equaling or longer than blade; blade rhombic-ovate, ovate, or ovate-lanceolate, 3-10 (-15) cm × 1.5-6 cm, base broadly cuneate, margins entire, plane or slightly undulate, apex acute or subobtuse. Inflorescences simple or compound terminal staminate spikes and axillary subglobose mostly pistillate clusters, erect or with reflexed or nodding tips, usually green to silvery green. Bracts of pistillate flowers lanceolate to ovate-lanceolate, shorter than tepals, apex attenuate. Pistillate flowers: tepals 5, obovate-lanceolate or spatulate-lanceolate, equal or subequal, 1.2-2 mm, apex mucronate; styles erect or spreading; stigmas 3. Staminate flowers: often terminal or in proximal glomerules; tepals 5, equal or subequal, 1.7-2.5 mm; stamens 5. Utricles ovoid to subglobose, 1.5-2.5 mm, membranaceous proximally, wrinkled and spongy or inflated distally, irregularly dehiscent or indehiscent. Seeds black, lenticular or subglobose, 0.7-1 mm diameter, smooth, shiny (Flora of North America, 2015) .

The striated, often reddish, stem with two sharp, long spines at the base of the petioles, and the fruit which opens by a line around the centre are distinguishing characteristics of this species.

反枝苋

学　名：***Amaranthus retroflexus*** L.

异　名：*Amaranthus bulgaricus* Kov.，*Amaranthus bullatus* Besser ex Spreng.，*Amaranthus chlorostachys* Willk.，*Amaranthus curvifolius* Spreng.，*Amaranthus delilei* Richt. & Loret，*Amaranthus johnstonii* Kov.，*Amaranthus recurvatus* Desf.，*Amaranthus retroflexus* var. *delilei*（Richt. & Loret）Thell.，*Amaranthus retroflexus* subsp. *delilei*（Richt. & Loret）Tzvelev，*Amaranthus retroflexus* var. *genuinus*（L.）Thell. ex Probst，*Amaranthus retroflexus* var. *rubricaulis* Thell.，*Amaranthus retroflexus* f. *rubricaulis* Thell. ex Probst，*Amaranthus retroflexus* var. *salicifolius* II. M. Johnst.，*Amaranthus rigidus* Schult. ex Steud.，*Amaranthus spicatus* Lam.，*Amaranthus strictus* Ten.，*Amaranthus tricolor* L.，*Galliaria retroflexa*（L.）Nieuwl.，*Galliaria scabra* Bubani

别　名：野苋菜、苋菜、西风谷、茵茵菜（东北）

英文名：redroot pigweed，carelessweed，common amaranth，redroot

形态特征

反枝苋为一年生草本植物。

植物茎直立，粗壮，单一或分枝，淡绿色，有时具紫色条纹，稍具钝棱，密生短柔毛。植株高20～80cm，有的高达2m多。

植物叶片菱状卵形或椭圆状卵形，长5～12cm，宽2～5cm，顶端锐尖或尖凹，有小凸尖，基部楔形，全缘或波状缘，两面及边缘有柔毛，下面毛较密。叶柄长1.5～5.5cm，淡绿色，有时淡紫色，有柔毛。

圆锥花序顶生及腋生，直立，直径2～4cm，由多数穗状花序形成。顶生花穗较侧生者长。苞片及小苞片钻形，长4～6mm，白色，背面有1龙骨状突起，伸出顶端成白色尖芒。花被片矩圆形或矩圆状倒卵形，长2～2.5mm，薄膜质，白色，有1淡绿色细中脉，顶端急尖或尖凹，具凸尖。雄蕊比花被片稍长。柱头3，有时2。

胞果扁卵形，长约1.5mm，环状横裂，薄膜质，淡绿色，包裹在宿存花被片内。

种子近球形，直径1mm，棕色或黑色，边缘钝。

花期7～8月。果期8～9月。

生物危害

反枝苋主要危害棉花、豆类、瓜类、薯类、蔬菜等多种旱作物。植株还可富集硝酸盐，家畜过量食用后会引起中毒。

反枝苋也可以与其他多种美洲苋属植物如绿穗苋、鲍氏苋、尾穗苋和刺苋等进行杂交，形成新的杂草品种。

此外，反枝苋还是桃蚜［*Myzus persicae*（Sulzer）］、黄瓜花叶病毒（CMV）、小地老虎（*Agrotis ypsilon* Rottemberg）、美国草牧盲蝽［*Lygus pratenszs*（Linnaeus）］、欧洲玉米螟［*Ostrinia nubilalis*（Hubner）］等的田间寄主。

| 地理分布 |

反枝苋原产于美洲热带。现广泛传播并归化于世界各地。19世纪中叶，在河北和山东发现。现分布于安徽、北京、甘肃、广东、广西、贵州、河北、河南、黑龙江、湖北、湖南、吉林、江苏、江西、辽宁、内蒙古自治区（以下简称内蒙古）、宁夏回族自治区（以下简称宁夏）、青海、山东、山西、陕西、上海、四川、台湾、天津、西藏（芒康）、新疆维吾尔自治区（以下简称新疆）、云南、浙江、重庆等地。

| 传播途径 |

人为传播和自然扩散。

| 管理措施 |

综合治理（Integrated Pest Management，IPM）或综合控制（Integrated Pest Control，IPC），但不包括植物卫生措施（Phytosanitary Measures）。

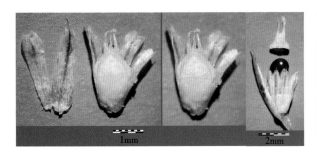 
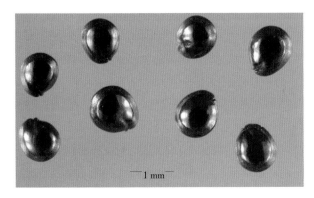 
 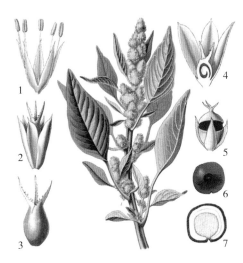

（图片选自www.unavarra.es，articles.extension.org，www.plantarium.ru，cals.arizona.edu，weedecology.css.cornell.edu，en.wikipedia.org，www.pfaf.org，commons.wikimedia.org，www.wildutah.us，newyork.plantatlas.usf.edu，image.baidu.com，www.forestryimages.org）

反枝苋（*Amaranthus retroflexus*）

【 www.efloras.org 】*Amaranthus retroflexus* Linnaeus, Sp. Pl. 2: 991. 1753.

Redroot pigweed, redroot amaranth, wild-beet amaranth, rough pigweed, common amaranth
Amaranthus retroflexus var. *salicifolius* I. M. Johnston

Plants Densely to moderately pubescent, especially distal parts of stem and branches.

Stems Erect, reddish near base, branched in distal part to simple 0.2-1.5（-2）m; underdeveloped or damaged plants rarely ascending to nearly prostrate.

Leaves: Petiole 1/2 to equaling blade; blade ovate to rhombic-ovate, 2-15 cm × 1-7 cm, base cuneate to rounded-cuneate, margins entire, plane or slightly undulate, apex acute, obtuse, or slightly emarginate, with terminal mucro.

Inflorescences Terminal and axillary, erect or reflexed at tip, green or silvery green, often with reddish or yellowish tint, branched, leafless at least distally, usually short and thick.

Bracts Lanceolate to subulate,（2.5-）3.5-5（-6）mm, exceeding tepals, apex acuminate with excurrent midrib.

Pistillate flowers: Tepals 5, spatulate-obovate, lanceolate-spatulate, not clawed, subequal or unequal,（2-）2.5-3.5（-4）mm, membranaceous, apex emarginate or obtuse, with mucro; style branches erect or slightly spreading,; stigmas 3.

Staminate flowers Few at tips of inflorescences; tepals 5; stamens（3-）4-5.

Utricles Broadly obovoid to broadly elliptic, 1.5-2.5 mm, shorter than or subequal to tepals, smooth or slightly rugose, especially near base and in distal part, dehiscence regularly circumscissile.

Seeds Black to dark reddish brown, lenticular to subglobose-lenticular, 1-1.3 mm, smooth, shiny.

【 www.efloras.org 】*Amaranthus retroflexus* Linnaeus, Sp. Pl. 2: 991. 1753.

Stem erect, light green, 20-80 cm tall, stout, branched or not, slightly obtusely angulate, densely pubescent. Petiole light green, 1.5-5.5 cm, hairy; leaf blade ovate-rhombic or elliptic, 5-12 cm × 2-5 cm, both surfaces shortly hairy, but densely hairy abaxially, base cuneate, margin entire and undulate, apex acute or notched, with a mucro. Complex thyrsoid structures terminal and axillary, erect, 2-4 cm in diam., including many spikes; terminal spikes longer than lateral ones. Bracts and bracteoles white, subulate, 4-6 mm, apex slenderly long pointed. Tepals white, oblong or oblong-obovate, 2-2.5 mm, membranous, with a green midvein, apex acute or notched, with a mucro. Stamens slightly longer than perianth. Stigmas 3, rarely 2. Utricles light green, ovoid, compressed, shorter than perianth, circumscissile. Seeds brown or black, subglobose, ca. 1 mm in diam., obtuse at margin. Fl. Jul-Aug, fr. Aug-Sep. $2n = 32, 34, 102$.

【 www.cabi.org 】*Amaranthus retroflexus*（redroot pigweed）

A. retroflexus is a monoecious, erect, finely hairy, freely-branching, herbaceous annual growing to 2 m tall; taproot pink or red, depth varies with soil profile; leaves alternate, egg-shaped or rhombic-ovate, cuneate at base, up to 10 cm long, margins somewhat wavy, veins prominent on underside, apex may be sharp, petiole shorter or longer than leaf; flowers numerous, small, borne in dense blunt spikes 1 to 5 cm long, densely crowded onto terminal panicle 5 to 20 cm long but may be smaller on upper axils; three spiny-tipped, rigid, awl-shaped bracteoles surround the flower, exceeding the perianth, length 4 to 8 mm, persistent; tepals five, much longer than fruit, usually definitely recurved at tips, obovate or highly spatulate, one pistil and five stamens; style branches erect or a bit recurved; fruit a utricle, membranous, flattened, 1.5 to 2 mm long, dehiscing by a transverse line at the middle, wrinkled upper part falling away; seed oval to egg-shaped, somewhat flattened, notched at the narrow end, 1 to 1.2 mm long, shiny black or dark red-brown.

【 www.discoverlife.org 】 *Amaranthus retroflexus* **Linnaeus, Sp. Pl. 2: 991. 1753.**

Amaranthus retroflexus var. *salicifolius* I. M. Johnston

Plants Densely to moderately pubescent, especially distal parts of stem and branches.

Stems Erect, reddish near base, branched in distal part to simple 0.2-1.5（-2）m; underdeveloped or damaged plants rarely ascending to nearly prostrate.

Leaves: Petiole 1/2 to equaling blade; blade ovate to rhombic-ovate, 2-15 cm × 1-7 cm, base cuneate to rounded-cuneate, margins entire, plane or slightly undulate, apex acute, obtuse, or slightly emarginate, with terminal mucro.

Inflorescences Terminal and axillary, erect or reflexed at tip, green or silvery green, often with reddish or yellowish tint, branched, leafless at least distally, usually short and thick.

Bracts Lanceolate to subulate,（2.5-）3.5-5（-6）mm, exceeding tepals, apex acuminate with excurrent midrib.

Pistillate flowers: Tepals 5, spatulate-obovate, lanceolate-spatulate, not clawed, subequal or unequal,（2-）2.5-3.5（-4）mm, membranaceous, apex emarginate or obtuse, with mucro; style branches erect or slightly spreading,; stigmas 3.

Staminate flowers Few at tips of inflorescences; tepals 5; stamens（3-）4-5. Utricles broadly obovoid to broadly elliptic, 1.5-2.5 mm, shorter than or subequal to tepals, smooth or slightly rugose, especially near base and in distal part, dehiscence regularly circumscissile.

Seeds Black to dark reddish brown, lenticular to subglobose-lenticular, 1-1.3 mm, smooth, shiny.

【 articles.extension.org 】 Redroot Pigweed（*Amaranthus retroflexus*）**, Smooth Pigweed**（**A. hybridus**）**, and Powell Amaranth**（*A. powellii*）

长芒苋

学　　名：**Amaranthus palmeri** S.Watson

异　　名：*Amaranthus palmeri* var. *glomeratus* Uline & W.L.Bray

别　　名：绿苋、野苋

英文名：Palmer amaranth，Carelessweed，dioecious amaranth，Palmer's amaranth，Palmer's pigweed

| 形态特征 |

长芒苋为一年生草本植物。

植物茎直立，粗壮，有纵棱，无毛或上部散生短柔毛，有分枝。植株高0.5～1.5m。

植物叶无毛。叶片卵形至菱状卵形。茎上部叶片则常为披针形，长（3）5～8cm，宽（1.5）2～5cm，先端钝、急尖或微凹，常具小突尖，基部楔形，边缘全缘。

花单性，雌雄异株。穗状花序生茎及分枝顶端，顶端常下垂，长7～30cm，宽1～1.2cm，生于叶腋者较为短，呈短圆柱状至头状。苞片钻状披针形，长4～6cm，先端芒刺状。雄花花被片5，极不等长，长圆形，先端急尖；最外面花被片长约5mm，其余花被片长3.5～4mm。雄蕊5枚，短于内轮花被片。雌花花被片5，极不等长，最外面一片倒披针形，长3～4mm，先端急尖，其余花被片匙形，长2～2.5mm，先端截形至微凹，上部边缘啮蚀状。花柱2或3。

果近球形，长1.5～2mm，包藏于宿存花被片内。果皮膜质，上部微皱，周裂。

种子近圆形，长1～1.2mm，深红褐色，有光泽。

花果期6～11月。

| 生物危害 |

长芒苋适应性强，常生于河岸低地、旷野、村落边及耕地中，产种量很大，竞争力强，易形成优势群落，威胁当地生物多样性。

作为一种旱地杂草，其植株高大，与农作物争夺水、肥、光照和生存空间，危害农田和果园，也可侵入湿地。

植株能富集亚硝酸盐。牲畜取食过量会引起中毒。

雌株成熟果序上的宿存苞片和花被片均具硬刺，可扎伤人。

| 地理分布 |

长芒苋原产于美国西南部。现广布北美洲、欧洲和亚洲。1985年，首次发现于北京的丰台区范庄子村（现槐房路附近）路边，此后沿道路蔓延侵入菜地。现主要分布于北京、天津、河北、辽宁、江苏、山东等地。

传播途径

人为传播或自然扩散。

管理措施

综合治理（Integrated Pest Management，IPM）或综合控制（Integrated Pest Control，IPC），但不包括植物卫生措施（Phytosanitary Measures）。

中国外来入侵 物种图鉴

中国外来入侵物种图鉴

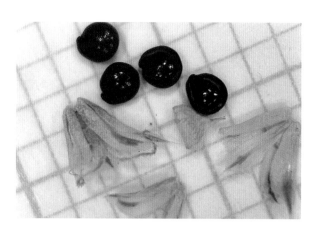
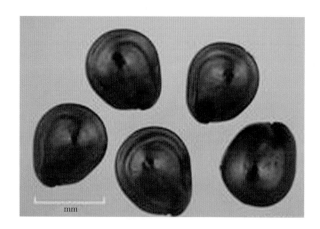

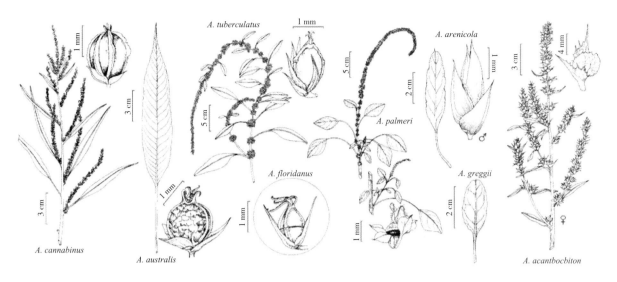

（长芒苋与Amaranthus属的其他物种比较）

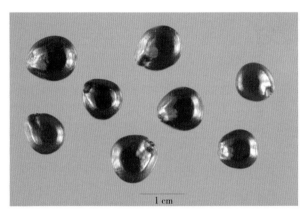

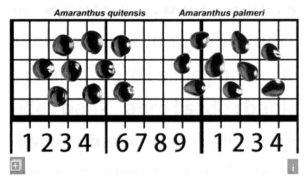

长芒苋与青苋（Amaranthus quitensis）种子比较

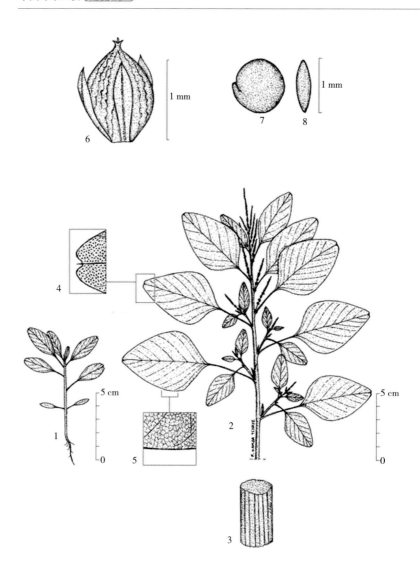

皱果苋（*Amaranthus viridis*）

（图片选自www.bivocational.org，bugwoodcloud.org，bucket2.glanacion.com，newfs.s3.amazonaws.com，www.mda.state.mn.us，extension.psu.edu，www.eeaoc.org.ar，asb.com.ar，cabezaprieta.org，en.wikipedia.org，articles.extension.org，www.actaplantarum.org，www.wnmu.edu，www.fireflyforest.com，www.uapress.arizona.edu，calphotos.berkeley.edu，flora.huh.harvard.edu，www.floravascular.com，plants.usda.gov，image.slidesharecdn.com，idao.cirad.fr）

长芒苋（*Amaranthus palmeri*）

【 www.efloras.org 】 *Amaranthus palmeri* S. Watson, Proc. Amer. Acad. Arts. 12: 274. 1877 (as *Amarantus*).

Palmer's amaranth

Plants glabrous or nearly so. Stems erect, branched, usually (0.3-) 0.5-1.5 (-3) m; proximal branches often ascending. Leaves: long-petiolate; blade obovate or rhombic-obovate to elliptic proximally, sometimes lanceolate distally, 1.5-7 cm × 1-3.5 cm, base broadly to narrowly cuneate, margins entire, plane, apex subobtuse to acute, usually with terminal mucro. Inflorescences terminal, linear spikes to panicles, usually drooping, occasionally erect, especially when young, with few axillary clusters, uninterrupted or interrupted in proximal part of plant. Bracts: of pistillate flowers with long-excurrent midrib, 4-6 mm, longer than tepals, apex acuminate or mucronulate; of staminate flowers, 4 mm, equaling or longer than outer tepals, apex long-acuminate. Pistillate flowers: tepals 1.7-3.8 mm, apex acuminate, mucronulate; style branches spreading; stigmas 2 (-3). Staminate flowers: tepals 5, unequal, 2-4 mm, apex acute; inner tepals with prominent midrib excurrent as rigid spine, apex long-acuminate or mucronulate; stamens 5. Utricles tan to brown, occasionally reddish brown, obovoid to subglobose, 1.5-2 mm, shorter than tepals, at maturity walls thin, almost smooth or indistinctly rugose. Seeds dark reddish brown to brown, 1-1.2 mm diam., shiny.

【 articles.extension.org 】 *Amaranthus palmeri* (Palmer amaranth)

Palmer amaranth is a tall, erect, branching summer annual, commonly reaching heights of 6-8 feet, and occasionally 10 feet or more. Stems and foliage are mostly smooth and lacking hairs (glabrous). Leaves have fairly long petioles and are arranged symmetrically around the stem, giving the plant a distinctly pointsettia-like appearance when viewed from above. Leaf blades are elliptical to diamond-shaped with pointed tips, and measure 0.6-3 inches long by 0.4-1.5 inches wide.

Male and female flowers are borne on separate plants (dioecious), and the small (<0.25 inch) flowers are clustered tightly in linear or sparingly branched terminal spikes up to 18 inches long. The perianth (whorl of petal-like structures) around each female flower bears small, rigid spines that give the female spikes a markedly bristly texture. In contrast, male inflorescences are fairly soft to the touch.

【 www.eppo.int 】 *Amaranthus palmeri* (Amaranthaceae)

A. palmeri is an annual plant with one central reddish-green stem up to 1.5 m tall with many lateral branches. Leaves are alternate, hairless, borne on long petioles that often exceed the length of the leaf blade, they are lanceolate in young plants and become ovate as the plant matures, with prominent whitish veins on the underside. The leaves often have a distinctive V-shaped chevron on the upper surface. Female and male flowers occur on separate plants, but are both 2 to 3 mm each, clustering in cylindrical inflorescences or spikes up to 60 cm on the central stem. The fruit is a thin-walled one-seeded utricule about 1.5 mm long. The top half of the fruit separates at maturity to expose the single, round, black to dark purple seed which is of 1 to 2 mm in diameter.

【 gobotany.newenglandwild.org 】 *Amaranthus palmeri* S. Wats.

【 plants.ces.ncsu.edu 】 *Amaranthus palmeri*

落葵薯

学　　名：***Anredera cordifolia***（Tenore）Steenis

异　　名：*Boussingaultia basselloides*，*Boussingaultia cordifolia* Ten.，*Boussingaultia cordasta* Spreng.，*Boussingaultia gracilis* Miers

别　　名：马德拉藤、藤三七、心叶落葵薯、洋落葵、川三七、热带皇宫菜、藤子三七、田三七

英文名：Madeira vine，basell-potatoes，bridal wreath，lamb's tails，mignonette vine，potato vine

| 形态特征 |

落葵薯为缠绕藤本植物。

植物根状茎粗壮。

植株缠绕藤状，可长达30m。

植物叶片卵形至近圆形，长2~6cm，宽1.5~5.5cm，顶端急尖，基部圆形或心形，稍肉质。腋生小块茎（珠芽）。叶具短柄。

总状花序具多花。花序轴纤细，下垂，长7~25cm。苞片狭，不超过花梗长度，宿存。花梗长2~3mm，花托顶端杯状，花常由此脱落。下面1对小苞片宿存，宽三角形，急尖，透明；上面1对小苞片淡绿色，比花被短，宽椭圆形至近圆形。花直径约5mm。花被片白色，渐变黑，开花时张开，卵形、长圆形至椭圆形，顶端钝圆，长约3mm，宽约2mm。雄蕊白色，花丝顶端在芽中反折，开花时伸出花外。花柱白色，分裂成3个柱头臂，每臂具1棍棒状或宽椭圆形柱头。

花期为6~10月。

果实、种子未见。

| 生物危害 |

落葵薯能以块根、珠芽、断枝高效率繁殖，生长迅速，极易扩散蔓延。由于其枝叶可密集覆盖植物而致使死亡，同时也对多种农作物有显著的化感作用。

| 地理分布 |

落葵薯原产于南美热带和亚热带地区；世界各地引种栽培，在温暖地区归化。20世纪70年代，我国从东南亚引种。现分布于重庆、四川、贵州、湖南、广西、广东、云南、香港、福建等地。为全球性入侵物种。

|传播途径|

落葵薯的块根、珠芽滚落后可长成新的植株,断枝也可繁殖。因此,这些无性繁殖材料可自然扩散或人为传播。

|管理措施|

综合治理(Integrated Pest Management,IPM)或综合控制(Integrated Pest Control,IPC),但不包括植物卫生措施(Phytosanitary Measures)。

中国外来入侵物种图鉴

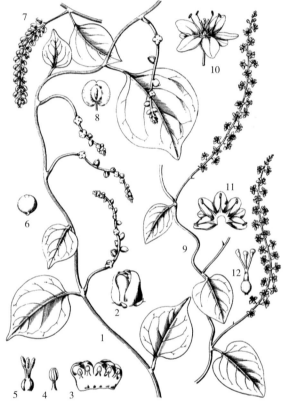

图版9 1—6.落葵Basella alba L.：1.植株一段；2.花；3.花解剖（示雄蕊着生）；4.雄蕊；5.雌蕊；6.种子。7—8.短序落葵薯Anredera scandens（L.）Moq.：7.果枝；8.果实。9—12.落葵薯A. cordifolia（Tenore）Steenis：9.植株一段；10.花；11.花解剖；12.雌蕊。（钱存源绘）

（图片选自www.dpi.nsw.gov.au，keyserver.lucidcentral. org，pukubook.jp，www1.lf1.cuni.cn，pic.sogou.com， image.baidu.com，frps.eflora.cn）

落葵薯（*Anredera cordifolia*）

【www.efloras.org】*Anredera cordifolia*（Tenore）Steenis, Fl. Malesiana, Ser. 1, Spermatoph. 5（3）: 303. 1957.

Boussingaultia cordifolia Tenore in Ann. Sci. Nat., Bot. sér. 3, 19: 355. 1853; *B. gracilis* Miers; *B. gracilis* f. *pseudobaselloides* Hauman; *B. gracilis* var. *pseudobaselloides*（Hauman）Bailey.

Vines twining, with thick, hardy rhizome. Leaves shortly petiolate; leaf blade ovate to subcordate, 2-6 cm × 0.5-5.5 cm, thinly fleshy, base rounded or cordate, apex acute, producing small axillary tubercles（bulbils）. Racemes many flowered; rachis pendent, thin, 7-25 cm. Bracts not longer than pedicel, narrow, persistent. Pedicel 2-3 mm, receptacular tip cupular, flowers shed from here. Lower bracteoles broadly triangular, hyaline, apex acute, persistent; upper bracteoles greenish white, flattened, orbicular to broadly elliptic, shorter than perianth. Flowers ca. 5 mm in diam., fragrant. Perianth white, inflexed, patent in anthesis; segments ovate or oblong to elliptic, 3 × 2 mm, apex blunt. Stamens white; filaments reflexed at apex in bud, spreading in anthesis. Style white, split to 3 stigmatic arms, each with 1 club-shaped or broadly elliptic stigma. Utricle and seed not seen. Fl. Jun-Oct.

【www.efloras.org】*Anredera cordifolia*（Tenore）Steenis, Fl. Males. Ser. 1. 5: 303. 1957. Madeira-vine

Boussingaultia cordifolia Tenore, Ann. Sci. Nat., Bot., sér. 3, 19: 355. 1853; *B. gracilis* Miers

Stems Twining to 5 m, often producing single or clustered axillary tubers.

Leaves: Petiole 6-12 mm; blade ovate to orbicular, 2-10 cm × 1-7 cm, base of larger leaf blades proximal to inflorescences cordate, base of small distal leaf blades tapering, apex obtuse to acute.

Inflorescences Racemes or in branched panicles of racemes, 10-35（-60）cm; single bract subtending each pedicel triangular-lanceolate, 1-2 mm × 0.3-0.4 mm; paired bracts subtending each flower persistent, triangular to obtuse, 0.6-1 mm × 0.5-0.6 mm, basally connate into cup.

Flowers Bisexual, usually functionally staminate; sepals basally adnate to petals, cream-white, not winged in fruit, ovate to elliptic, 1.2-2.3 mm × 1.1-2 mm, apex obtuse; petals basally connate, cream-white, ovate to elliptic, 2.1-3 mm × 1.4-2 mm, apex obtuse, spreading at anthesis; stamens fleshy; filaments basally connate and dilated, 1.9-3.5 mm; anthers early deciduous, 0.7-0.9 mm; pistils 0.4-0.6 mm; styles 1-1.5 mm, basally connate for 1/2-2/3 their length; stigmas clavate to capitate; pedicel 1-3 mm.

Utricles Rarely producing viable seeds, style bases persistent, globose, 0.8-1.1 mm.

$2n = 24$.

Flowering late summer-fall. Disturbed areas, fencerows, roadsides; 0-500 m; introduced; Calif., Fla., La., Tex.; Mexico; West Indies; Central America; Eurasia; Africa; Pacific Islands; Australia; native to South America.

Anredera cordifolia is sometimes cultivated as an ornamental, and it escapes and naturalizes in subtropical to subtemperate regions（Sperling C R, 1987）. In many floristic treatments, it has been confused with *A. baselloides*（Kunth）Baillon, a species restricted to Ecuador and Peru.

Fruit- and seed-set are rare in cultivated material; tubers are the main means by which Madeira-vine is propagated（Sperling C R, 1987）.

【www.cabi.org】 *Anredera cordifolia* (Madeira vine)

Broadleaved, Herbaceous, Perennial, Succulent, Vegetatively propagated Vine / climber, Woody plant

A. cordifolia is a perennial evergreen climbing vine or liana that grows from fleshy rhizomes. Stems are slender, climbing to 3-6 m in height in a single growing season, often reddish in colour. Oval or heart-shaped leaves are bright green and shiny, 2-13 cm long and 1-11 cm wide, broadly ovate, often involute, sometimes lanceolate, scarcely succulent to succulent according to degree of exposure, margins often turned inwards, base subcordate or cordate; apex obtuse, subsessile or with a petiole 1- (2) cm long, commonly with small irregular tubers in their axils. The potato-like tubers, produced on aerial stems covered in warts, are specific and typical in identifying the plant, but can grow to 25 cm in diameter. Masses of fragrant, cream flowers occur on simple or 2-4-branched racemes, pendent to 18cm cm long excluding the common peduncle, up to 30 cm including it, with numerous small, white, fragrant flowers. Pedicels are 2-3 mm long, bracts 1.5-1.8 mm long and lanceolate-subulate. Lower bracteoles are 0.5-1 mm long and cupulate, with upper bracteoles 2-2.5 mm long and suborbicular. The five tepals are 2-3 mm long and elliptic-oblong to broadly elliptic. Filaments are narrow-triangular, widely divergent, bending outwards near base, with a single style shorter than the stamens and clavate.

【keys.lucidcentral.org】 *Anredera cordifolia* (Madeira Vine)

A long-lived (perennial), twining or climbing plant growing over taller plants and trees up to 30 m tall.

The stems are hairless (glabrous) and grow in a twining fashion. Younger stems are green or reddish in colour and round in cross-section.

They become rope-like in appearance and turn greyish-brown in colour as they mature. Distinctive greyish-brown or greenish-coloured warty tubers (1-10 cm long, but usually 2-3 cm long) often form at the joints (nodes) along the older stems. These wart-like tubers are very characteristic.

The leaves are alternately arranged, slightly fleshy (semi-succulent) in nature, hairless (glabrous) and sometimes have a glossy appearance. They are borne on leaf stalks (petioles) 5-20 mm long and are more or less heart-shaped (cordate) or broadly egg-shaped with broad end at base (ovate). These leaves (2-15 cm long and 1.5-10 cm wide) either taper to a blunt point or have a somewhat rounded tip (acute or obtuse apex).

Plants produce masses of drooping flower clusters (6-30 cm long) which arise from the forks (axils) of the upper leaves. Each flower cluster (raceme) bears numerous small, white or cream-coloured, fragrant flowers (about 5 mm across). These star-shaped flowers have five 'petals' (sepals or perianth segments) and are borne on short stalks (pedicels) 2-3 mm long. They also have five stamens and an ovary topped with a three-branched style and three tiny club-shaped stigmas. The petals (2-3 mm long) are fleshy, persistent, turn dark brown or black in colour with age. This plant does not produce fruit in Africa.

钻形紫菀

学　　名：***Aster subulatus***（Michx.）Hort. ex Michx.

异　　名：*Aster flexicaulis* Raf.，*Aster linifolius* Torr. & A.Gray，*Aster squamatus*（Spreng.）Hieron. ex Sodiro（misapplied），*Aster subulatus* Michx.，*Aster subulatus*（Michx.）Hort. ex Michx.，*Aster subulatus* var. *euroauster* Fernald & Grisc.，*Chrysocoma linifolia* Steud.，*Erigeron linifolius* Bertero ex DC.，*Mesoligus subulatus*（Michx.）Raf.，*Symphyotrichum subulatum*（Michx.）G.L.Nesom，*Tripolium subulatum*（Michx.）DC.

别　　名：钻叶紫菀、剪刀菜、燕尾菜、土柴胡

英 文 名：annual saltmarsh aster，aster，aster weed，aster-weed，bushy starwort，eastern annual saltmarsh aster，saltmarsh aster，sea aster，slender aster，slim aster，small saltmarsh aster，wild aster

| 形态特征 |

钻形紫菀为一年生草本植物。

植物茎直立，无毛而富肉质，上部稍有分枝。植株高30～150cm。

植物基生叶倒披针形，花后凋落。茎中部叶线状披针形，先端尖或钝，有时具钻形尖头，全缘，无柄，无毛。

头状花序小排成圆锥状。总苞钟状，总苞片3～4层，外层较短，内层较长，线状钻形，无毛。舌状花细狭，淡红色，长与冠毛相等或稍长。管状花多数，短于冠毛。

瘦果长圆形或椭圆形，长1.5～2.5mm，有5纵棱，冠毛淡褐色。

花果期9～11月。

| 生物危害 |

钻形紫菀喜生于潮湿土壤中，在沼泽或含盐的土壤中也可以生长，常沿河岸、沟边、洼地、路边、海岸蔓延，侵入农田危害棉花、花生、大豆、甘薯、水稻等作物。也常侵入浅水湿地，影响湿地生态系统及景观。

| 地理分布 |

钻形紫菀原产于北美洲。现广布于世界温暖地区。1827年，在我国澳门发现。1947年，在湖北武昌发现。现分布于安徽、澳门、北京、福建、广东、广西、贵州、河北、河南、湖北、湖南、江苏、江西、辽宁、山东、上海、四川、台湾、天津、香港、云南、浙江、重庆等地。

| 传播途径 |

钻形紫菀可产生大量瘦果，果具冠毛，随风扩散。也可以人为传播。

| 管理措施 |

综合治理（Integrated Pest Management，IPM）或综合控制（Integrated Pest Control，IPC），但不包括植物卫生措施（Phytosanitary Measures）。

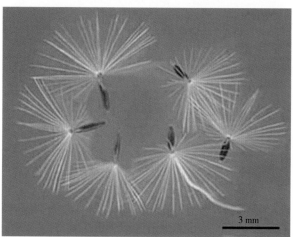

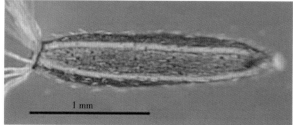

（图片选自keyserver.lucidcentral.org，tupian.baike.com，www.yeehua.net，hasbrouck.asu.edu，www.backyardnature.net，npsot.org www.osaka-c.ed.jp，pests.agridata.cn，calphotos.berkeley.edu，www.smmflowers.org，mikawanoyasou.org）

钻形紫菀（*Aster subulatus*）

【plants.jstor.org】 *Aster subulatus*

Plants 30-120（-150）cm. Stems simple. Heads（10-）30-100（-150）, in elongate, pyramidal, much branched arrays. Involucres（5-）6-7（-8.2）mm. Phyllaries 20-30, subulate to lanceolate, narrow, green zones narrowly to broadly lanceolate, extending phyllary length. Ray florets 16-30 in 2 series; laminae white（drying white or lavender）,（1.3-）mm1.5-2.6（-3.1）mm × 0.2-0.5 mm, shorter to slightly longer than pappi, drying in 0-1 coil（apices often deeply lobed）. Disc florets 4-10（-13）; corollas 3.8-4.6（-4.9）mm. Cypselae（1.2-）1.5-2.5 mm; pappi（3.5-）4-5.5 mm. $2n = 10$.

【plantnet.rbgsyd.nsw.gov.au】 *Aster subulatus* Michx.

Annual, biennial or short-lived perennial, 0.3-1.8 m high, erect, almost glabrous; stems branched above, rigid, smooth, reddish.

Leaves linear-lanceolate to narrow-lanceolate, 1-15 cm long, 3-10 mm wide, apex acute, margins entire; stem-clasping.

Heads numerous, in loose leafy panicles; peduncles branched, with numerous bracts 2-10 mm long. Heads cylindrical, 2-4 mm diam.; involucral bracts 3- or 4-seriate, often reddish with a green centre, glabrous. Ray florets white, pink or blue; ligule 3-5 mm long. Disc florets few.

Achenes 1.5-2 mm long, narrow, 4-5 ribbed, sparsely hairy; pappus whitish. Weed of seasonally wet or poorly drained land and damp areas such as swamp edges and roadside drains.

【nyc.books.plantsofsuburbia.com】 *Symphyotrichum subulatum*（*Aster subulatus*）

An annual herb 10-100 cm tall, rather fleshy, from a taproot.

Leaves Alternate, linear, entire, to 20 cm long, 1 cm wide.

Flowers Greenish, tubular, of two types; rays short, inconspicuous, pale blue, numerous, disk flowers 5-15, yellow; in narrow heads with basal bracts（nvolucres）0.5-0.8 cm tall, bracts overlapping, narrow, usually green with purple tips; inflorescence branched, heads solitary or few along a branch.

Fruit Dry, 1-seeded, white-plumed achenes.

【www.discoverlife.org】 *Aster subulatus*

Symphyotrichum subulatum（Michaux）G. L. Nesom, Phytologia. 77: 293. 1995.

Aster subulatus Michaux, Fl. Bor.-Amer. 2: 111. 1803

Annuals,（10-）30-150 cm; tap-rooted.

Stems 1, erect（often with purple or purplish brown areas）, glabrous or glabrate, sometimes strigillose in leaf axils.

Leaves Thin（green to dark green）, margins often strigilloso-ciliolate, faces glabrous; basal withering by flowering, long-petiolate（petiole bases sheathing）, sparsely ciliate, blades ovate to oblanceolate, 10-90 mm × 6-14 mm, bases attenuate to cuneate, rounded, margins entire or serrulate or crenulate, apices rounded, obtuse, or acute; proximal cauline withering by flowering, petiolate, subpetiolate, or sessile, blades narrowly lanceolate or subulate, 20-100（-200）mm × 1.5-10（-20）mm, bases attenuate, margins subentire, entire, or serrulate, apices acute to acuminate; distal sessile, blades narrowly lanceolate to subulate, 5-113 mm × 0.5-5.5 mm, apices acuminate.

Heads（10-）30-100（-150）, in open, diffuse, paniculiform arrays.

Peduncles (0.2-) 0.5-4 cm, bracts 4-8 (-17).

Involucres Cylindric to turbinate, 5-7 (-8.2) mm.

Phyllaries In 3-5 series, broadly or narrowly lanceolate to subulate, unequal, bases indurate, margins hyaline, often purple-tinged, entire, green zones lanceolate (usually narrow, sometimes broad and covering most of distal portion), apices acute, faces glabrous.

Ray florets 16-30 (-54) in 1-3 series; corollas white, pink, or lavender, laminae 1.3-7 mm × 0.2-1.3 mm.

Disc florets 4-10 (-13); corollas yel-low, sometimes tinged with purple, 3.4-5.2 mm, throats narrowly funnelform, lobes ± spreading to erect, nar-rowly triangular, 0.3-0.7 mm, glabrous.

Cypselae Light brown to purple, narrowly obovoid to fusiform, some-times ± compressed, (1.2-) 1.5-2.7 (-3) mm, 5-nerved, faces sparsely strigillose.

pappi White, (3-) 3.5-5.5 mm.

Varieties 5 (5 in the flora): North America, Mexico, West Indies, Bermuda, Central America, South America; widely introduced worldwide.

Five varieties of *Symphyotrichum subulatum* are recognized for North America based on differences in chromosome number, ray lamina color and size, array shapes, number of series of ray florets, number of disc and ray florets, and other, more cryptic characters (Sundberg S D, 2004). These varieties were treated as species by G. L. Nesom (1994b, 2005d). Variety ligulatum is apparently an obligate outcrosser and is the least variable variety (Sundberg). Other varieties are self-compatible, which could facilitate the fixation of mutations in populations.

The five varieties are nearly entirely allopatric, and intermediates between pairs of varieties are not uncommon where they approach one another. Populations that are intermediate in ray lamina size between vars. *ligulatum* and *parviflorum* are widespread in southern Texas, New Mexico, Arizona, and northern Mexico. Intermediates between vars. elongatum and parviflorum and between vars. elongatum and subulatum occur in Florida. Despite these observations, hybridization experiments and chromosome number differences suggest that the varieties are mostly reproductively isolated (Sundberg S D, 1986, 2004).

In older floras the name Aster exilis Elliott has been applied to Symphyotrichum subulatum vars. ligulatum and parviflorum. The status of this name is uncertain; the type specimen has been lost and the description of the plant is inadequate for determining the taxon to which the name should be applied (Nesom G L, 1994b; Sundberg S D, 2004).

三叶鬼针草

学　名：***Bidens pilosa*** L.

异　名：*Bidens abadiae* DC.，*Bidens adhaerescens* Vell.，*Bidens africana* Klatt，*Bidens alausensis* Kunth，*Bidens alba*（L.）DC.，*Bidens arenaria* Gand.，*Bidens arenicola* Gand.，*Bidens aurantiaca* Colenso，*Bidens barrancae* M. E. Jones，*Bidens bimucronata* Turcz.，*Bidens bonplandii* Sch. Bip.，*Bidens brachycarpa* DC.，*Bidens calcicola* Greenm.，*Bidens californica* DC.，*Bidens cannabina* Lam.，*Bidens caracasana* DC.，*Bidens caucalidea* DC.，*Bidens chilensis* DC.，*Bidens ciliata* Hoffmanns. ex Fisch. & C. A. Mey.，*Bidens daucifolia* DC.，*Bidens deamii* Sherff，*Bidens decussata* Pav. ex DC.，*Bidens decussata* Pav. ex Steud.，*Bidens dichotoma* Desf. ex DC.，*Bidens exaristata* DC.，*Bidens hirsuta* Nutt. 1841 not Sw. 1788，*Bidens hirta* Jord，*Bidens hispida* Kunth，*Bidens hybrida* Thuill.，*Bidens inermis* S.Watson，*Bidens leucantha*（L.）Willd.，*Bidens leucantha* Poepp. ex DC.，*Bidens leucanthemus*（L.）E. H. L. Krause，*Bidens minor*（Wimm. & Grab.）Vorosch.，*Bidens minuscula* H.Lév. & Vaniot，*Bidens montaubani* Phil.，*Bidens odorata* Cav.，*Bidens orendainae* M.E.Jones，*Bidens orientalis* Velen. ex Bornm.，*Bidens paleacea* Vis.，*Bidens pinnata* Noronha，*Bidens pumila*（Retz.）Steud.，*Bidens ramosissima* Sherff，*Bidens reflexa* Link，*Bidens rosea* Sch. Bip.，*Bidens scandicina* Kunth，*Bidens striata* Schott ex Sweet，*Bidens sundaica* Blume，*Bidens taquetii* H. Lév. & Vaniot，*Bidens trifoliata* Norona，*Bidens valparadisiaca* Colla，*Bidens viciosoi* Pau，*Ceratocephalus pilosus* Rich. ex Cass.，*Coreopsis alba* L.，*Coreopsis corymbifolia* Buch.-Ham. ex DC.，*Coreopsis leucantha* L.，*Coreopsis leucorrhiza* Lour.，*Coreopsis multifida* DC.，*Coreopsis odorata* Poir.，*Coreopsis odorata* Lam.，*Glossogyne chinensis* Less.，*Kerneria dubia* Cass.，*Kerneria pilosa*（L.）Lowe，*Kerneria tetragona* Moench

别　名：虾钳草、蟹钳草（广东，广西）、对叉草、粘人草、粘连子（云南）、一包针、引线包（江苏、浙江）、豆渣草、豆渣菜（四川、陕西）、盲肠草（福建、广东、广西）、王八叉、小狗叉（山东等地）

英文名：blackjack，beggar tick，bur marigold，cobbler's pegs，duppy needles，farmer's friend，needle grass，spanish needle，stick tight

| 形态特征 |

三叶鬼针草为一年生草本植物。

植物茎直立，下部略带淡紫色，钝四棱形，无毛或上部被极稀疏的柔毛，基部直径可达6mm。植株高30~60cm。

茎上部叶互生，较小，羽状分裂；中、下部叶对生，长11~19cm，2回羽状深裂，裂片披针形或卵状披针形，先端尖或渐尖，边缘具不规则的细尖齿或钝齿，两面略具短毛，有长柄。下部叶较小，3裂或不分裂，通常在开花前枯萎。

中部叶具柄，长1.5~5cm，无翅，叶三出，小叶3枚，很少为具5（~7）小叶的羽状复叶，两侧小叶椭圆形或卵状椭圆形，长2~4.5cm，宽1.5~2.5cm，先端锐尖，基部近圆形或阔楔形，有时偏斜，不对称，具短柄，边缘有锯齿；顶生小叶较大，长椭圆形或卵状长圆形，长3.5~7cm，先端渐尖，基部渐狭或近圆形；具柄，柄长1~2cm，边缘有锯齿，无毛或被极稀疏的短柔毛；上部叶小，3裂或不分裂，条状披针形。

头状花序直径6~10mm，有梗，长1.8~8.5cm。总苞杯状，基部被短柔毛，苞片7~8枚，条状披针形，上部稍宽，开花时长3~4mm，结果时长5~6mm，背面褐色，具黄色边缘，内层较狭。苞片线状椭圆形，先端尖或钝，被有细短毛。花托托片椭圆形，先端钝，长4~12mm，花杂性，边缘舌状花黄色，通常有1~3朵不发育。中央管状花黄色，两性，全育，长约4.5mm，裂片5枚；雄蕊5，聚药；雌蕊1，柱头2裂。

瘦果黑色，条形，略扁，具棱，长7~13mm，宽约1mm，上部具稀疏瘤状突起及刚毛，顶端冠毛芒状，3~4枚，长2~5mm，具倒刺毛。

花期8~9月。果期9~11月。

| 生物危害 |

三叶鬼针草是常见的旱田、桑园、茶园和果园杂草，影响作物产量。该植物也是棉蚜（*Aphis gossypii* Glover）等害虫的中间寄主。

| 地理分布 |

三叶鬼针草原产于热带美洲。现广布于亚洲和美洲的热带及亚热带地区。

1857年在我国香港被报道。现分布于我国安徽、澳门、北京、福建、广东、广西、贵州、海南、河北、河南、湖北、湖南、江苏、江西、山东、山西、四川、台湾、天津、西藏、香港、云南、浙江、重庆等地。

| 传播途径 |

三叶鬼针草随进口农作物和蔬菜带入中国，由于瘦果冠毛芒刺状具倒钩，可附着于人畜和货物而到处传播。

| 管理措施 |

综合治理（Integrated Pest Management，IPM）或综合控制（Integrated Pest Control，IPC），但不包括植物卫生措施（Phytosanitary Measures）。

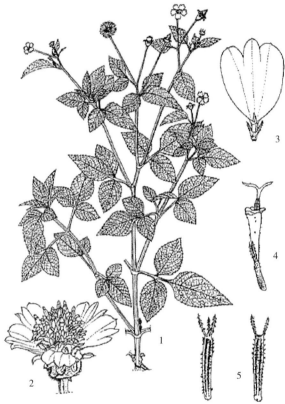
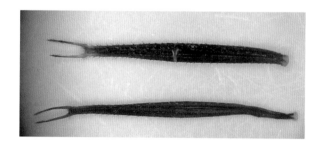
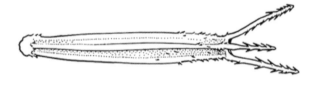

（图片选自publish.plantnet-project.org，keyserver.lucidcentral.org，pic.sogou.com，upload.wikimedia.org，newfs.s3.amazonaws.com，fioridisicilia.altervista.org，www.wnmu.edu，nathistoc.bio.uci.edu，www.westafricanplants.senckenberg.de，www.flowerspictures.org，www.pariscotejardin.fr，www.photomazza.com，www.tisanes-indigenes.re，alienplantsbelgium.be，luirig.altervista.org，i1.treknature.com，data.kew.org，database.prota.org）

三叶鬼针草（*Bidens pilosa*）

【 www.efloras.org 】 *Bidens pilosa* Linnaeus, Sp. Pl. 2: 832. 1753.

Bidens chilensis Candolle; *B. pilosa* var. *minor* (Blume) Sherff; *B. pilosa* f. *radiata* Schultz Bipontinus; *B. pilosa* var. *radiata* (Schultz Bipontinus) J. A. Schmidt; *B. pilosa* f. *rubiflora* S. S. Ying;*B. sundaica* Blume var. *minor* Blume; *Kerneria tetragona* Moench, nom. illeg. superfl.

Annuals. Stems 30-180 cm tall, glabrous or very sparsely pubescent in upper part. Petiole 10-30 (-70) mm; leaf blade either ovate to lanceolate, 30-70 (-120) mm × 12-18 (-45) mm, or pinnately 1-lobed, primary lobes 3-7, ovate to lanceolate, (10-) 25-80 mm × (5-) 10-40 mm, both surfaces pilosulose to sparsely hirtellous or glabrate, bases truncate to cuneate, ultimate margin serrate or entire, usually ciliate, apices acute to attenuate. Synflorescence of solitary capitula or capitula in lax corymbs. Capitula radiate or discoid; peduncles 10-20 (-90) mm; calycular bracts (6 or) 7-9 (-13) , appressed, spatulate to linear, (3-) 4-5 mm, abaxially usually hispidulous to puberulent, margins ciliate; involucres turbinate to campanulate, 5-6 × 6-8 mm; phyllaries (7 or) 8 or 9 (-13) , lanceolate to oblanceolate, 4-6 mm. Ray florets absent or (3-) 5-8; lamina whitish to pinkish, 5-15 mm. Disk florets 20-40 (-80) ; corollas yellowish, (2-) 3-5 mm. Outer achenes red-brown, ± flat, linear to narrowly cuneate, (3-) 4-5 mm, faces obscurely 2-grooved, sometimes tuberculate-hispidulous, margin antrorsely hispidulous, apex truncate or somewhat attenuate; inner achenes blackish, ± equally 4-angled, linear-fusiform, 7-16 mm, faces 2-grooved, tuberculate-hispidulous to sparsely strigillose, margin antrorsely hispidulous, apex attenuate; pappus absent, or of 2 or 3 (-5) erect to divergent, retrorsely barbed awns (0.5-) 2-4 mm. Fl. year-round. $2n = 24, 36, 48, 72$.

【 www.efloras.org 】 *Bidens pilosa* Linnaeus, Sp. Pl. 2: 832. 1753.

Bident poilu

Bidens alba (Linnaeus) de Candolle; *B. alba* var. *radiata* (Schultz-Bipontinus) R. E. Ballard; *B. odorata* Cavanilles;*B. pilosa* var. *radiata* (Schultz-Bipontinus) Schultz-Bipontinus

Annuals [perennials], (10-) 30-60 (-180+) [-250] cm.

Leaves: Petioles 10-30 (-70) mm; blades either ovate to lanceolate, 30-70 (-120) mm × 12-18 (-45) mm, or 1-pinnately lobed, primary lobes 3-7, ovate to lanceolate [linear], (10-) 25-80+ × (5-) 10-40+ mm [blades 2 (-3) -pinnatisect], bases truncate to cuneate, ultimate margins serrate or entire, usually ciliate, apices acute to attenuate, faces pilosulous to sparsely hirtellous or glabrate.

Heads Usually borne singly, sometimes in open, ± corymbiform arrays.

Peduncles 10-20 (-90) mm.

Calyculi (6-) 7-9 (-13) ± appressed, spatulate to linear bractlets (3-) 4-5 mm, margins ciliate, abaxial faces usually hispidulous to puberulent.

Involucres Turbinate to campanulate, 5-6 mm × (6-) 7-8 mm.

Phyllaries (7-) 8-9 (-13) , lanceolate to oblanceolate, 4-6 mm.

Ray florets 0 or (3-) 5-8+; laminae whitish to pinkish [yellowish], 2-3 or 7-15+ mm.

Disc florets 20-40 (-80+) ; corollas yellowish, (2-) 3-5 mm.

Cypselae: Outer red-brown, ± flat, linear to narrowly cuneate, (3-) 4-5+ mm, margins antrorsely hispidulous, apices ± truncate or somewhat attenuate, faces obscurely 2-grooved, sometimes tuberculate-hispidulous; inner blackish, ± equally 4-angled, linear-fusiform, 7-16 mm, margins antrorsely hispidulous, apices ± attenuate, faces 2-grooved, tuberculate-hispidulous to sparsely strigillose;

pappi 0, or of 2-3（-5）, erect to divergent, retrorsely barbed awns（0.5-）2-4 mm. **2*n*** = 24, 36, 48, 72.

【www.cabi.org】*Bidens pilosa*（blackjack）

B. pilosa seedlings have lanceolate（strap-shaped）cotyledons, 25 mm long, and purple-tinged hypocotyls. The first true leaf is similar to later leaves. Finot et al.（1996）describe the morphology of dry seed, unfolded cotyledons, first true leaf or leaf pair unfolded and two to five true leaves unfolded. Original drawings and photographs accompany each description.

The plant is an erect annual herb, 20-150 cm tall（in tall plants sometimes the branches straggling）, very variable, reproducing by seeds. Main root pivotant. Stems square, glabrous or minutely hairy, green or with brown strips. Dark green, opposite leaves on stems and branches, 4-20 cm long, up to 6 cm wide, the lower leaves simple, ovate and serrate, the upper leaves trifoliolate or imparipinnate with 2-3 pairs of pinnae and a single terminal leaflet. Petioles are 2-5 cm long.

The inflorescence is an isolated or grouped pedunculated capitula, emerging from the leaf axil. Heads borne singly at the ends of long, slender, nearly leafless branches; narrow, discoid, the disk 4-6 mm wide at anthesis; ray florets, absent or 4-7 per head, white or pale-yellow, 2-8 mm long, disk florets, 35-75 per head, yellow.

Achenes（commonly referred to as 'seeds'）linear, black or dark brown, 1-1.5 cm long, flat, 4-angled, sparsely hairy. Pappus with 2-3（-5）yellowish barbed awns, 1-2 mm long. The achenes are the dispersal units; dispersion is aided by the awns as they readily attach to animal skin, machinery and clothing.

【en.wikipedia.org】*Bidens pilosa*

Bidens pilosa is a species of flowering plant in the aster family. It is native to the Americas but it is known widely as an introduced species of other regions, including Eurasia, Africa, Australia, and the Pacific Islands. It is a tall branched weed with thin yellow flowers that develop into a cluster of barbed seeds. Its many common names include black-jack, beggar-ticks, cobbler's pegs, and Spanish needle. The seeds are like short, stiff hairs. They get stuck in feathers, fur, or socks, etc. This bur is widespread throughout the warmer regions of the world. Its little black seeds hook onto clothes or horses and thereby the bur spreads itself around. It is susceptible to hand weeding if small enough, even then must be bagged, and thick mulches may prevent it from growing. Each seed has two to four barbed spines. A weed of gardens, woodlands, and waste areas, a person who brushes against it will end up covered in the burs and need to pick them off one by one. Although this plant is considered a weed in some parts of the world, in other parts it is a source of food or medicine. For example, it is reportedly widely eaten in Africa,and in Vietnam, during the Vietnam War soldiers adopted the herb as a vegetable, which lead to it being known as the "soldier vegetable".

大狼杷草

学　名：***Bidens frondosa*** Linn.

异　名：*Bidens melanocarpa* Wiegand，*Bidens frondosa* var. *anomala*，*Bidens frondosa* var. *caudata* Sherff，*Bidens frondosa* var. *minor* Hook.，*Bidens frondosa* var. *pallida* (Wiegand) Wiegand，*Bidens frondosa* var. *stenodonta*，*Bidens melanocarpa* var. *pallida* Wiegand

别　名：接力草、外国脱力草（上海）、鬼叉、鬼针、鬼刺、夜叉头、大狼把草、外国脱力草、仙鹤草、狼把草、婆婆针、紫茎鬼针草

英文名：devil's beggar-ticks

| 形态特征 |

大狼杷草为一年生草本植物。

植物茎直立，分枝，被疏毛或无毛，常带紫色。植株高20~60cm。

植物叶对生，具柄，为一回羽状复叶，小叶3~5枚，披针形，长3~10cm，宽1~3cm，先端渐尖，边缘有粗锯齿，通常背面被稀疏短柔毛，至少顶生者具明显的柄。

头状花序单生茎端和枝端，连同总苞苞片直径12~25mm，高约12mm。总苞钟状或半球形，外层苞片5~10枚，通常8枚，披针形或匙状倒披针形，叶状，边缘有缘毛，内层苞片长圆形，长5~9mm，膜质，具淡黄色边缘。无舌状花或舌状花不发育，极不明显。筒状花两性，花冠长约3mm，冠檐5裂。

瘦果扁平，楔形，长5~10mm，近无毛或者为糙伏毛，顶端芒刺2枚，长约2.5mm，有倒刺毛。

| 生物危害 |

大狼杷草适应性强，喜于湿润的土壤上生长，常生长在荒地、路边和沟边，具有较强的繁殖能力，易形成优势群落，排挤本地植物。在低洼的水湿处及稻田的田埂上生长较多，在稻田缺水的条件下，可大量侵入田中，与农作物竞争养分，降低作物产量。

| 地理分布 |

大狼杷草原产于北美洲。现广泛归化。1926年，在江苏采到标本。现主要分布于北京、河北、辽宁、吉林、黑龙江、上海、江苏、浙江、安徽、福建、江西、山东、河南、湖北、湖南、广东、广西、海南、重庆、四川、云南、台湾等地。

| 管理措施 |

综合治理（Integrated Pest Management，IPM）或综合控制（Integrated Pest Control，IPC），但不包括植物卫生措施（Phytosanitary Measures）。

中国外来入侵物种图鉴

外来入侵植物

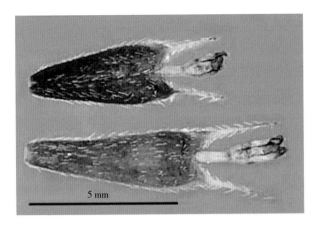

（图片选自baike.baidu.com、en.wikipedia.org、www.cabi.org、en.wikibooks.org、florafinder.com、luirig.altervista.org、newfs.s3.amazonaws.com、s3.amazonaws.com、www.fnanatureserach.org、de.academic.ru、www.opsu.edu、www.minnesotawildflowers.info、www.wildflower.org、media.eol.org、eol.org、cn.bing.com、www.discoverlife.org、upload.wikimedia.org、mikawanoyasou.org、plants.usda.gov、chestofbooks.com）

大狼杷草（*Bidens frondosa*）

【www.efloras.org】*Bidens frondosa* Linnaeus, Sp. Pl. 2: 832. 1753.

Bidens frondosa var. *anomala* Porter ex Fernald; *B. frondosa* var. *caudata* Sherff; *B. frondosa* var. *pallida* (Wiegand) Wiegand; *B. frondosa* var. *stenodonta* Fernald & H. St. John; *B. melanocarpa* Wiegand; *B. melanocarpa* var. *pallida* Wiegand.

Annuals, 20-120 cm tall. Leaves petiolate; petiole 10-40 (-60) mm; blade deltate to ovate-lanceolate overall, 30-80 (-150) mm × 20-60 (-100) mm, 3 (-5)-foliolate, leaflets petiolulate, lanceolate to ovate-lanceolate, (15-) 35-60 (-120) mm × (5-) 10-20 (-30) mm, both surfaces glabrous or hirtellous, bases cuneate, margins dentate to serrate, sometimes ciliate, apices acuminate to attenuate. Capitula radiate or discoid, usually solitary, sometimes in 2s or 3s or in lax corymbs; peduncles 10-40 (-80) mm; calycular bracts (5-) 8 (-10), ascending to spreading, spatulate or oblanceolate to linear, sometimes ± leaflike, 5-20 (-60) mm, abaxially glabrous or hirtellous, margins usually ciliate; involucres campanulate to hemispheric or broader, 6-9 mm × 7-12 mm; phyllaries 6-12, oblong or ovate to ovate-lanceolate, 5-9 mm. Ray florets 0 or 1-3+; lamina golden yellow, 2-3.5 mm. Disk florets 20-60 (-120+); corollas ± orange, 2.5-3+ mm. Achenes blackish to brown or straw-colored, ± obcompressed, obovate to cuneate, outer 5-7 mm, inner 7-10 mm, faces usually 1-veined, sometimes tuberculate, glabrous or sparsely hirtellous, margin antrorsely or retrorsely barbed, apices ± truncate to concave; pappus of 2 ± erect to spreading, antrorsely or retrorsely barbed awns 2-5 mm. Fl. Aug-Sep. $2n = 24, 48, 72$.

【www.efloras.org】*Bidens frondosa* Linnaeus, Sp. Pl. 2: 832. 1753.

Bident feuillu

Bidens frondosa var. *anomala* Porter ex Fernald; *B. frondosa* var. *caudata* Sherff; *B. frondosa* var. *pallida* (Wiegand) Wiegand; *B. frondosa* var. *stenodonta* Fernald & H. St. John; *B. melanocarpa* Wiegand

Annuals, (10-) 20-60 (-180) cm.

Leaves: Petioles 10-40 (-60) mm; blades deltate to lance-ovate overall, 30-80 (-150+) mm × 20-60 (-100+) mm, 3 (-5)-foliolate, leaflets petiolulate, lanceolate to lance-ovate, (15-) 35-60 (-120) mm × (5-) 10-20 (-30) mm, bases cuneate, margins dentate to serrate, sometimes ciliate, apices acuminate to attenuate, faces glabrous or hirtellous.

Heads Usually borne singly, sometimes in 2s or 3s or in open, corymbiform arrays.

Peduncles 10-40 (-80+) mm.

Calyculi Of (5-) 8 (-10) ascending to spreading, spatulate or oblanceolate to linear, sometimes ± foliaceous bractlets or bracts 5-20 (-60) mm, margins usually ciliate, abaxial faces glabrous or hirtellous.

Involucres Campanulate to hemispheric or broader, 6-9 mm × 7-12 mm.

Phyllaries 6-12, oblong or ovate to lance-ovate, 5-9 mm.

Ray florets 0 or 1-3+; laminae golden yellow, 2-3.5 mm.

Disc florets 20-60 (-120+); corollas ± orange, 2.5-3+ mm.

Cypselae Blackish to brown or stramineous, ± obcompressed, obovate to cuneate, outer 5-7 mm, inner 7-10 mm, margins antrorsely or retrorsely barbed, apices ± truncate to concave, faces usually 1-nerved, sometimes tuberculate, glabrous or sparsely hirtellous;

pappi Of 2 ± erect to spreading, antrorsely or retrorsely barbed awns 2-5 mm.

$2n = 24, 48, 72$.

【www.cabi.org】 *Bidens frondosa* (beggarticks)

Annual, Broadleaved, Herbaceous, Seed propagated plant

The following description has been modified from the Flora of North America (2014). Annual, (10-) 20-60 (-180) cm high. Leaves: petioles 10-40 (-60) mm long; blades deltate to lance-ovate overall, 30-80 (-150+) mm × 20-60 (-100+) mm long, 3 (-5) -foliolate, leaflets petiolulate, lanceolate to lance-ovate, (15-) 35-60 (-120) mm long × (5-) 10-20 (-30) mm across, bases cuneate, margins dentate to serrate, sometimes ciliate, apices acuminate to attenuate, faces glabrous or hirtellous. Heads usually borne singly, sometimes in 2s or 3s or in open, corymbiform arrays. Peduncles 10-40 (-80+) mm long. Calyculi (subsidiary circle of small bracts outside the involucral phyllaries) of (5-) 8 (-10) ascending to spreading, spatulate or oblanceolate to linear, sometimes ± foliaceous bractlets or bracts 5-20 (-60) mm, margins usually ciliate, abaxial faces glabrous or hirtellous. Involucres campanulate to hemispheric or broader, 6-9 mm × 7-12 mm. Phyllaries (bracts surrounding the capitulum) 6-12, oblong or ovate to lance-ovate, 5-9 mm. Ray florets 0 or 1-3+; laminae golden yellow, 2-3.5 mm. Disc florets 20-60 (-120+) mm long; corollas ± orange, 2.5-3+ mm. Cypselae (fruits) blackish to brown or stramineous, ± obcompressed, obovate to cuneate, outer 5-7 mm, inner 7-10 mm, margins antrorsely or retrorsely barbed, apices ± truncate to concave, faces usually 1-nerved, sometimes tuberculate, glabrous or sparsely hirtellous; pappi of 2 ± erect to spreading, antrorsely or retrorsely barbed awns 2-5 mm.

【en.wikipedia.org】 *Bidens frondosa*

Bidens frondosa is annual herb is usually about 20 to 60 (8-20 feet) centimeters tall, but it can reach 1.8 meters (72 inches or 6 feet) at times. The stems are square in cross-section and may branch near the top. The leaves are pinnate, divided into a few toothed triangular or lance-shaped leaflets usually up to 6 or 8 centimeters long, sometimes up to 12. The inflorescence is often a solitary flower head, but there may be pairs or arrays of several heads. The head contains many orange disc florets. It often lacks ray florets but some heads have a few small yellow rays. The fruit is a flat black or brown barbed cypsela up to a centimeter long which has two obvious hornlike pappi at one end. The barbed pappi on the fruit help it stick to animals, facilitating seed dispersal.

【近似种】狼杷草

学　　名：*Bidens tripartita* Linn.

异　　名：*Bidens acuta*（Wiegand）Britton，*Bidens orientalis* Velen.，*Bidens shimadai* Hayata，*Bidens bullata* L. syn of subsp. *Bullatus*，*Bidens hirtus* Godr. syn of subsp. *Bullatus*，*Bidens repens* D. Don syn of var. *repens*，*Bidens trifida* Buch.-Ham. ex Roxb. syn of var. *repens*

别　　名：乌阶、乌杷、郎耶草、狼反映草、小鬼叉、豆渣草、针包草、引钱包、引线包、狼耻草、切才曼巴、叉子草、老蟹叉、田边菊、鬼叉

英文名：three-lobed beggar-ticks，threelobe beggarticks，three-lobe beggarticks

| 形态特征 |

狼杷草为一年生草本植物。

植物植株高20~150cm。

植物茎圆柱或具钝棱而稍呈四方形，绿色或带紫色，基部直径2~7mm，无毛，绿色或带紫色，上部分枝或有时自基部分枝。

植物叶对生，无毛，叶柄有狭翅，椭圆形或长椭圆状披针形，边缘有锯齿。上部叶较小，披针形，3深裂或不裂；茎中、下部的叶片羽状分裂或深裂，裂片3~5，卵状披针形至狭披针形，边缘疏生不整齐大锯齿，顶端裂片通常比下方的大，叶柄有翼。

茎中部叶片长椭圆状披针形，长4~13cm，不分裂或近披针形，顶生裂片较大，两端渐狭，边缘具锯齿；叶具柄，柄长0.8~2.5cm，有狭翅。

头状花序花序单生茎端及枝端，直径1~3cm，高1~1.5cm，具较长的花序梗。总苞盘状，外层苞片5~9枚，线形或匙状倒披针形，长1~3.5cm，先端钝，叶状，具缘毛，内层苞片长椭圆形或卵状披针形，长6~9mm，膜质，褐色，有纵条纹，具透明或淡黄色的边缘。托片条状披针形，约与瘦果等长，背面有褐色条纹，边缘透明。无舌状花，全为筒状两性花。花冠长4~5mm，冠檐4裂。花药基部钝，顶端有椭圆形附器，花丝上部增宽。

瘦果扁，楔形或倒卵状楔形，长6~11mm，宽2~3mm，边缘有倒刺毛，顶端芒刺通常2~3枚，极少4枚，长2~4mm，两侧有倒刺毛。

花、果期为8~10月。

| 管理措施 |

综合治理（Integrated Pest Management，IPM）或综合控制（Integrated Pest Control，IPC），但不包括植物卫生措施（Phytosanitary Measures）。

外来入侵植物

中国外来入侵**物种图鉴**

[178]

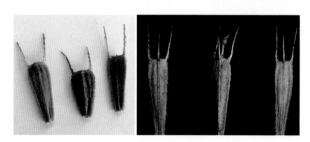
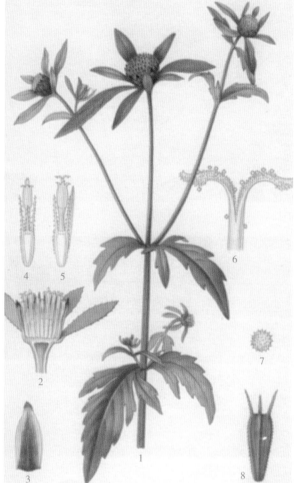

（图片选自 www.floralimages.co.uk, bugwoodcloud.org, canope.ac-besancon.fr, www2.arnes.si, botany.cz, s3.amazonaws.com, plantevaernonline.dlbr.dk, www.habitas.org.uk, biopix.com, www.skolvision.se, www.floraiwww.syngenta.fr, www.2test.nl, taliae.actaplantarum.org, www.florafinder.com, www.ipmdss.dk, biodiversite.ville-larochesuryon.fr, luirig.altervista.org, www.kuleuven-kulak.be, www.nahuby.sk, piantemagiche.it, abiris.snv.jussieu.fr, www.discoverlife.org, www.minnesotawildflowers.info, www.botanic.jp, img.botanicayjardines.com, en.wikipedia.org）

狼杷草（*Bidens tripartita*）

【近似种】三叶鬼针草　大狼杷草　狼杷草

	三叶鬼针草 （Bidens pilosa）	大狼杷草 （Bidens frondosa）	狼杷草 （Bidens tripartita）
茎	四棱形	近圆形	近圆形
成株茎颜色	绿色	常紫色	多为紫色
叶	枝梢叶对生或互生，三裂或不裂	叶对生，具柄，为一回羽状复叶，小叶3～5片，披针形	羽状复叶对生，小叶3～5片，披针形
花冠顶端	5裂	5裂	4裂
花序	黄色的舌状花	无舌状花或舌状花不发育	无舌状花，花序下有发达的叶状苞片
瘦果	细棒状，顶端有3～4个短刺	扁平，楔形，顶端有2枚芒刺	扁平，楔形或倒卵状楔形，顶端通常有2～3个短刺，少有4个短刺

小蓬草

学　　名：***Conyza canadensis*** (L.) Cronquist

异　　名：*Conyza parva*，*Erigeron canadensis* L.，*Erigeron canadensis* L. (basionym)，
Erigeron pusillus Nutt.，*Trimorpha canadensis* (L.) Lindm.

别　　名：小白酒草、加拿大蓬飞草、小飞蓬、飞蓬

英文名：Canadian fleabane，horseweed

| 形态特征 |

小蓬草为一年生草本植物。

植物具纤维状根，根纺锤状。

植物茎直立，具纵条纹，疏被长硬毛，上部分枝。植株高0.5~2.0m，全体绿色。

植物叶密集，基部叶花期常枯萎。下部叶倒披针形，长6~10cm，宽1~1.5cm，顶端尖或渐尖，基部渐狭成柄，边缘具疏锯齿或全缘。中部和上部叶较小，线状披针形或线形，近无柄或无柄，全缘或少有具1~2个齿，两面或仅上面被疏短毛，边缘常被上弯的硬缘毛。

头状花序多数，小，直径3~4mm，排列成顶生多分枝的大圆锥花序。花序梗细，长5~10mm。总苞近圆柱状，长2.5~4mm。总苞片2~3层，淡绿色，线状披针形或线形，顶端渐尖，外层约短于内层之半，背面被疏毛，内层长3~3.5mm，宽约0.3mm，边缘干膜质，无毛。花托平，直径2~2.5mm，具不明显的突起。雌花多数，舌状，白色，长2.5~3.5mm，舌片小，稍超出花盘，线形，顶端具2个钝小齿。两性花淡黄色，花冠管状，长2.5~3mm，上端具4或5个齿裂，管部上部被疏微毛。

瘦果线状披针形，长1.2~1.5mm，稍扁压，被贴微毛。冠毛污白色，1层，糙毛状，长2.5~3mm。

花期5~9月。

| 生物危害 |

小蓬草可产生大量瘦果，蔓延极快，对秋收作物、果园和茶园危害严重。

该植物还通过分泌化感物质，抑制邻近其他植物的生长。还是棉铃虫和棉蝽象的中间宿主。其叶汁和捣碎的叶对皮肤有刺激作用。

| 地理分布 |

小蓬草原产于北美洲。现广布于世界各地。1860年，在山东烟台发现。现分布于安徽、澳门、北京、福建、甘肃、广东、广西、贵州、海南、河北、河南、黑龙江、湖北、湖南、吉林、江苏、江西、辽宁、内蒙古、宁夏、青海、山东、山西、陕西、四川、台湾、天津、西藏、香港、新疆、云南、浙江、重庆。我国各地均有分布，是在我国分布最广的外来入侵物种之一。

传播途径

小蓬草瘦果可人为传播或自然扩散。

管理措施

综合治理（Integrated Pest Management，IPM）或综合控制（Integrated Pest Control，IPC），但不包括植物卫生措施（Phytosanitary Measures）。

外来入侵植物

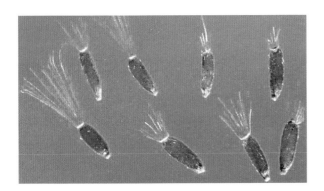

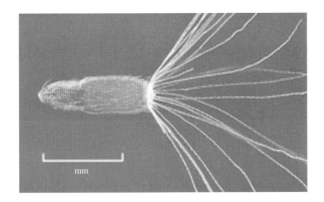

（图片选自ograsradgivaren.slu.se, healthyhomegardening.com, extension.umass.edu, pic.sogou.com, delawarewildflowers.org, www.kansasnativeplants.com, www.kuleuven-kulak.be, www.aphotoflora.com, www.dorsetnature.co.uk, nyc.books.plantsofsuburbia.com, plantevaernonline.dlbr.dk, upload.wikimedia.org, wnmu.edu, pfaf.org）

小蓬草（*Conyza canadensis*）

【www.efloras.org】*Conyza canadensis* (Linnaeus) Cronquist, Bull. Torrey Bot. Club. 70: 632. 1943.

Vergerette du Canada

Erigeron canadensis Linnaeus, Sp. Pl. 2: 863. 1753 (as canadense); *Conyza canadensis* var. *glabrata* (A. Gray) Cronquist; *C. canadensis* var. *pusilla* (Nuttall) Cronquist; *C. parva* Cronquist; *E. canadensis* var. *pusillus* (Nuttall) B. Boivin.

Plants Erect, (3-) 50-200 (-350+) cm, branched mostly distally.

Leaves: Faces usually glabrate (proximal margins ± ciliolate, hairs usually stiff, spreading and hispid on nerves, hairs erect); proximal blades oblanceolate to linear, 20-50 (-100+) mm × 4-10 (-15+) mm, toothed to entire; distal similar, smaller, entire.

Heads Usually in paniculiform, sometimes corymbiform arrays.

Involucres 3-4 mm.

Phyllaries Usually glabrous, sometimes sparsely strigose (margins chartaceous to scarious); outer greenish to stramineous, lanceolate to linear, shorter; inner stramineous to reddish, lance-attenuate to linear.

Receptacles 1-1.5 (-3) mm diam. in fruit.

Pistillate florets 20-30 (-45+); corollas ± equaling or surpassing styles, laminae 0.3-1 mm.

Disc florets 8-30+.

Cypselae Uniformly pale tan to light gray-brown, 1-1.5 mm, faces sparsely strigillose.

pappi of 15-25, white bristles 2-3 mm.

$2n$ = 18.

【www.cabi.org】*Conyza canadensis* (Canadian fleabane)

Annual, Biennial, Broadleaved, Herbaceous, Seed propagated plant

C. canadensis is an erect annual with a long taproot and one or more stems arising from a basal rosette, it is usually about 1 m high but may be much taller. Leaves are up to 10 cm long and about 1 cm wide with some shallow teeth, clear green (not greyish as in other common Conyza species), almost glabrous on the surfaces, but with some scattered hairs. Leaf margins ciliate and with longer conspicuous hairs towards the leaf base. Flower heads are very numerous on short pedicels, only 2-3 mm in diameter when fresh (broader in pressed specimens), involucral bracts about 5 mm long, glabrous. Disc florets yellow, contrasting with distinct white ray florets which are 0.5 to 1 mm long, the latter distinguishing C. canadensis from other common weedy Conyza species. Seeds 1.0-1.3 mm long with 10-25 off-white pappus hairs, 2-4 mm long (Holm et al., 1997).

【en.wikipedia.org】(*Erigeron canadensis*) *Conyza canadensis*

Erigeron canadensis is an annual plant growing to 1.5 m (60 inches) tall, with sparsely hairy stems. The leaves are unstalked, slender, 2-10 centimetres (0.79-3.94 inches) long and up to 1 cm (0.4 inches) across, with a coarsely toothed margin. They grow in an alternate spiral up the stem and the lower ones wither early. The flowers are produced in dense inflorescences 1 cm in diameter. Each individual flower has a ring of white or pale purple ray florets and a centre of yellow disc florets. The fruit is a cypsela tipped with dirty white down.

Erigeron canadensis can easily be confused with *Conyza sumatrensis,* which may grow to a height of 2

m, and the more hairy Erigeron bonariensis which does not exceed 1 m (40 inches). Erigeron canadensis is distinguished by bracts that have a brownish inner surface and no red dot at the tip, and are free (or nearly free) of the hairs found on the bracts of the other species.

【www.efloras.org】*Conyza canadensis* (**Linnaeus**) **Cronquist, Bull. Torrey Bot. Club. 70: 632. 1943.**

Erigeron canadensis Linnaeus, Sp. Pl. 2: 863. 1753 (as canadense); *Conyza canadensis* var. *glabrata* (A. Gray) Cronquist; *C. canadensis* var. *pusilla* (Nuttall) Cronquist; *C. parva* Cronquist; *E. canadensis* var. *pusillus* (Nuttall) B. Boivin

Plants erect, (3-) 50-200 (-350+) cm, branched mostly distally. Leaves: faces usually glabrate (proximal margins ± ciliolate, hairs usually stiff, spreading and hispid on nerves, hairs erect); proximal blades oblanceolate to linear, 20-50 (-100+) mm × 4-10 (-15+) mm, toothed to entire; distal similar, smaller, entire. Heads usually in paniculiform, sometimes corymbiform arrays. Involucres 3-4 mm. Phyllaries usually glabrous, sometimes sparsely strigose (margins chartaceous to scarious); outer greenish to stramineous, lanceolate to linear, shorter; inner stramineous to reddish, lance-attenuate to linear. Receptacles 1-1.5 (-3) mm diam. in fruit. Pistillate florets 20-30 (-45+); corollas ± equaling or surpassing styles, laminae 0.3-1 mm. Disc florets 8-30+. Cypselae uniformly pale tan to light gray-brown, 1-1.5 mm, faces sparsely strigillose; pappi of 15-25, white bristles 2-3 mm. $2n = 18$.

Conyza canadensis is thought to be native to North America and is now widely adventive, e.g., in South America, Europe, Asia, and Africa. Plants with stems glabrous and phyllaries red-tipped are sometimes treated as var. *pusilla*; similar plants with stems glabrous and phyllaries stramineous (not red-tipped) are sometimes treated as var. *glabrata*.

苏门白酒草

学　　名：***Conyza sumatrensis***（Retz.）Walker

异　　名：*Conyza albida* Willd. ex Sprengel，*Conyza floribunda*（H. B. & K.），*Coyza bonariensis* var. *microcephala*（Cabrera）Cabrera，*Erigeron floribundus*（Sch. Bip.），*Erigeron sumatrensis* Retz.

别　　名：洋蒿、野蒿、茵陈、茵陈蒿、竹叶艾

英文名：tall fleabane，broad-leaved fleabane，fleabane; Guernsey fleabane

| 形态特征 |

苏门白酒草为一年生或二年生草本植物。

植物具纤维状根，纺锤形，直或弯。

植物茎粗壮，直立。基部径4~6mm，具条棱，绿色或下部红紫色。中部或中部以上有长分枝，被较密灰白色上弯糙短毛，杂有展开的疏柔毛。植株高80~150cm。

植物叶密集，基部叶花期凋落。下部叶倒披针形或披针形，长6~10cm，宽1~3cm。顶端尖或渐尖，基部渐狭成柄，边缘上部每边常有4~8个粗齿，基部全缘。中部和上部叶渐小，狭披针形或近线形，具齿或全缘，两面特别下面被密糙短毛。

头状花序多数，径5~8mm，在茎枝端排列成大而长的圆锥花序。花序梗长3~5mm。总苞卵状短圆柱状，长4mm，宽3~4mm。总苞片3层，灰绿色，线状披针形或线形，顶端渐尖，背面被糙短毛，外层稍短或短于内层之半，内层长约4mm，边缘干膜质。花托稍平，具明显小窝孔，径2~2.5mm。雌花多层，长4~4.5mm，管部细长，舌片淡黄色或淡紫色，极短细，丝状，顶端具2细裂。两性花6~11个，花冠淡黄色，长约4mm，檐部狭漏斗形，上端具5齿裂，管部上部被疏微毛。

瘦果线状披针形，长1.2~1.5mm，扁压，被贴微毛。冠毛1层，初时白色，后变黄褐色。

花果期为5~10月。

| 生物危害 |

苏门白酒草为常见的区域性恶性杂草，危害性同小蓬草（*C. canadensis*（L.）Cronq.）。但植株更高大，可产生更多瘦果。

| 地理分布 |

苏门白酒草原产于南美洲。现在已成为一种热带和亚热带地区广泛分布的杂草。19世纪中期引入中国。现发生于云南、贵州、四川、广西、广东、海南、江西、浙江、福建、台湾、西藏等地。

| 传播途径 |

瘦果可人为传播或自然扩散。

|管理措施|

综合治理（Integrated Pest Management，IPM）或综合控制（Integrated Pest Control，IPC），但不包括植物卫生措施（Phytosanitary Measures）。

中国外来入侵 物种图鉴

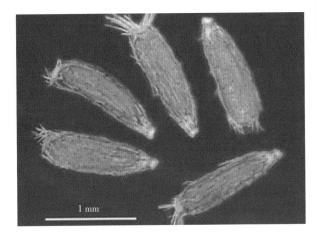
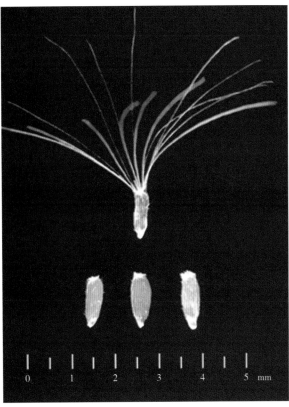
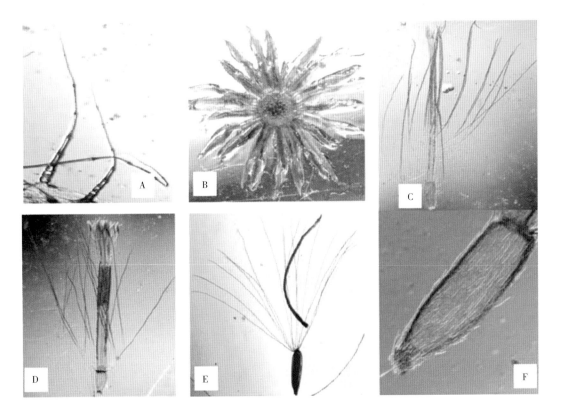

（图片选自asb.com.ar, agpest.co.nz, www.nzpcn.org.nz, www7a.biglobe.ne.jp, www.iewf.org, www.virboga.de, www.naturanelmondo.com, www.kuleuven-kulak.be, kaede.nara-edu.ac.jp, www.sarinalandcare.org.au, flore-bis.lecolebuissonniere.eu, www.british-wild-flowers.co.uk, invasoras.pt, files.stadsplantenbreda.webnode.com, www.kuleuven-kulak.be, www.researchgate.net, mikawanoyasou.org）

苏门白酒草（*Conyza sumatrensis*）

【efloras.org】*Erigeron sumatrensis* Retzius, Observ. Bot. 5: 28. 1788.

Conyza sumatrensis（Retzius）E. Walker.

Herbs, annual or biennial, 80-150 cm tall; roots fusiform. Stems erect, thick, branched above middle, densely leafy, densely gray-white strigose, sparsely hirsute. Leaves: basal withered at anthesis, lower cauline petiolate, blade oblanceolate or lanceolate, 6-10 cm × 1-3 cm, surfaces densely strigose, especially abaxially, base attenuate, margin usually coarsely 4-8-serrate per side, apex acute or acuminate, mid and upper reduced, blade narrowly lanceolate to linear, margin serrate or entire. Capitula 5-8 mm in diam., numerous, in large and long paniculiform synflorescences; peduncles 3-5 mm. Involucre campanulate to urceolate, ca. 4 × 3-4 mm; phyllaries 3-seriate, gray-green, linear-lanceolate or linear, abaxially scabrous, apex acuminate, outer short or ca. 1/2 as long as inner, inner ca. 4 mm, margin scarious. Ray florets numerous, 4-4.5 mm, lamina yellowish or purplish, short, filiform, apex 2-denticulate; disk florets 6-11, yellowish, ca. 4 mm, tube sparsely puberulent. Achenes linear-lanceoloid, compressed, 1.2-1.5 mm, strigillose. Pappus white, later yellowish brown. Fl. May-Oct.

【www.cabi.org】*Conyza sumatrensis*（tall fleabane）

Annual，Biennial，Broadleaved，Herbaceous，Seed propagated plant

An erect annual or short-lived perennial, up to 3 m tall, usually about 1-1.5 m. For full description and illustration see Kostermans et al.（1987）, Reutelingsperger（2000）.

【en.wikipedia.org】*Conyza sumatrensis*

When fully grown（in summer or autumn）, *Conyza sumatrensis* reaches one to two metres in height. Flowers are white rather than purple-pink. Its leaves are like dandelion leaves, but longer, thinner and more like primrose leaves in colour and texture. Its seeding heads are like dandelions, but straw coloured and smaller. In certain countries the plant has started to exhibit resistance to herbicides.

【invasoras.pt】*Conyza sumatrensis*（tall fleabane）

【近似种】香丝草

学　　名：*Conyza bonariensis*（L.）Cronq.

异　　名：*Conyza ambigua* DC.，*Conyza bonariensis* var. *leiotheca*（S. F. Blake）Cuatrec.，*Conyza crispus*（Pourr.）Rupr.，*Conyza linifolia*（Willd.）Tackh，*Erigeron bonariensis* L.，*Erigeron crispus* Pourr.，*Erigeron crispus* subsp. *naudinii*（Bonnet）Bonnier，*Erigeron linifolius* Willd.，*Leptilon bonariense*（L.）Small，*Leptilon linifolium*（Willd.）Small

别　　名：野塘蒿、野地黄菊、蓑衣草

英文名：hairy fleabane，Argentine fleabane，fleabane

| 形态特征 |

香丝草为一年生或二年生草本植物。

植物具纤维根，纺锤状，常斜升。

植物茎直立或斜升。中部以上常分枝，常有斜上不育的侧枝，密被贴短毛，杂有开展的疏长毛。植株高20～50cm。

植物叶密集，基部叶花期常枯萎。下部叶倒披针形或长圆状披针形，长3～5cm，宽0.3～1cm，顶端尖或稍钝，基部渐狭成长柄，通常具粗齿或羽状浅裂。中部和上部叶具短柄或无柄，狭披针形或线形，长3～7cm，宽0.3～0.5cm。中部叶具齿，上部叶全缘，两面均密被贴糙毛。

头状花序多数，径8～10mm。在茎端排列成总状或总状圆锥花序，花序梗长10～15mm。总苞椭圆状卵形，长约5mm，宽约8mm。总苞片2～3层，线形，顶端尖，背面密被灰白色短糙毛，外层稍短或短于内层之半，内层长约4mm，宽0.7mm，具干膜质边缘。花托稍平，有明显的蜂窝孔，径3～4mm。雌花多层，白色，花冠细管状，长3～3.5mm，无舌片或顶端仅有3～4个细齿。两性花淡黄色，花冠管状，长约3mm，管部上部被疏微毛，上端具5齿裂。

瘦果线状披针形，长1.5mm，扁压，被疏短毛。冠毛1层，淡红褐色，长约4mm。

花果期为5～10月。

| 生物危害 |

香丝草可产生大量瘦果。瘦果借冠毛随风扩散，蔓延极快，对秋收作物、果园和茶园危害严重，为一种常见杂草。

该种还可以分泌化感物质，抑制邻近其他植物的生长。

| 地理分布 |

香丝草原产于南美洲。现广布于热带、亚热带。在我国分布于中部、东部、南部及西南部各省区。

| 传播途径 |

瘦果可人为传播或自然扩散。

| 管理措施 |

综合治理（Integrated Pest Management，IPM）或综合控制（Integrated Pest Control，IPC），但不包括植物卫生措施（Phytosanitary Measures）。

（图片选自www.cropscience.bayer.com，www.agromonitoreo.com.ar，weeds.brisbane.qld.gov.au，upload.wikimedia. org，hasbrouck.asu.edu，www7a.biglobe.ne.jp，www.floravascular.com，www.friendsofqueensparkbushland.org.au， www.freenatureimages.eu，www.alabamaplants.co，cdn1.arkive.org，it.wikipedia.org，idtools.org，invasoras.pt，agpest. co.nz，www.floradecanarias.com，tribes.eresmas.net）

香丝草（*Conyza bonariensis*）

【近似种】苏门白酒草　小蓬草　香丝草

	小蓬草（*Conyza canadensis*）	苏门白酒草（*Conyza sumatrensis*）	香丝草（*Conyza bonariensis*）
头状花序直径	3～4 mm	4～8 mm	8～15 mm
总苞叶及苞片	苞片无茸毛，但在苞片边缘及中脉处有纤毛。苞片顶端无红点，内侧棕色。雌花具舌片	苞片密被茸毛，但在苞片端部无长毛。苞片顶端具红点，内侧淡红棕色。雌花舌片极短细，细丝状	苞片密被茸毛，在苞片端部具长毛。苞片顶端具红点，内侧淡红棕色。雌花无舌片

小蓬草（*Conyza canadensis*）、苏门白酒草（*Conyza sumatrensis*）及香丝草（*Conyza bonariensis*）

（图片选自invasoras.pt）

一年蓬

学　名：**Erigeron annuus** Pers.

异　名：*Erigeron annuum*（L.）Pers.，*Aster annuus* L.，*Stenactis annua*（L.）Cass. ex Less.，*Erigeron septentrionalis*（Fernald & Wiegand）Holub

别　名：女菀、野蒿、牙肿消、牙根消、千张草、墙头草、长毛草、地白菜、油麻草、白马兰、千层塔、治疟草、瞌睡草、白旋覆花。

英文名：annual fleabane，daisy fleabane

| 形态特征 |

一年蓬为一年生或二年生草本植物。

植物茎粗壮，直立。上部绿色，有分枝，被较密的短硬毛，毛上弯；下部被开展的长硬毛；基部径6mm。植株高30～150cm。

植物基部叶长圆形或宽卵形，少有近圆形，长4～17cm，宽1.5～4cm，或更宽。顶端尖或钝，基部狭成具翅的长柄，边缘具粗齿。下部叶与基部叶同形，但叶柄较短。中部和上部叶较小，长圆状披针形或披针形，长1～9cm，宽0.5～2cm，顶端尖，具短柄或无柄，边缘有不规则的齿或近全缘，最上部叶线形，全部叶边缘被短硬毛，两面被疏短硬毛，或有时近无毛。

头状花序数个或多数，排列成疏圆锥花序，长6～8mm，宽10～5mm。总苞半球形。总苞片3层，草质，披针形，长3～5mm，宽0.5～1mm，近等长或外层稍短，淡绿色或多少褐色，背面密被腺毛和疏长节毛。外围的雌花舌状，2层，长6～8mm，管部长1～1.5mm，上部被疏微毛，舌片平展，白色，或有时淡天蓝色，线形，宽0.6mm，顶端具2小齿，花柱分枝线形。中央的两性花管状，黄色，管部长约0.5mm，檐部近倒锥形，裂片无毛。

瘦果披针形，长约1.2mm，扁压，被疏贴柔毛。冠毛异形，雌花的冠毛极短，膜片状连成小冠，两性花的冠毛2层，外层鳞片状，内层为10～15条长约2mm的刚毛。

花期为6～9月。

| 生物危害 |

一年蓬可产生大量具冠毛的瘦果。瘦果可借冠毛随风扩散，蔓延极快，对秋收作物、桑园、果园和茶园危害严重，亦可入侵草原、牧场、苗圃造成危害，也常入侵山坡湿草地、旷野、路旁、河谷或疏林下，排挤本土植物。

该植物还是害虫地老虎的宿主。

| 地理分布 |

一年蓬原产于北美洲。现广布北半球温带和亚热带地区。1827年，在我国澳门发现。目前我国除内蒙古、宁夏和海南外，各地均有采集记录。

| 传播途径 |

瘦果可人为传播或自然扩散。

| 管理措施 |

综合治理（Integrated Pest Management，IPM）或综合控制（Integrated Pest Control，IPC），但不包括植物卫生措施（Phytosanitary Measures）。

（图片选自www.botanickafotogalerie.cz，www.minnesotawildflowers.info，www.missouriplants.com，www.flowerspictures.org，newfs.s3.amazonaws.com，www.undkraut.de，www.monde-de-lupa.fr，www.discoverlife.org，delawarewildflowers.org，gobotany.newenglandwild.org，en.wikipedia.org，luirig.altervista.org，chestofbooks.com，wildebloemen.info，clopla.butbn.cas.cz）

一年蓬（*Erigeron annuus*）

【www.efloras.org】 *Erigeron annuus* （Linnaeus） Persoon, Syn. Pl. 2: 431. 1807.

Herbs, annual, [10-]30-100[-150] cm tall. Stems erect, branched in upper part, sparsely hispid, strigose above (hairs spreading). Leaves: surfaces strigose-hirsute or sometimes glabrate, eglandular; basal withered at anthesis, winged petiolate, blade elliptic or broadly ovate, rarely spatulate, 4-17 cm × 1.5-4 cm or more, base attenuate, margin coarsely serrate, scabrous, apex acute or obtuse; lower cauline similar, shortly petiolate, mid and upper shortly petiolate or sessile, blade oblong-lanceolate or lanceolate, 1-9 cm × [0.3-]0.5-2 cm, margin irregularly serrate to subentire, apex acute, uppermost linear. Capitula 5-50+, in loose paniculiform or corymbiform synflorescences, 6-8 mm × 10-15 mm. Involucre hemispheric; phyllaries 2- or 3 (or 4) -seriate, greenish or ± brownish, abaxially sparsely hirsute, minutely glandular, subequal or outer shorter, lanceolate, 3-5 mm × 0.5-1 mm, herbaceous. Ray florets 80-125, 2-seriate, [4-]6-8[-10] mm, tube 1-1.5 mm, sparsely hairy above, lamina white or sometimes bluish, linear, ca. 5 × 0.6 mm, flat, tardily coiling; disk florets yellow, 2-2.8 mm, sparsely hairy, lobes glabrous. Achenes lanceolate, flattened, [0.8-]1.2 mm, sparsely strigillose. Pappus 2-seriate, outer of scales or setae, inner absent in ray florets, in disk florets of [8-]10-15 long bristles. Fl. Jun-Sep.

【www.efloras.org】 *Erigeron annuus* （Linnaeus） Persoon, Syn. Pl. 2: 431. 1807.

Eastern daisy fleabane, vergerette annuelle

Aster annuus Linnaeus, Sp. Pl. 2: 875. 1753; *Erigeron annuus* var. *discoideus* （Victorin & J. Rousseau） Cronquist

Annuals, (10-) 60-150 cm; fibrous-rooted or taprooted.

Stems Erect, sparsely piloso-hispid (hairs spreading), sometimes strigose distally, eglandular.

Leaves Basal (usually withering by flowering) and cauline; basal blades mostly lanceolate to oblanceolate or ovate, 15-80 mm × 3-20 mm, margins coarsely serrate to nearly entire, faces sparsely strigoso-hirsute, eglandular; cauline lanceolate to oblong, little reduced proximal to midstem.

Heads ca. 5-50 + in loosely paniculiform or corymbiform arrays.

Involucres 3-5 mm × 6-12 mm.

Phyllaries In 2-3 (-4) series, sparsely villous or hirsuto-villous, minutely glandular.

Ray florets 80-125; corollas white, 4-10 mm, laminae tardily coiling.

Disc corollas 2-2.8 mm.

Cypselae 0.8-1 mm, 2-nerved, faces sparsely strigose;

pappi: Outer minute crowns of setae or narrow scales, inner 0 (rays) or of 8-11 bristles (disc).
$2n = 27$.

【en.wikipedia.org】 *Erigeron annuus*

Erigeron annuus is a herbaceous plant with alternate, simple leaves, and green, sparsely hairy stems, which can grow between 30 and 150 centimeters (about 1 to 5 feet) in height. Leaves are numerous and large relative to other species of *Erigeron*, with lower leaves, especially basal leaves, coarsely toothed or cleft, a characteristic readily distinguishing this species from most other *Erigeron*. Upper leaves are sometimes, not always toothed, but may have a few coarse teeth towards the outer tips. The flower heads are white with yellow centers, with the white to pale lavender rays, borne spring through fall depending on the individual plant. Ray florets number 40-100.

【gobotany.newenglandwild.org】*Erigeron annuus*（L.）Pers.

【florafinder.com】*Erigeron annuus*

【wizzley.com】Annual Fleabane or Eastern Daisy Fleabane, *Erigeron annuus*

假臭草

学　　名：***Praxelis clematidea***（Grisebach.）King et Robinson

异　　名：*Chrysocoma pauciflora* Vell. Conc.，*Eupatorium catarium* Veldkamp，*Eupatorium clematideum* Griseb.，*Eupatorium pauciflorum* auct. non Kunth，*Eupatorium urticifolium* auct. non L.，*Eupatorium urticifolium* var. *clematideum*（Griseb.）Hieron.，*Eupatorium urticifolium* var. *clematideum*（Griseb.）Hieron. ex Kuntze，*Eupatorium urticifolium* var. *clematideum* Chodat，*Eupatorium urticifolium* var. *nana* Hiero，*Eupatorium urticifolium* var. *nanum* Hieron.

别　　名：猫腥菊

英文名：praxelis

| 形态特征 |

假臭草为亚灌木或一年生草本植物。

植株高0.3～1m，多分枝。全株被长柔毛。

植物茎直立或上斜，单一或于下部分枝，散生贴伏的短柔毛和腺状短柔毛。

植物叶对生，卵形或长椭圆状卵形，具腺点。边缘齿状，先端急尖，基部圆楔形。叶长1.5～5.5cm，宽1～3.5cm，具3出脉或不明显的5出脉。叶柄长0.3～2cm。上部叶较小，通常披针形。

头状花序生于茎、枝端，花序有长梗，排成疏松的伞房花序。花序梗的毛长约0.2mm。总苞半球形或宽钟状，直径3～6mm。小花25～30mm，蓝紫色。

瘦果黑色或黑褐色，长1～1.5mm，具白色冠毛，具3～5棱。

花果期全年。

| 生物危害 |

假臭草所到之处，其他低矮草本逐渐被排挤。在我国华南地区的果园中，它能迅速覆盖整个果园地面。由于其对土壤肥力吸收能力强，便极大地消耗土壤中的养分，对土壤的可耕性造成严重破坏，从而极大地影响了作物生长。

此外，假臭草还能分泌一种有毒恶臭物质，影响家畜觅食。

| 地理分布 |

假臭草原产于南美。现散布于东半球热带地区。20世纪80年代在香港发现。现分布于澳门、福建、广东、广西、海南、台湾、香港、云南等地。

| 传播途径 |

瘦果可人为传播或自然扩散。

| 管理措施 |

综合治理（Integrated Pest Management，IPM）或综合控制（Integrated Pest Control，IPC），但不包括植物卫生措施（Phytosanitary Measures）。

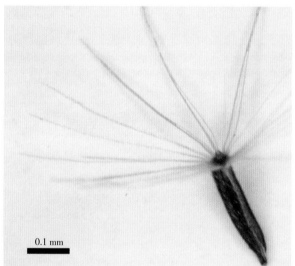

（图片选自www.hear.org，cdn.xl.thumbs.canstockphoto.com，kplant.biodiv.tw，npuir-3d.npust.edu.tw，e.share.photo.xuite.net，cn.bing.com，keyserver.lucidcentral.org，pic.sogou.com，www.natureloveyou.sg，weeds.dpi.nsw.gov.au）

假臭草（*Praxelis clematidea*）

【www.efloras.org】*Praxelis clematidea* R. M. King & H. Robinson, Phytologia. 20: 194. 1970.

Eupatorium clematideum Grisebach, Abh. Königl. Ges. Wiss. Göttingen 24: 172. 1879, not (Wallich ex Candolle) Schultz Bipontinus (1866); *E. urticifolium* Linnaeus f. var. *clematideum* Hieronymus ex Kuntze; *E. catarium* Veldkamp.

Subshrubs or annual herbs, to 0.6 m tall. Stems erect or ascending, bright green, simple or poorly branched at base, leafy throughout except at base, pubescent throughout, hairs simple, eglandular. Leaves opposite, rank-smelling; petiole 3-7 mm; blade ovate, 20-35 mm × 12-25 mm, pubescent below, hairs long, simple, eglandular along venation, stipitate glandular and gland-dotted between veins, base attenuate, margin coarsely serrate, apex acute. Synflorescence terminal, corymbose, capitula pedunculate; peduncles 4-7 mm, pubescent; involucre narrowly campanulate, 4-5 mm in diam.; phyllaries 2- or 3-seriate, with scattered short simple eglandular hairs at base, glabrous apically, margin ciliate, apex long attenuate; receptacle conical, epaleate. Florets 35-40; corollas bright lilac-blue, ca. 4.5 mm; corolla lobes long papillose on inner surface, usually glabrous outside or with few simple eglandular hairs; anther appendages longer than wide, apex acute; style base not swollen, glabrous, bright lilac-blue; style branches coarsely papillose. Achenes 2-2.5 mm, 3-5-ribbed, ribs pale, setuliferous or glabrous, body black, with scattered setulae; pappus setae 3.5-4.5 mm, coarsely barbellate, off-white. Fl. and fr. often year-round.

【keyserver.lucidcentral.org】*Praxelis clematidea* (Griseb.) R.M. King & H. Robinson

Distinguishing Features

A short-lived or long-lived herbaceous plant usually growing 20-80 cm tall.

Its stems are hairy and bear leaves in pairs.

Its leaves have deeply toothed margins with about 5-8 teeth on each side.

The leaves emit a foul odour when crushed. its small flower-heads are purplish, blue or lilac in colour and its 'seeds' are topped with a ring bristles (3-4 mm long).

Stems and Leaves The branched stems are rounded (i.e. terete) or angular and covered in long hairs (i.e. they are hirsute).

The leaves are oppositely arranged along the stems and borne on stalks (i.e. petioles) 3-20 mm long. The leaf blades (2.5-6 cm long and 1-4 cm wide) are egg-shaped in outline (i.e. ovate) to somewhat diamond-shaped (i.e. rhomboid) and have pointed tips (i.e. acute apices). They are hairy (i.e. hirsute), particularly underneath, and have deeply toothed (i.e. coarsely dentate) margins with about 5-8 large teeth on each side. The leaves emit a foul odour when crushed.

Flowers and Fruit The small flower-heads (i.e. capitula) are borne in dense clusters at the tips of the branches (i.e. in terminal corymbs or cymes). Each flower-head (7-10 mm long and 4.5 mm wide) is borne on a hairy stalk (i.e. hirsute pedicel) 2-10 mm long. These flower-heads consist of numerous (25-50) tiny flowers (i.e. tubular florets) surrounded by about 20 green or yellowish-green coloured bracts (i.e. an involucre). The elongated (i.e. lanceolate or linear) bracts are sometimes purplish towards their pointed tips (i.e. acuminate or acute apices) and are hairless or have a few close-lying hairs (i.e. glabrous to appressed strigose). The tiny flowers (i.e. tubular florets) are purplish, blue or lilac in colour and consist of four or five petals that are fused into tiny tube (i.e. corolla tube) 3.5-4.8 mm long. Flowering occurs mostly during summer and autumn (i.e. from January to May), but some flowers may be present throughout the year.

The 'seeds' (i.e. achenes) are black in colour (2-3 mm long) and are topped with a ring (i.e. pappus) of 15-40 bristles (3-4 mm long).

【 keys.trin.org.au 】*Praxelis clematidea*

Stem Stems 30-100 cm tall with the leafy twigs clothed in hairs.

Leaves Leaf blades 20-80 mm × 12-52 mm, clothed in hairs and gland-dotted especially on the lower surface. Petioles 4-20 mm long. Leaves and twigs emit an unpleasant odour resembling cat's urine when crushed.

Flowers Each flower head subtended by bracts about 5 mm long. Flowers 25-30 per flower head. Each flower about 2 mm diam. Calyx, i.e. pappus, about 4 mm long. Corolla 4-4.5 mm long, lower half white, upper half purple. Anthers fused together, each anther about 1.5 mm long, filaments about 2 mm long. Ovary about 2 mm long. Style + stigma about 7 mm long.

Fruit Fruits 2.5-3 mm long and the pappus persistent, 4-4.5 mm long. Embryo about 2 mm long, the cotyledons about 1 mm long and the radicle also about 1 mm long. Seeds black, 2.5-3 mm long with a pale tuft of finely barbed bristles, 3-4 mm long.

Seedlings Cotyledons ca. 5 × 4 mm, petiole ca. 5 mm long. First leaf hairy, toothed 5-9 mm × 2-3 mm. Tenth leaf stage, blades 5.4-6 cm × 2.5 cm, toothed, petioles 12 mm long. Growth bud hairy. Whole seedling hairy.

【 www.natureloveyou.sg 】*Praxelis clematidea*

刺苍耳

学　名：*Xanthium spinosum* L.
异　名：*Xanthium spinosum* var. *ambrosioides*（Hook. & Arn.）Love & Dans.，*Xanthium spinosum* var. *heterocephalum* Widder，X*anthium spinosum* var. *inerme* Bel
英文名：bathurst burr，dagger cocklebur，daggerweed，prickly burweed，spiny burweed，spiny clotburr，spiny cocklebur，thorny burweed

| 形态特征 |

刺苍耳为一年生草本植物。

植物茎直立，上部多分枝，节上具三叉状棘刺。植株高30~60cm。

植物叶狭卵状披针形或阔披针形，长3~8cm，宽6~30mm，边缘3~5浅裂或不裂，全缘，中间裂片较长，长渐尖，基部楔形，下延至柄，背面密被灰白色毛。叶柄细，长5~15mm，被茸毛。

花单性，雌雄同株。雄花序球状，生于上部。总苞片一层。雄花管状，顶端裂。雄蕊5。雌花序卵形，生于雄花序下部。总苞囊状，长8~14mm，具钩刺，先端具2喙，内有2朵无花冠的花。花柱线形，柱头2深裂。

总苞内有2个长椭圆形瘦果。

花期8~9月，果期9~10月。

| 生物危害 |

刺苍耳为我国检疫性有害生物。其全株有毒，以果实最毒。鲜叶比干叶毒。嫩枝比老叶毒。一般中毒症状出现较晚，常于误食后二日发病。患者表现为上腹胀闷，恶心呕吐、腹痛，有时腹泻、乏力、烦躁。重者则造成肝损伤，出现黄疸，毛细血管渗透性增高而出血，甚至导致昏迷、惊厥，以及呼吸、循环功能或肾功能衰竭，乃至死亡。

本种可入侵农田，危害白菜、小麦、大豆等旱地作物。对牧场危害也比较严重。

| 地理分布 |

刺苍耳为原产于南美洲。现在欧洲中、南部，亚洲和北美归化。1974年，在北京丰台区发现。现分布于安徽、北京、河北、河南、辽宁、内蒙古、宁夏、新疆等地。

| 传播途径 |

果实具钩刺，常随人和动物传播，或混在作物种子中散布。

| 管理措施 |

综合治理（Integrated Pest Management，IPM）或综合控制（Integrated Pest Control，IPC），包括植物卫生措施（Phytosanitary Measures）。

（图片选自www.unavarra.es，www.bayer-agri.fr，keys.lucidcentral.org，species.wikimedia.org，botany.cz，canope.ac-besancon.fr，waste.ideal.es，acorral.es，i39.servimg.com，upload.wikimedia.org）

刺苍耳（*Xanthium spinosum*）

【www.efloras.org】*Xanthium spinosum* Linnaeus, Sp. Pl. 2: 987. 1753.

Acanthoxanthium spinosum（Linnaeus）Fourreau; *Xanthium cloessplateaum* D. Z. Ma; *X. spinosum* var. *inerme* Bel.

Herbs, annual, 10-60（-120）cm; nodal spines usually in pairs, simple or 2- or 3-partite, 15-30 mm. Petiole 1-15（-25）mm; leaf blade ± ovate to lanceolate or lanceolate-linear, 4-8（-12）cm × 1-3（-5）cm, often pinnately 3（-7）-lobed, abaxially gray to white, densely strigose. Burs 10-12（-15）mm. Fl. Jul-Oct. $2n = 36$.

【www.efloras.org】*Xanthium spinosum* Linnaeus, Sp. Pl. 2: 987. 1753.

Spiny cocklebur, clotbur, lampourde épineuse

Xanthium ambrosioides Hooker & Arnott; *X. spinosum* var. *inerme* Bel

Plants 10-60（-120+）cm; nodal spines usually in pairs, simple or 2-3-partite, 15-30+ mm.

Leaves: petioles 1-15（-25+）mm; blades ± ovate to lanceolate or lance-linear, 4-8（-12+）cm × 1-3（-5+）cm, often pinnately 3（-7+）-lobed, abaxial faces gray to white, densely strigose.

Burs 10-12（-15+）mm.

$2n = 36$.

【www.cabi.org】*Xanthium spinosum*（bathurst burr）

Annual，Broadleaved，Herbaceous，Seed propagated plant

X. spinosum is a much branched annual herb, generally erect and somewhat woody, often 0.3-0.6 m in height, but sometimes up to 1 m tall and 1.5 m across. Stems are striate, yellowish or brownish grey and finely pubescent. True leaves are lanceolate, entire, irregularly toothed or lobed, mostly three-lobed with the center lobe much longer than the other two, 3-8 cm long, 0.6-2.6 cm wide. They are hairy（glabrous or strigose）and a dull grey-green colour above, and paler and downy（silvery-tomentulose）beneath, with a conspicuous white midrib, and each on short petioles approximately 1 cm long. Each leaf base is armed at the axil with three-pronged yellow spines usually up to 2.5 cm long, often opposite in pairs.

Flower heads are in axillary clusters or solitary. Flowers are inconspicuous, greenish, and monoecious; male flowers in almost globular heads in axils of upper most leaves, and female flowers in axils of lower leaves, developing into a burr. The burr is two-celled, oblong, nearly egg-shaped, slightly flattened, 10-13 mm long, 4 mm wide, pale yellowish to brown covered with yellowish hairs, more or less striate, glandular, covered with numerous slender, hooked, glabrous spines up to 3 mm long from more or less thickened bases, with the two apical beaks short and straight. Each burr contains two flattened, thick-coated, dark brown or black seeds, about 1 cm long, the lower germinating first.

【keyserver.lucidcentral.org】*Xanthium spinosum* L.

Distinguishing Features

An upright and much-branched herbaceous plant usually growing 30-100 cm tall.

Its stems are armed with yellowish three-pronged spines（15-50 mm long）in the leaf forks.

Its leaves are usually irregularly lobed, with dark green and shiny upper surfaces and pale green lower surfaces covered in downy hairs.

Male flowers are borne in dense clusters near the tips of the stems, while separate female flowers are borne in the leaf forks.

Its stalkless 'burrs' (8-15 mm long) are covered in numerous small hooked spines (2-3 mm long).

Stems and Leaves The stems are greenish-yellow when young and are covered with fine hairs (i.e. finely pubescent). They are armed with spines that occur singly or in pairs at the base of each leaf stalk (i.e. in the leaf axils). These spines are usually three-pronged from near their bases and may appear to be several spines at first glance. They are yellow or greenish-white in colour with prongs 15-50 mm long.

The alternately arranged leaves (2-10 cm long and 6-30 mm wide) are borne on stalks (i.e. petioles) up to 30 mm long. The lower leaves are usually irregularly three-lobed, or occasionally with five lobes, with the middle lobe much larger than the others. However, on upper leaves the side lobes may be insignificant or absent, thereby giving the leaf blade an elongated (i.e. lanceolate) shape. The leaf upper surfaces are dark green and shiny with prominent whitish-coloured veins, while their undersides are pale green or whitish in colour with a dense covering of downy hairs.

Flowers and Fruit Separate male and female (i.e. unisexual) flower-heads are produced on different parts of the same plant (i.e. this species is monoecious). Male flower-heads consist of numerous tiny flowers (i.e. florets) that are arranged in dense rounded clusters. These male flower-heads are borne at the tips of the stems, and are yellowish or creamy-white in colour. The greenish-coloured female flower-heads are borne singly or in small clusters in the upper leaf forks (i.e. axils), usually below the male flower-heads. Flowering occurs from late spring through to early autumn, but is most abundant during summer.

The fruit (8-15 mm long and 4-6 mm wide) is greenish when young, later becoming yellowish or straw-coloured, then eventually brownish as it matures. It is an oval-shaped (i.e. ellipsoid) 'burr' containing two seeds. These 'burrs' are stalkless (i.e. sessile), finely hairy, and covered in numerous small hooked spines (2-3 mm long). They also have two small, straight, spines or 'beaks' at the tip (1-2 mm long), which may be difficult to distinguish from the hooked spines. These fruit are mostly formed during late summer and autumn. The brown or black seeds (about 10 mm long) are flattened, and one of each pair is slightly larger than the other.

【en.wikipedia.org】*Xanthium spinosum*

Xanthium spinosum is an annual herb producing a slender stem up to 1 metre (3 feet 3 inches) tall or slightly taller. It is lined at intervals with very long, sharp, yellowish spines which may exceed three centimeters in length and may divide into two or three separate spines.

The leaves are divided into linear or lance-shaped lobes, the middle much longer than the others, and are arranged alternately all along the stem. Each is up to 10 or 12 centimeters long and dark green or grayish on top with a white underside.

The plant produces male and female flower heads, the female heads developing into burs one or 1.5 centimeters long and covered in thin spines.

【wiki.bugwood.org】*Xanthium spinosum*

Appearance *Xanthium spinosum* is an erect, rigid, much-branched annual herb, 11.8-39.4 in. (30-100 cm) tall and up to 59.1 inches (150 cm) or more wide.

Stems are striate, yellowish or brownish gray, and finely pubescent.

Foliage Leaves are lanceolate, entire, toothed or lobed, 1.2-3.1 inches (3-8 cm) long, 0.24-1.02 inches (6-26 mm) wide, glabrous or strigose above, and silvery-tomentulose beneath. They are dull gray-green above with a conspicuous white midrib and short petioles 0.4 inches (1 cm). Each leaf base is armed at the axil with yellow three-pronged spines 0.8-2 inches (2-5 cm) long, often opposite in pairs.

Flowers The monoecious flowers may be solitary or found in axillary clusters. Greenish flowers are inconspicuous.

Fruits Fruits are yellowish egg shaped burs covered with hooked glabrous spines. Each fruit contains two seeds.

圆叶牵牛

学　　名：*Ipomoea purpurea*（L.）Roth

异　　名：*Ipomoea diversifolia* Lindl.，*Ipomoea glandulifera* Ruiz & Pav.，*Ipomoea hirsutula* J. Jacq.，*Ipomoea hispida* Zucc.，*Ipomoea purpurea* var. *diversifolia*（Lindl.）O'Donell，*Ipomoea purpurea* var. *purpurea*，*Pharbitis diversifolius* Lindl.，*Pharbitis hispida*（Zucc.）Choisy，*Pharbitis nil* var. *diversifolia*（Lindl.）Choisy，*Pharbitis purpurea* Asch. in Schweinf.，*Pharbitis purpurea*（L.）Voigt

别　　名：牵牛花、喇叭花、紫花牵牛、圆叶旋花、小花牵牛

英 文 名：tall morning glory，common morning glory，tall morning-glory

形态特征

圆叶牵牛为多年生攀援草本植物。

植物茎缠绕，多分枝。茎长2~3m，被短柔毛和倒向的长硬毛。

植物叶互生。叶片宽卵圆形，顶端渐尖，基部心形，全缘。叶柄长5~9cm。

花腋生，1~5朵，总花梗与叶柄近等长。苞片线形，长6~7mm。萼片5，长圆形，长1~1.6cm，基部被开展的长硬毛。花冠漏斗状，直径4~6cm，紫色、淡红色或白色，无毛。雄蕊5，内藏，不等长，花丝基部被短柔毛。雌蕊内藏。子房无毛，3室，每室2胚珠。柱头头状，3裂。

蒴果近球形，直径9~10mm，无毛，3瓣裂。

种子黑色或禾秆色，卵球状，三棱形，表面粗糙，无毛或种脐处疏被柔毛。

花期5~10月。果期8~11月。

生物危害

圆叶牵牛属于旱田、果园及苗圃杂草，可缠绕和覆盖其他植物，导致后者生长不良。

地理分布

圆叶牵牛原产于南美洲。现已经在世界各地广泛栽培和归化。主要是由于人为引种而引起传播和扩散。1890年，在我国已有栽培记录。现分布于安徽、北京、福建、甘肃、广东、广西、贵州、海南、河北、河南、湖北、湖南、吉林、江苏、江西、辽宁、内蒙古、宁夏、青海、山东、山西、陕西、上海、四川、台湾、天津、西藏、香港、新疆、黑龙江、云南、浙江、重庆等地。

传播途径

人为传播或自然扩散。

|管理措施|

综合治理（Integrated Pest Management，IPM）或综合控制（Integrated Pest Control，IPC），但不包括植物卫生措施（Phytosanitary Measures）。

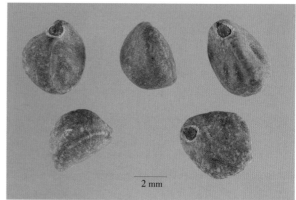

（图片选自www.viarural.com.ar，bugwoodcloud.org，newfs.s3.amazonaws.com，www.aanbodpagina.nl，pics.davesgarden.com，keys.lucidcentral.org，http2.mlstatic.com，www.lubera.com，facultystaff.richmond.edu，www.conabio.gob.mx，keyserver.lucidcentral.org，climbers.lsa.umich.edu，commons.wikimedia.org，botanix.org）

圆叶牵牛（*Ipomoea purpurea*）

【www.efloras.org】*Ipomoea purpurea* (Linnaeus) Roth, Bot. Abh. 27. 1787.

Convolvulus purpureus Linnaeus, Sp. Pl. ed. 2. 1: 219. 1762; *Ipomoea chanetii* H. Léveillé; *I. hispida* Zuccarini; *Pharbitis hispida* Choisy; *P. purpurea* (Linnaeus) Voigt.

Herbs annual, twining; axial parts short pubescent and long retrorse hirsute. Stems 2-3 m. Petiole 2-12 cm; leaf blade circular-ovate or broadly ovate, 4-18 cm × 3.5-16.5 cm, ± strigose, base cordate, margin entire or ± 3-lobed, apex acute or ± abruptly acuminate. Inflorescences 1-5-flowered; pedun-cle 4-12 cm; bracts linear, 6-7 mm, spreading hirsute. Pedicel recurved before and after anthesis, 1.2-1.5 cm. Sepals subequal, 1.1-1.6 cm, spreading hirsute abaxially in basal 1/2; outer 3 oblong, apex acuminate; inner 2 linear-lanceolate. Corolla red, reddish purple, or blue-purple, with a fading to white center, funnelform, 4-6 cm, glabrous. Stamens included, unequal; filaments pubescent basally. Pistil included; ovary glabrous, 3-loculed. Stigma 3-lobed. Capsule subglobose, 9-10 mm in diam., 3-valved. Seeds black or straw colored, ovoid-trigonous, glabrous or hilum sparsely pilose. $2n = 30, 32$.

【www.cabi.org】*Ipomoea purpurea* (tall morning glory)

Annual, Herbaceous, Seed propagated Vine / climber plant

Herbaceous vine, twining, 2-3 m in length. Stems cylindrical, slender, pilose or hirsute. Leaves alternate; blades simple, 2-10 cm × 2-10 cm, cordiform or deeply trilobed, the lobes ovate orlanceolate, chartaceous, strigulose on both surfaces, the apex acuminate, the base cordiform, the margins entire or slightly sinuate, ciliate; upper and lower surface with the veins slightly prominent; petioles 2.5-6 cm long, slender, strigulose, sulcate. Flowers solitary or in simple dichasia, axillary; peduncles longer than the petioles; bracts subulate, approximately 3 mm long, not forming an involucre. Calyx green, of 5 subequal sepals, 8-16 mm long, chartaceous, oblong lanceolate, the outer ones slightly broader than the inner ones, acute at the apex, hirsute outside on the basal portion; corolla blue, purple, pink, or with lines (forming a star) of these colours on a white background, 4-4.5 cm long, the throat white, limb with shallow, rounded lobes; stamens and stigmas pink, not exserted. Capsule, 9-10 mm in diameter, glabrous, the pericarp thin, with the chartaceous sepals persistent at the base; seeds 4 per fruit, pyriform, 3-4 mm long, black, glabrous (Acevedo-Rodríguez, 2005).

【climbers.lsa.umich.edu】*Ipomoea purpurea* (L.) Roth

Vegetative Plant Description: This annual plant has simple, alternate, pinnately veined leaves with entire margins. Leaves are ovate with cordate bases and acute or acuminate tips, ranging from 1-11 cm in length and 1-12 cm in width. The stems are branched or simple, loosely pubescent to tomentose with short appressed trichomes. They are sparsely hirsute to glabrate. The petioles range from 1-14 cm long.

Flower Description: Inflorescences with 1-5 flowers are borne in a cymose cluster. The calyx is 5-lobed, densely hirsute at the base, and more glabrous toward the tips. The sepals are sub-equal. The outer sepals are narrowly ovate to elliptic ranging from 0.8-1.5cm long and 2.5-4.5mm wide. The inner sepals are ovate, 0.8 to 1.5cm long and 2.5 to 3mm wide. The glabrous, funnelshaped corollas range from white to blue to purple, however white is always found within the corolla. The flowers vary from 2.5-5cm long and 2.4-5cm wide. The five epipetalous stamens range from 8-10mm long with anthers that are 1.5-2mm long. The 3-locular glabrous ovary is superior. The shape varies from ovoid to conic. The glabrous style ranges from 14-22mm long and there are 3 stigmas.

Fruit Type and Description: The fruit is a subglobose to ovoid capsule that is approximately 1cm in diameter and up to 2.5cm long. It is 6-valvate and contains 3-6 seeds.

Seed Description: The seed surface is granular, dull brown to black in color, and densely covered with small, brown hairs. The seeds are from 4 to 5.7 mm long. They are wedge shaped, with a horseshoe-shaped scar.

Distinguished by:*Ipomoea coccinea* grows to a similar height and can grow tangled in amongst other *Ipomoea* species. Once in flower, *I. coccinea* and *I. purpurea* are easily distinguished by dark orange to red salverform corollas of *I. coccinea,* however both *I. purpurea* and *I. hederacea* have light blue to purple funnelform corollas. *Ipomoea purpurea* and *Ipomoea hederacea* are distinguishable by sepal characteristics: *I. purpurea*'s sepals are ovate-lanceolate with an acute to abruptly acuminate apex. The lobes are shorter to slightly longer than the body. In contrast, *I. hederacea*'s sepals are lanceolate, long-attenuate, and caudate with lobes that are much longer than the body. Species in the genus *Hedera* have two leaf types: palmately lobed juvenile leaves and unlobed cordate adult leaves. The green-yellow flowers produced in late autumn are borne in 3-5 cm diameter umbels unlike *Ipomoea hederacea*（purple, white, or rosy large funnelform corollas）. It is very easy to distinguish *Hedera* from *Ipomoea* because *Hedera* grows using adventitious roots. Fruits of species from the genus *Hedera* are small black berries, while *I. purpurea*'s are capsules. *Ipomoea lacunosa* typically has a white corolla with pink to purple anthers. Flowers of *I. lacunose* are usually fewer per inflorescence than those of *I. purpurea* .

【keys.lucidcentral.org】*Ipomoea purpurea*（L.）**Roth**

Ipomoea purpurea is a herbaceous annual twining climber. Stems of Ipomoea purpurea are hairy and may be trailing or twinning. The leaf blade is ovate（egg-shaped in outline with broad end at base）, entire or 3-lobed, acuminate（gradually tapering to a sharp point）at the apex, cordate（heart-shaped）at the base, glabrous or pubescent（hairy）.

Flowers of *I. purpurea* are solitary or in few-flowered cymes. The stalk of the inflorescence（peduncle）is up to 12 cm long. Sepals finely pubescent all over; corolla is white, pink or magenta, and white below.

【gobotany.newenglandwild.org】*Ipomoea purpurea*（L.）**Roth**

【近似种】提琴叶牵牛

学　　名：*Ipomoea pandurata*（L.）G. F. W. Mey.

异　　名：*Convulvulus candicans*，*C. panduratus* L.

别　　名：野生马铃薯藤

英文名：man of the earth，wild potato vine，manroot，wild sweet potato，wild rhubarb，big-root morning-glory，man-root，wild-potato，wild-sweet-potato-vine，Indian potato

| 形态特征 |

提琴叶牵牛为多年生藤本植物。

植物具块状储藏根，垂直，长可达50cm。

植物茎长达3.5m。分枝或不分枝，有毛或无毛。

植物叶互生。心形，边缘锯齿或全缘。正面绿色，反面浅绿色，长15cm，宽11cm，有毛或无毛。叶柄长9cm。

聚伞花序腋生，小花2～10朵。花梗2cm，无毛。花冠白色，花瓣内部底侧为暗红色。雄蕊5枚，不均等，贴生于花管中。花丝顶端无毛，基部稍微膨胀、有毛。花药为白色或粉红色，长8mm。子房上位，绿色，圆锥形，长1～2mm，2室，每室2个胚珠；中轴胎座，子房底部有绿色的蜜腺。柱头宽2～3mm，变干后为棕色。萼片5，不等长，分离，相互重叠，有皱折，最大的长2.5cm，宽1.6cm，光滑无毛，椭圆形或卵圆形，紫色。

蒴果具2～4粒浅褐色种子，表面有毛。

花期5～9月。

| 生物危害 |

提琴叶牵牛为我国检疫性有害生物。可对湿地植物造成危害。

| 传播途径 |

通过种子传播。

| 管理措施 |

综合治理（Integrated Pest Management，IPM）或综合控制（Integrated Pest Control，IPC），包括植物卫生措施（Phytosanitary Measures）。

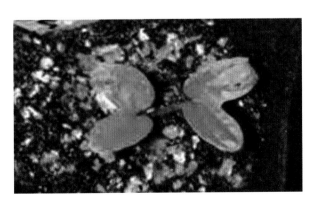

中国外来入侵**物种图鉴**

外来入侵植物

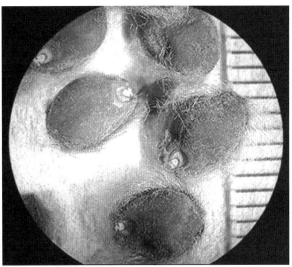
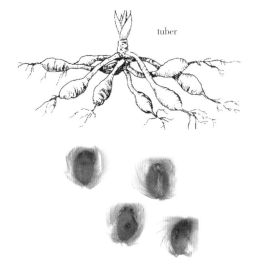
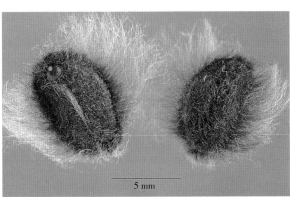

（图片选自pics.davesgarden.com，oak.ppws.vt.edu，www.florafinder.com，luirig.altervista.org，www.discoverlife.org，www.oardc.ohio-state.edu，meltonwiggins.com，climbers.lsa.umich.edu，www.namethatplant.net，www.exot-nutz-zier.d，www.missouriplants.com，src.sfasu.edu，www.pacificbulbsociety.org，www.asergeev.com，www.eattheweeds.com，s3.amazonaws.com，extension.missouri.edu，lowres-picturecabinet.com，s3-eu-west-1.amazonaws.com）

提琴叶牵牛（*Ipomoea pandurata*）

垂序商陆

学　名：**Phytolacca americana** L.
异　名：*Phytolacca decandra* L.，*Phytolacca rigida* Small，*Phytolacca thyrsiflora* FENZL
别　名：十蕊商陆、美商陆、美洲商陆、美国商陆、洋商陆、见肿消
英文名：pokeweed，American pokeweed，common pokeweed，Pigeonberry，Pokeberry

| 形态特征 |

垂序商陆为多年生草本植物。

植物根粗壮，肥大，倒圆锥形。

植物茎直立，圆柱形，有时带紫红色。植株高1～2m。

植物叶片椭圆状卵形或卵状披针形，长9～18cm，宽5～10cm，先端急尖，基部楔形。叶柄长1～4cm。

总状花序，顶生或侧生，长5～20cm，花序较纤细。花梗长6～8mm。花白色，微带红晕，直径约6mm。花被片5，雄蕊、心皮及花柱通常均为10，心皮合生。

果序下垂。浆果扁球形，熟时紫黑色。

种子肾圆形，直径约3mm，表面光滑。

花期为6～8月。果期8～10月。

| 生物危害 |

垂序商陆环境适应性强，生长迅速，在营养条件较好时，植株高可达2m，易形成单优群落，主茎有的能达到3cm粗，可与其他植物竞争养分。其茎具有多数开展的分枝，叶片宽阔，能覆盖其他植物体，导致其他植物生长不良甚至死亡。

该种具有较为肥大的肉质直根，消耗土壤肥力。

垂序商陆全株有毒，根及果实毒性最强，对人和牲畜均有毒害作用。由于其根酷似人参，常被人误做人参服用，人取食后会造成腹泻症状。

| 地理分布 |

垂序商陆原产于北美。现已经在世界各地引种或归化。1935年，在杭州采到此物种标本，之后在我国各地广泛逸生。现主要分布于北京、天津、河北、山西、辽宁、上海、江苏、浙江、安徽、福建、江西、山东、河南、湖北、湖南、广东、广西、重庆、四川、贵州、云南、陕西、甘肃、新疆、台湾、香港等地。

| 传播途径 |

人为传播或自然扩散。种子可通过鸟类进行远距离传播。

|管理措施|

综合治理（Integrated Pest Management，IPM）或综合控制（Integrated Pest Control，IPC），但不包括植物卫生措施（Phytosanitary Measures）。

中国外来入侵 物种图鉴

（图片选自bugwoodcloud.org，www.efferus.no，meltonwiggins.com，www.sbs.utexas.edu，upload.wikimedia.org，www.missouriplants.com，www.maltawildplants.com，plants.ces.ncsu.edu，media.eol.org，www.ortobotanico.unina.it，en.wikipedia.org，luirig.altervista.org，www.asergeev.com，www.vaplantatlas.org，pic.baike.soso.com，actaplantarum.org，www.wildflower.org，www.floravascular.com）

垂序商陆（*Phytolacca americana*）

【www.efloras.org】*Phytolacca americana* Linnaeus, Sp. Pl. 1: 441. 1753.

Phytolacca decandra Linnaeus.

Herbs perennial, 1-2 m tall. Root obconic, thick. Stems erect, sometimes reddish purple, terete. Petiole 1-4 cm; leaf blade elliptic-ovate or ovate-lanceolate, 9-18 cm × 5-10 cm, base cuneate, apex acute. Racemes terminal or lateral, 5-20 cm. Pedicel 6-8 mm. Flowers ca. 6 mm in diam. Tepals 5, white, slightly red. Stamens, carpels, and styles 10; carpels connate. Infructescence pendent. Berry purple-black when mature, oblate. Seeds reniform-auricular, ca. 3 mm. Fl. Jun-Aug, fr. Aug-Oct. $2n = 18^*, 36^*$.

【www.efloras.org】*Phytolacca americana* Linnaeus, Sp. Pl. 1: 441. 1753.

Pokeweed, poke, pokeberry

Plants to 3 (-7) m.

Leaves: Petiole 1-6 cm; blade lanceolate to ovate, to 35 × 18 cm, base rounded to cordate, apex acuminate.

Racemes Open, proximalmost pedicels sometimes bearing 2-few flowers, erect to drooping, 6-30 cm; peduncle to 15 cm; pedicel 3-13 mm.

Flowers: Sepals 5, white or greenish white to pinkish or purplish, ovate to suborbiculate, equal to subequal, 2.5-3.3 mm; stamens (9-) 10 (-12) in 1 whorl; carpels 6-12, connate at least in proximal 1/2; ovary 6-12-loculed.

Berries Purple-black, 6-11 mm diam.

Seeds Black, lenticular, 3 mm, shiny.

$2n = 36$.

【en.wikipedia.org】*Phytolacca americana*

Perennial herbaceous plant which can reach a height of 10 feet (3 meters), but is usually 4 ft (1.2 m) to 6 ft (2 m). However, the plant must be a few years old before the root grows large enough to support this size. The stem is often red as the plant matures. There is an upright, erect central stem early in the season, which changes to a spreading, horizontal form later in the season with the weight of the berries. Plant dies back to roots each winter. Stem has a chambered pith.

Leaves: The leaves are alternate with coarse texture with moderate porosity. Leaves can reach sixteen inches in length. Each leaf is entire. Leaves are medium green and smooth with what some characterize as an unpleasant odor.

Flowers: The flowers have 5 regular parts with upright stamens and are up to 0.2 inches (5 mm) wide. They have white petal-like sepals without true petals, on white pedicels and peduncles in an upright or drooping raceme, which darken as the plant fruits. Blooms first appear in early summer and continue into early fall.

Fruit: A shiny dark purple berry held in racemose clusters on pink pedicels with a pink peduncle. Pedicels without berries have a distinctive rounded five part calyx. Fruits are round with a flat indented top and bottom. Immature berries are green, turning white and then blackish purple.

Root: Thick central taproot which grows deep and spreads horizontally. Rapid growth. Tan cortex, white pulp, moderate number of rootlets. Transversely cut root slices show concentric rings. No nitrogen fixation ability.

【www.missouribotanicalgarden.org】*Phytolacca americana*

Phytolacca americana, commonly known as pokeweed, common poke or scoke, is a vigorous, herbaceous perennial that typically grows to 4-10 inches tall with a spread to 3-5 inches wide. This plant features (a) showy reddish-purple stems, (b) large, alternate, lanceolate green leaves (each 5-10 feet long spreading to 2-4 feet wide) which exude an unpleasant (some say fetid) aroma when bruised, (c) apetulous, bisexual, summer flowers (to 3/4 feet wide) which bloom July to September in slender racemes to 8 feet long, each flower composed of five showy petal-like greenish-white sepals, 10 stamens and a pistil composed of united carpels, (d) grape-like fruits (each to 1/4 feet across) which emerge green but mature to a deep reddish-purple, and (e) very large taproots which will grow to 12 feet long and 4 feet thick.

Flower racemes are typically erect when in bloom but begin to droop as the fruit develops. Berries typically appear about 30 days after the flowers bloom. As this pokeweed continues to bloom from mid-summer to fall, it will begin to simultaneously display flowers and fruits in various stages of development.

【近似种】商陆

学　　名：*Phytolacca acinosa* Roxb.

异　　名：*Phytolacca acinosa*，*Phytolacca acinosa* f. *insularis*（Nakai）M.Kim，*Phytolacca acinosa* var. *esculenta*（Van Houtte）Maxim. Variety，*Phytolacca acinosa* var. *kaempferi*（A.Gray）Makino variety，*Phytolacca esculenta* Van Houtte，*Phytolacca insularis* Nakai，*Phytolacca kaempferi* A. Gray，*Phytolacca latbenia*（Moq.）H. Walter，*Phytolacca pekinensis* Hance，*Pircunia esculenta*（van Houtte）Moq.，*Pircunia latbenia* Moq.，*Sarcoca acinosa*（Roxb.）V. Skalick，*Sarcoca esculenta*（van Houtte）V. Skalick，*Sarcoca latbenia*（Moq.）Skalický，*Sarcoca latbenia*（Moquin）V. Skalick

别　　名：章柳、山萝卜、见肿消、倒水莲、金七娘、猪母耳、白母鸡

英文名：Indian pokeweed，Indian poke，Indian poke eng，Indian pokeweed eng，Asiatische Kermesbeere

| 形态特征 |

商陆为多年生草本植物。

植物根肥大，肉质，倒圆锥形，外皮淡黄色或灰褐色，内面黄白色。

植物茎直立，圆柱形，有纵沟，肉质，绿色或红紫色，多分枝。植株高0.5~1.5m，全株无毛。

植物叶片薄纸质，椭圆形、长椭圆形或披针状椭圆形，长10~30cm，宽4.5~15cm，顶端急尖或渐尖，基部楔形，渐狭，两面散生细小白色斑点（针晶体），背面中脉凸起。叶柄长1.5~3cm，粗壮，上面有槽，下面半圆形，基部稍扁宽。

总状花序，顶生或与叶对生，圆柱状，直立，通常比叶短，密生多花。花序梗长1~4cm；花梗基部的苞片线形，长约1.5mm，上部2枚小苞片线状披针形，均膜质；花梗细，长6~10（13）mm，基部变粗。花两性，直径约8mm。花被片5，白色、黄绿色，椭圆形、卵形或长圆形，顶端圆钝，长3~4mm，宽约2mm，大小相等，花后常反折。雄蕊8~10，与花被片近等长，花丝白色，钻形，基部成片状，宿存，花药椭圆形，粉红色。心皮通常为8，有时少至5或多至10，分离。花柱短，直立，顶端下弯。柱头不明显。

果序直立。

浆果扁球形，直径约7mm，熟时黑色。

种子肾形，黑色，长约3mm，具3棱。

| 生物危害 |

除了红根有剧毒外，商陆具有经济意义，如可以作绿肥、用作聚锰植物（改良土壤），亦可用于提取生物农药（防治钉螺）等。

| 地理分布 |

美国（威斯康星州）、欧洲、朝鲜、日本、印度、锡金、不丹、缅甸、越南。我国除东北、内蒙古、青海、新疆外均有分布，主要分布于河南、湖北、山东、浙江、江西等地。

| 传播途径 |

根和种子均可繁殖,因此也可以传播扩散。

| 管理措施 |

综合治理(Integrated Pest Management,IPM)或综合控制(Integrated Pest Control,IPC),但不包括植物卫生措施(Phytosanitary Measures)。

（图片选自pestid.msu.edu，unkraeuter.info，flowers2.la.coocan.jp，www.botanische-spaziergaenge.at，asperupgaard.dk，zhiwutong.com，warehouse1.indicia.org.uk，www.alpinegardensociety.net，www.anniesannuals.com，www.tuinadvies.be，plantis.info，commons.wikimedia.org，flora-emslandia.de，pflanzenbestimmung.info，img.fotocommunity.com，digituin.tuinadvies.be，luirig.altervista.org，flora.nhm-wien.ac.at）

商陆（*Phytolacca acinosa*）

垂序商陆　商陆

	垂序商陆（*Phytolacca americana*）	商陆（*Phytolacca acinosa*）
茎	茎直立，圆柱形，有时带紫红色	茎直立，圆柱形，有纵沟，肉质，绿色或红紫色，多分枝
花序	花序、果序下垂，雄蕊10枚，心皮10，心皮合生	花、果序直立，雄蕊8～10枚，心皮5～10，通常8，分离
浆果	呈球形	呈扁球形

光荚含羞木

学　　名：*Mimosa bimucronata*（DC.）Kuntze
异　　名：*Acacia bimucronata* DC.，*Mimosa sepiaria* Benth.，*Mimosa stuhlmannii* Harms，*Mimosa thyrsoidea* Griseb.
别　　名：簕仔树、光叶含羞草
英文名：Thorny Mimosa，Giant Sensitive Plant

| 形态特征 |

光荚含羞木为落叶灌木植物。

植物小枝圆柱状，小枝无刺，密被黄色茸毛。植株高3～6m。

植物叶为二回羽状复叶，羽片6～7对，长2～6cm。叶轴无刺，被短柔毛。小叶12～16对，线形，长5～7mm，宽1～1.5mm，革质，先端具小尖头，除边缘疏具缘毛外，其余无毛。

头状花序，球形，花白色。花萼杯状。花瓣长约2mm，基部连合。雄蕊8枚，花丝长4～5mm。

荚果带状，劲直，长3.5～4.5cm，宽约6mm，无刺毛，褐色，通常有5～7个荚节。成熟时荚节脱落而残留荚缘。

花果期几乎全年。

| 生物危害 |

光荚含羞木常生于村边、溪流边、果园及荒地中，适应性强，具有较强的抗逆性，生长迅速，栽后当年就能长到2m左右；具有较强的竞争能力，能在短时间内形成单优群落，排挤本地物种，造成严重的生态或经济损害。该种入侵性很强，在我国已侵入自然保护区内，威胁当地生物多样性。

| 地理分布 |

光荚含羞木原产热带美洲。20世纪50年代由广东中山县（现为中山市）旅美华侨从美国引入我国，现主要分布于我国福建、江西、湖南、广东、广西、海南、云南、香港、澳门等地。

| 传播途径 |

人为传播或自然扩散。

| 管理措施 |

综合治理（Integrated Pest Management，IPM）或综合控制（Integrated Pest Control，IPC），但不包括植物卫生措施（Phytosanitary Measures）。

L.— F Bl., G Hülse, H S. von

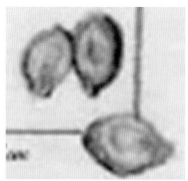 

光荚含羞木（*Mimosa bimucronata*）与粒糙叶含羞木（*Mimosa scabrella* Benth.）种子

（图片选自pic.sogou.com，tropical.theferns.info，www.fpcn.net，cn.bing.com，www.ufrgs.br，upload.wikimedia.org，baike.sogou.com，lh3.googleusercontent.com，image.slidesharecdn.com，www.clickmudas.com.br）

光荚含羞木（*Mimosa bimucronata*）

【www.efloras.or】*Mimosa bimucronata* (Candolle) O. Kuntze, Revis. Gen. Pl. 1: 198. 1891.

Acacia bimucronata Candolle, Prodr. 2: 469. 1825; *Mimosa sepiaria* Bentham.

Shrubs, deciduous, 3-6 m tall. Branchlets unarmed in distal parts, in lower parts armed by recurved prickles to 1 cm, densely yellow tomentose. Pinnae 4-9 pairs, 1.5-8 cm; rachis unarmed, pubescent; leaflets 12-16 pairs, linear, 5-7 mm × 1-1.5 mm, leathery, glabrous to puberulent with ciliate margin, main vein near upper side, apex mucronate. Heads globose, forming a spreading panicle with compound, spreading lower branches. Flowers white, scented. Calyx cup-shaped, minute. Petals oblong, 2.5-4 mm, connate at base. Stamens 8; filaments 4-5 mm. Ovary initially glabrous. Legume brown, straight, strap-shaped, 3.5-4.5 × ca. 0.6 cm, unarmed, finely reticulate veined, usually with 4-8 segments. Seeds olivaceous, ovoid, compressed, ca. 4.5 mm. $2n = 26*$

【powo.science.kew.org】*Mimosa bimucronata* (DC.) Kuntze

Habit Shrub or small tree up to 10 m high; stems varying from densely pubescent or puberulous to almost glabrous, and also ± sparsely armed with scattered straight or slightly recurved prickles 2-10 mm. long.

Leaves Leaves unarmed, petiole 0.3-1.7 cm long; rhachis 1.3-9.5 cm long, with 3-9 pairs of pinnae; leaflets 10-30 pairs (the lowest pair very reduced, ± equal, subulate), 4-12 mm × 0.8-2.6 mm, linear-oblong, venation basal and pinnate, prominent beneath; margins ± ciliate, not setulose.

Flowers Flowers whitish, in subglobose pedunculate heads 0.7-1.7 cm in diam., clustered 1-4 together along the leafless branches of a terminal panicle; clusters leafless or with pinnate bracts up to 0.3 mm long.

Calyx Calyx 0.3-0.5 mm long.

Corolla Corolla c. 1.5-2.5 mm long.

Stamens Stamens 8 (10).

Fruits Pods brown, 2-6 cm × 0.5-0.7 cm, without bristles or prickles, glabrous or almost so, breaking up transversely into segments c. 5-7 mm. long, the margins persisting as an empty frame.

【plantsoftheworldonline.org】*Mimosa bimucronata* (DC.) Kuntze

Leaves Leaves unarmed, petiole 0.3-1.7 cm long; rhachis 1.3-9.5 cm long, with 3-9 pairs of pinnae; leaflets 10-30 pairs (the lowest pair very reduced, ± equal, subulate), 4-12 mm × 0.8-2.6 mm, linear-oblong, venation basal and pinnate, prominent beneath; margins ± ciliate, not setulose.

Flowers Flowers whitish, in subglobose pedunculate heads 0.7-1.7 cm in diam., clustered 1-4 together along the leafless branches of a terminal panicle; clusters leafless or with pinnate bracts up to 0.3 mm long.

Calyx Calyx 0.3-0.5 mm long.

Corolla Corolla c. 1.5-2.5 mm long.

Stamens Stamens 8 (10).

Fruits Pods brown, 2-6 cm × 0.5-0.7 cm, without bristles or prickles, glabrous or almost so, breaking up transversely into segments 0.5-7 mm long, the margins persisting as an empty frame.

【近似种】含羞草

学　　名：*Mimosa pudica* Linn.

异　　名：*Mimosa hispidula* Kunth，*Mimosa pudica* var. *pudica*

别　　名：感应草、知羞草、呼喝草、怕丑草、见笑草、夫妻草、害羞草

英文名：bashful mimosa，humble plant，sensitive plant，shame plant，tickle-me plant，touch-me-not

形态特征

含羞草为披散、亚灌木状草本植物。

植物茎圆柱状，具分枝，有散生、下弯的钩刺及倒生刺毛。植株可高可达1m。

植物托叶披针形，长5～10mm，有刚毛。羽片和小叶触之即闭合而下垂；羽片通常2对，指状排列于总叶柄之顶端，长3～8cm；小叶10～20对，线状长圆形，长8～13mm，宽1.5～2.5mm，先端急尖，边缘具刚毛。

头状花序圆球形，直径约1cm，具长总花梗，单生或2～3个生于叶腋；花小，淡红色，多数；苞片线形；花萼极小；花冠钟状，裂片4，外面被短柔毛；雄蕊4枚，伸出于花冠之外；子房有短柄，无毛；胚珠3～4颗，花柱丝状，柱头小。

荚果长圆形，长1～2cm，宽约5mm，扁平，稍弯曲；荚缘波状，具刺毛，成熟时荚节脱落，荚缘宿存。

种子卵形，长3.5mm。

花期3～10月。果期5～11月。

生物危害

含羞草为入侵物种。含羞草碱o-β-D-葡萄糖甙有微毒，可致皮肤细胞的毛囊衰败，从而引起头发、眉毛变黄，甚至脱落。还会引起白内障和生长抑制。

地理分布

含羞草原产热带美洲，已广布于世界热带地区。目前在我国台湾、福建、广东、广西、云南等地有分布。

传播途径

人为传播或自然扩散。

管理措施

综合治理（Integrated Pest Management，IPM）或综合控制（Integrated Pest Control，IPC），但不包括植物卫生措施（Phytosanitary Measures）。

中国外来入侵物种图鉴

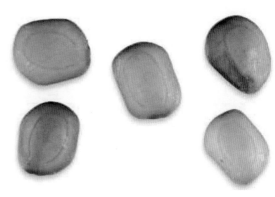

（图片选自zhidao.baidu.com，baike.baidu.com，pic.sogou.com，www.cabi.org，upload.wikimedia.org，stuartxchange.com）

含羞草（*Mimosa pudica*）

【近似种】巴西含羞草

学　　名：*Mimosa invisa* Mart. ex Colla

（1）巴西含羞草（原变种）

学　　名：*Mimosa invisa* Mart. ex Colla var. *invisa*

| 形态特征 |

植物亚灌木状草本。

植物茎直立、攀援或平卧，长达60cm，五棱柱状，沿棱上密生钩刺，其余被疏长毛，老时毛脱落。

植物叶为二回羽状复叶，长10~15cm；总叶柄及叶轴有钩刺4~5列；羽片（4）7~8对，长2~4cm；小叶（12）20~30对，线状长圆形，长3~5mm，宽约1mm，被白色长柔毛。

头状花序花时连花丝直径约1cm，1个或2个生于叶腋，总花梗长5~10mm；花紫红色，花萼极小，4齿裂；花冠钟状，长2.5mm，中部以上4瓣裂，外面稍被毛；雄蕊8枚，花丝长为花冠的数倍。子房圆柱状，花柱细长。

荚果长圆形，长2~2.5cm，宽4~5mm，边缘及荚节有刺毛。

花果期3~9月。

| 生物危害 |

巴西含羞草（原变种）为侵害性物种，覆盖植物。含有皂素，牲畜误食会引起中毒。

| 地理分布 |

巴西含羞草（原变种）原产于巴西。目前在我国广东有分布。

| 传播途径 |

人为传播或自然扩散。

| 管理措施 |

综合治理（Integrated Pest Management，IPM）或综合控制（Integrated Pest Control，IPC），但不包括植物卫生措施（Phytosanitary Measures）。

（图片选自luirig.altervista.org，www.jungleseeds.co.uk，www.plant.csdb.cn，keyserver.lucidcentral.org，www.foodmate.net，idao.cirad.fr）

巴西含羞草（原变种）（*Mimosa invisa* Mart. ex Colla var. *invisa*）

【近似种】巴西含羞草

学　　名：*Mimosa invisa* Mart. ex Colla

（2）无刺含羞草（变种）

学　　名：*Mimosa invisa* Mart. ex Colla var. *inermis* Adelb.

别　　名：无刺巴西含羞草

英文名：Spineless mimosa

| 形态特征 |

与巴西含羞草（原变种）近似，但区别在于：该变种植株茎上无钩刺，荚果边缘及荚节上无刺毛。

| 生物危害 |

无刺含羞草（变种）为侵害性物种，可作胶园覆盖植物，但全株有毒，牲畜误食后能中毒致死。

| 地理分布 |

无刺含羞草（变种）原产于印度尼西亚爪哇岛。目前在我国广东、海南、云南等省有种植。

| 传播途径 |

人为传播或自然扩散。

| 管理措施 |

综合治理（Integrated Pest Management，IPM）或综合控制（Integrated Pest Control，IPC），但不包括植物卫生措施（Phytosanitary Measures）。

（图片选自baike.baidu.com，tupian.baike.com，pic.sogou.com）

无刺含羞草（变种）（*Mimosa invisa* Mart. ex Colla var. *inermis* Adelb.）

五爪金龙

学　名：*Ipomoea cairica*（L.）Sweet
异　名：*Convolvulus adansonii* Desrousseaux，*Convolvulus cairicus* L.，*Convolvulus pendulus*
（R. Br.）Spreng.，*Convolvulus repens* Willdenow，*Convolvulus repens* Vahl，
Convolvulus reptans Linnaeus，*Ipomoea palmata* Forssk.，*Ipomoea pendula* R. Br.，
Ipomoea repens Roth，*Ipomoea reptans* Poiret，*Ipomoea stipulacea* Jacq.，*Ipomoea*
subdentata Miquel，*Ipomoea tuberculata*（Desr.）Roem. & Schult.
别　名：假土瓜藤、黑牵牛、牵牛藤、上竹龙、五爪龙
英文名：five-fingered morningglory，cairo morningglory（USA），coast morningglory
（Australia）

| 形态特征 |

五爪金龙为多年生草质藤本植物。

植物茎缠绕，灰绿色，常有小瘤状突起，有时平滑。茎长达5m。

植物叶互生。叶片指状，5深裂几达基部，直径5~9cm。裂片椭圆状披针形，先端近钝但有小锐尖，两面均无毛，边缘全缘或最下一对裂片有时再分裂。叶柄长2~4cm。

花序有花1~3朵，腋生。总花梗短。萼片5，不等大，长4~9mm，边缘薄膜质，外轮萼片较大，先端钝，并具小凸尖。花冠漏斗状，粉红色至紫红色，长5~7cm，径4.5~5cm，顶端5浅裂；雄蕊5，内藏。子房3室，花柱长，柱头2裂，头状。

蒴果近球形，直径约1cm。

种子黑褐色，长约5mm，密被绒毛。

花果期几乎全年。

| 生物危害 |

五爪金龙分布于海拔100~600m的荒地、海岸边的矮树林、灌丛、人工林、山地次生林等生境。常缠绕在其他乔灌木上，覆盖其林冠，使其无法得到足够的阳光而慢慢枯死。目前在我国南方已成为园林中的常见有害杂草。

| 地理分布 |

一般认为五爪金龙原产热带亚洲或非洲，也有学者认为原产于热带美洲，现已广泛栽培或归化于泛热带。

根据Dunn&Tutcher（1912年）记载，该种当时已在我国香港归化，攀于乔木和灌丛上。现主要分布于江苏、福建、广东、广西、海南、贵州、云南、台湾、香港、澳门等地。

| 传播途径 |

人为传播或自然扩散。

| 管理措施 |

综合治理（Integrated Pest Management，IPM）或综合控制（Integrated Pest Control，IPC），但不包括植物卫生措施（Phytosanitary Measures）。

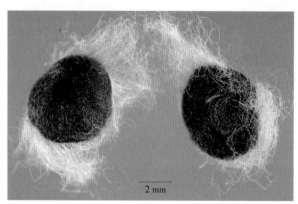

（图片选自asb.com.ar，keyserver.lucidcentral.org，gardenbreizh.org，upload.wikimedia.org，www.botanic.jp，pic.sogou.com，en.wikipedia.org，plant.cila.cn）

五爪金龙（*Ipomoea cairica*）

【www.efloras.org】 *Ipomoea cairica* (Linnaeus) Sweet, Hort. Brit., ed. 1. 287. 1827.

Herbs perennial, twining, with a tuberous root; axial parts glabrous. Stems to 5 m, thinly angular, ± tuberculate or smooth. Petiole 2-8 cm, base with leafy pseudostipules; leaf blade palmately 5-parted to base; lobes entire or minutely undulate, apex acute or obtuse, mucronulate, basal pair usually again lobed or parted; middle lobe larger, ovate, ovate-lanceolate, or elliptic, (2.5-) 4-5 cm × (0.5-) 2-2.5 cm. Inflorescences 1- or several flowered; peduncle 2-8 cm; bracts and bracteoles early deciduous, squamiform, small. Pedicel 0.5-2 cm, sometimes verruculose. Sepals unequal; outer 2 sepals 4-6.5 mm; inner ones 5-9 mm, glabrous, abaxially ± verruculose, margin paler, scarious. Corolla pink, purple, or reddish purple, with a darker center, rarely white, funnelform, (2.5-) 5-7 cm. Stamens included, unequal. Ovary glabrous. Stigma 2-lobed. Capsule ± globose, ca. 1 cm. Seeds black, ca. 5 mm, densely tomentose, margin with longer hairs. $2n = 30^*$.

【keyserver.lucidcentral.org】 *Ipomoea cairica*

Distinguishing Features

A rampant climber or creeper with hairless slender stems.

Its very distinctive leaves have 5-7 finger-like lobes.

Its large purple, purplish-pink or whitish tubular flowers (4-6 cm long and 5-8 cm across) have a darker centre.

Its small capsules (10-12 mm across) turn brown as they mature and contain four seeds. these seed are partly covered in long silky hairs.

Stems and Leaves The slender stems are hairless (i.e. glabrous), grow in a twining habit, and sometimes produce roots at the joints (i.e. nodes). The alternately arranged leaves (3-10 cm long and 3-10 cm wide) are divided into five or seven narrow lobes, like the fingers of a hand (i.e. they are palmately lobed). These leaves are hairless (i.e. glabrous) and borne on stalks (i.e. petioles) 2-6 cm long.

Flowers and Fruit The funnel-shaped (i.e. tubular) flowers are purple to pinkish-purple (occasionally white) with a darker purple centre. They are borne singly or in small clusters on short stalks originating in the leaf forks (i.e. axils). These flowers (4-6 cm long and 5-8 cm across) have five petals that are fused into a tube (i.e corolla tube) and five small sepals (4-7 mm long). Flowering occurs throughout most of the year.

The fruit capsules are more or less globular (i.e. sub-globose) in shape and turn from green to brown in colour as they mature. These capsules (10-12 mm across) contain four large brown seeds (about 6 mm across) that are slightly three-angled in shape. The seeds have smooth surfaces interspersed with dense tufts of long silky hairs.

【近似种】五爪金龙（原变种）

学　　名：*Ipomoea cairica*（Linn.）Sweet var. *cairica*

| 形态特征 |

五爪金龙为多年生缠绕草本，全体无毛，老时根上具块根。

植物茎细长，有细棱，有时有小疣状突起。

植物叶掌状5深裂或全裂，裂片卵状披针形、卵形或椭圆形，中裂片较大，长4～5cm，宽2～2.5cm，两侧裂片稍小，顶端渐尖或稍钝，具小短尖头，基部楔形渐狭，全缘或不规则微波状，基部1对裂片通常再2裂；叶柄长2～8cm，基部具小的掌状5裂的假托叶（腋生短枝的叶片）。

聚伞花序腋生，花序梗长2～8cm，具1～3花，或偶有3朵以上；苞片及小苞片均小，鳞片状，早落；花梗长0.5～2cm，有时具小疣状突起；萼片稍不等长，外方2片较短，卵形，长5～6mm，外面有时有小疣状突起，内萼片稍宽，长7～9mm，萼片边缘干膜质，顶端钝圆或具不明显的小短尖头；花冠紫红色、紫色或淡红色、偶有白色，漏斗状，长5～7cm；雄蕊不等长，花丝基部稍扩大下延贴生于花冠管基部以上，被毛；子房无毛，花柱纤细，长于雄蕊，柱头2球形。

蒴果近球形，高约1cm，2室，4瓣裂。

种子黑色，长约5mm，边缘被褐色柔毛。

【近似种】纤细五爪金龙

学　　名：*Ipomoea cairica*（Linn.）Sweet var. *gracillima*（Coll. et Hemsl.）C. Y. Wu

| 形态特征 |

与五爪金龙区别在于：茎较纤细，叶较小而裂片较狭，中裂片长2.5～3.3cm，宽0.5～1cm，花较小，长2.5～3.5cm。

喀西茄

学　　名：***Solanum aculeatissimum*** Jacquin
异　　名：*Solanum angustispinosum* De Wild.，*Solanum khasianum* C. B. Clarke
别　　名：苦颠茄、苦天茄、刺天茄
英文名：Dutch eggplant，love-apple

| 形态特征 |

喀西茄为草本至亚灌木植物。

植株直立，植株高达1~2m。

全株多混生腺毛及直刺。茎、叶、花梗及花萼均被硬毛、腺毛及基部宽扁的直刺，刺长0.2~1.5cm。

植物叶互生。叶片宽卵形，长6~15cm，先端渐尖，基部戟形，5~7浅裂，裂片边缘具不规则齿裂及浅裂，上面沿叶脉毛密，侧脉疏被直刺。叶柄长3~7cm。

蝎尾状聚伞花序腋外生，花单生或2~4聚生。花萼钟状，裂片长圆状披针形，长约5mm，具长缘毛。花冠筒淡黄色，隐于萼内，长约1.5mm。冠檐白色，裂片披针形，长约1.4cm，具脉纹，反曲。花药顶端延长，长6~7mm，顶孔向上。子房被微绒毛。

浆果球形，直径2~3cm。幼果具绿色斑纹，成熟时淡黄色。宿萼被毛及细刺，后渐脱落。

种子褐黄色，近倒卵圆形，直径2~2.8mm。

花期为3~8月。果期为11~12月。

| 生物危害 |

喀西茄常分布于海拔100~2 300m的路边、沟边、灌丛、荒地、草坡或疏林生境，已入侵到我国自然保护区内。为一种大型具刺杂草或亚灌木。

全株含有毒生物碱，未成熟果实毒性较大，人和家畜误食会引起中毒。

| 地理分布 |

喀西茄原产于南美洲热带地区。19世纪末，在贵州南部首次发现。现主要分布于上海、江苏、浙江、福建、江西、湖北、湖南、广东、广西、海南、重庆、四川、贵州、云南、西藏、台湾、香港等地。

| 传播途径 |

借种子人为传播或自然扩散。

| 管理措施 |

综合治理（Integrated Pest Management，IPM）或综合控制（Integrated Pest Control，IPC），但不包括植物卫生措施（Phytosanitary Measures）。

（图片选自commons.wikimedia.org，upload.wikimedia.org，www.plantnames.unimelb.edu.au，plantgenera.org）

喀西茄（*Solanum aculeatissimum*）

【 www.efloras.org 】 *Solanum aculeatissimum* Jacquin, Collectanea. 1: 100. 1787.

Solanum cavaleriei H. Léveillé & Vaniot; *S. khasianum* C. B. Clarke.

Herbs to subshrubs, erect, 1-2（-3）m tall, copiously armed, minutely tomentose with simple, many-celled, mostly glandular hairs, often with a pinkish cast. Stems and branches terete, erect, loosely pilose with many-celled, simple and stellate hairs to 2 mm, armed with recurved flat prickles 1-5 mm × 2-10 mm and sometimes straight spines. Leaves sometimes unequal paired; petiole, stout, 3-7 cm, copiously prickly; leaf blade broadly ovate, 6-15 cm × 4-15 cm, with coarse, many-celled simple hairs and straight prickles on both surfaces, mixed with sparse, stellate hairs abaxially, base truncate to subhastate, margin 5-7-lobed or -parted, with angular or dentate sharp lobes, apex acute or obtuse. Inflorescences extra-axillary, short, 1-4-flowered scorpioid racemes; peduncle obsolete or to 1 cm. Pedicel 5-10 mm, pilose. Calyx campanulate, ca. 5.5 cm; lobes oblong-lanceolate, 5 mm × 1.5 mm, hairy and sometimes prickly abaxially. Corolla white; lobes lanceolate, ca. 4 × 14 mm, pubescent as on calyx. Filaments 1-2 mm; anthers lanceolate, acuminate, 6-7 mm. Ovary glabrous or minutely stipitate glandular. Style 6-7 mm. Berry pale yellow, globose, 2-3 cm in diam. Seeds light brown, lenticular, 2-2.8 mm in diam. Fl. Mar-Aug, fr. Nov-Dec.

【 uses.plantnet-project.org 】 *Solanum aculeatissimum*

Description

Perennial herb or small shrub up to 120（-200）mm tall; stems densely covered with yellowish, slender prickles up to 18 mm long, hairs mostly simple, some hairs stellate. Leaves alternate, simple; stipules absent; petiole 1-9.5 cm long; blade broadly ovate, 2.5-16 cm × 2.5-19 cm, base unequal, cuneate to cordate, apex usually acute, margin coarsely toothed or more or less deeply lobed, prickles present on main veins. Inflorescence a cyme, inserted above the leaf axil, 2-3.5 cm long, 2-7-flowered. Flowers bisexual or functionally male, regular, 5（-6）-merous; calyx bell- or cup-shaped, 4-10 mm in diameter, lobes unequal, triangular, acuminate, reflexed; corolla stellate, up to 22 mm in diameter, white to mauve or purple; stamens alternate with corolla lobes, filaments 1-2 mm long, anthers lanceolate, c. 6 mm long, opening with apical pores; ovary superior, with stalked glands and short hairs near base, style glabrous, 6-10 mm long. Fruit a globose berry 2-3 cm in diameter, glabrous, green marbled white or cream, yellowish when ripe, many-seeded. Seeds ovoid, compressed, 2.5-4.5 mm in diameter, margin thickened. Seedling with epigeal germination; cotyledons thin, leafy.

黄花刺茄

学　　名：***Solanum rostratum*** Dunal

异　　名：*Solanum cornutum* auct., non Lam., *Solanum heterandrum* Pursh

别　　名：刺萼龙葵、刺茄、尖嘴茄

英 文 名：prickly nightshade，beaked nightshade，beaked sandbur，buffalo berry，buffalo bitter apple，buffalo bur，buffalobur（USA），hedgehog bush，horned nightshade，horse nettle，kansas thistle（USA），pincushion nightshade，sandbur，spiny nightshade

| 形态特征 |

黄花刺茄为一年生草本植物。

植物茎直立，基部稍木质化，自中下部多分枝，密被长短不等淡黄色的刺，刺长0.5~0.8cm，并有具柄的星状毛。植株高30~70cm。

植物叶互生。叶片卵形或椭圆形，长8~18cm，宽4~9cm，不规则羽状深裂，其中部分裂片羽状半裂，裂片椭圆形或近圆形，先端钝，上面疏被5~7分叉星状毛，背面密被5~9分叉星状毛，两面脉上具疏刺，刺长3~5mm。叶柄长0.5~5cm，密被刺及星状毛。

蝎尾状聚伞花序腋外生，具3~10花。花后花轴伸长变成总状花序，长3~6cm，果期长达16cm。花两性。花萼筒钟状，长7~8mm，宽3~4mm，密被刺及星状毛，萼片5，线状披针形，长约3mm，密被星状毛。花冠黄色，辐状，直径2~3.5cm，5裂，裂片外面密被星状毛。雄蕊5，花药黄色，异形，下方1枚最长，长9~10mm，后期常带紫色，内弯曲成弓形，其余4枚长6~7mm。

浆果球形，直径1~1.2cm，完全被增大的带刺及星状毛宿萼包被。果皮薄，与萼合生，自顶端开裂后种子散出。

种子多数，黑色，直径2.5~3mm，具网纹。

花果期6~9月。

| 生物危害 |

黄花刺茄为我国检疫性有害生物。其适应性强，耐瘠薄、干旱，常生长于荒地、草原、河滩和过度放牧的牧场，也能侵入农田、果园中。竞争能力强，可严重抑制其他植物生长，常形成大面积单一群落，破坏当地生物多样性。全株密被刺毛，可伤害家畜，影响放牧和羊毛产量。

植株有毒，人误食后，可引起严重的肠炎和出血。果实含有神经毒素茄碱，牲畜误食后可致其死亡。

该种还是马铃薯甲虫（*Leptinotarsa decemlineata*（Say））和马铃薯卷叶病毒（PLRV）的野外寄主。

| 地理分布 |

黄花刺茄原产于墨西哥北部和美国西南部，除佛罗里达州外已经遍布美国，现已入侵到加拿大、俄罗斯、韩国、南非、澳大利亚等国家或地区。该种种子通过风、水流，或以刺萼扎入动物皮毛或人的衣服等方式传播。

我国早在1982年在辽宁省朝阳县就有报道，后来又相继在吉林省白城市、河北省张家口市、北京市密云区等地发现该物种。2005—2007年，在新疆境内也发现了乌鲁木齐县和石河子市两个分布区。2009年，该种被发现入侵到内蒙古。现主要分布于北京、河北、山西、内蒙古、辽宁、吉林、江苏、云南、新疆、香港等地。

| 传播途径 |

黄花刺茄仅由种子传播，以人为或自然方式进行。

| 管理措施 |

综合治理（Integrated Pest Management，IPM）或综合控制（Integrated Pest Control，IPC），包括植物卫生措施（Phytosanitary Measures）。

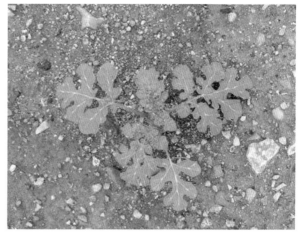

（图片选自soilcropandmore.info，bugwoodcloud.org，www.walterreeves.com，courses.missouristate.edu，www.minnesotawildflowers.info，fireflyforest.net，flora.nhm-wien.ac.at，commons.wikimedia.org，calphotos.berkeley.edu，www.cabi.org，wiki.bugwood.org，www.flora.sa.gov.au）

黄花刺茄（*Solanum rostratum*）

【www.cabi.org】 *Solanum rostratum* (prickly nightshade)

Individual plants reach 1-1.5 m tall, with once- or twice-pinnatifid leaves, and abundant prickles on the stems and leaves. It produces yellow flowers with pentagonal corollas 2-3.5 cm in diameter and weakly bilaterally symmetric (Whalen, 1979). In its native range *S. rostratum* is pollinated by medium- to large-sized bees including bumblebees (Bowers 1975). Flowers bear two sets of anthers that are unequal in size and may be distinctly coloured (Vallejo-Marín et al., 2009). The fruit, a berry, is enclosed by a prickly calyx and the seeds are released when the berries dry and split while still attached to the plant [Taken from Vallejo-Marin, 2014].

【en.wikipedia.org】 *Solanum rostratum*

It is an annual, self-compatible herb that forms a tumbleweed. Individual plants reach 1-1.5 m (3.3-4.9 feet) tall, have once or twice pinnatified leaves (see image of leaf), and abundant spines on the stems and leaves. It produces yellow flowers with pentagonal corollas 2-3.5 cm (0.79-1.38 inches) in diameter and weakly bilaterally symmetric (see flower-closeup image). In its native range *S. rostratum* is pollinated by medium- to large-sized bees including bumblebees. *Solanum rostratum* flowers exhibit heteranthery, i.e. they bear two sets of anthers of unequal size, possibly distinct colouration, and divergence in ecological function between pollination and feeding. The fruit, a berry, is enclosed by a prickly calyx. The seeds are released when the berries dry and dehisce (split apart) while still attached to the plant.

【wiki.bugwood.org】 *Solanum rostratum*

Appearance *Solanum rostratum* is an annual herbaceous plant that can grow up to 2 feet. (0.6 m) tall.

Foliage Stems and leaves have extremely sharp spines, so take care when handling. Leaves are alternate, bright green, petiolate, 4-5 inches. (10.2-12.7 cm) long, and pinnately lobed.

Flowers Flowers have five petals and are yellow, and 1 in. (2.5 cm) across. Calyx tube is 1 in. (2.5 cm) long, and spiny. Flowering occurs from May to October.

Fruit Fruits are dry berries up to 0.4 in. (1 cm) in diameter covered in sharp spines and contain several wrinkled, flat, black seeds.

【gobotany.newenglandwild.org】 *Solanum rostratum* Dunal

【www.fireflyforest.com】 *Solanum rostratum* – Buffalobur Nightshade

【近似种】北美刺龙葵

学　　名：*Solanum carolinense* L.

异　　名：*Solanum carolinense* f. *albiflorum*（Kuntze）Benke，*Solanum carolinense* var. *albiflorum* Kuntze，*Solanum carolinense* var. *floridanum*（Dunal）Chapm.，*Solanum carolinense* var. *pohlianum* Dunal，*Solanum floridanum* Raf. 1840，*Solanum floridanum* Shuttlew. ex Dunal 1852，*Solanum godfreyi* Shinners，*Solanum pleei* Dunal

别　　名：所多马之果、魔鬼马铃薯、魔鬼番茄、水牛荨麻、野番茄

英文名：horsenettle，apple of Sodom，ball nettle，ball nightshade，bullnettle，Carolina horsenettle，Carolina nettle，devil's potato，devil's tomato，sand brier，wild tomato

| 形态特征 |

北美刺龙葵为多年生草本植物。

植物茎（下胚轴）矮，绿色偏紫。覆盖有短而硬、微下垂、披散的毛。茎绿色，老后变为紫色。茎在近顶端分支，并有分散、坚硬、尖锐的刺。茎直立，分支或不分支，具毛，多刺。植株高30～100cm。

植物叶片轮生，椭圆形或卵形。叶片长1.9～14.4cm，宽0.4～8cm。叶片上表面光滑绿色，下表面浅绿色，两面光滑。叶边缘有短腺毛。叶中部的导管在上表面凹下，在下表面成脊状微微突起。沿主脉有锯齿状叶片，其上表面暗绿色，下表面浅绿色。老叶上表面覆有稀疏的未分支的星形毛。叶柄上表面扁平，覆有星形毛。第一对叶片上表面有稀疏毛，其次的叶片边缘呈波浪形或深裂，表面有毛和刺。

花星形5裂，长约2.5cm，生于上部枝条末端和边缘的分支上，丛生，一簇上可长有几朵或多朵白色、紫色或蓝色的花。萼片长2～7mm，表面常具有小刺。花瓣卵形、分裂，直径可达3cm。花药直立，长度为6～8mm。

果实为浆果，多汁，球形，直径为9～15mm。果实成熟时光滑，黄色到橘色，表面有皱纹。果实含有大量种子。种子直径为1.5～2.5mm。

花期为5～9月。果期夏末至秋季。

| 生物危害 |

北美刺龙葵为我国检疫性有害生物。植物蔓延快、生活力强。主要侵害种植花卉的花园及蔬菜和果园。植物全株有毒，牲畜误食会引起中毒。

| 地理分布 |

美国。

| 传播途径 |

种子和地下根茎均能繁殖。但仅由种子传播。

| 管理措施 |

综合治理（Integrated Pest Management，IPM）或综合控制（Integrated Pest Control，IPC），包括植物卫生措施（Phytosanitary Measures）。

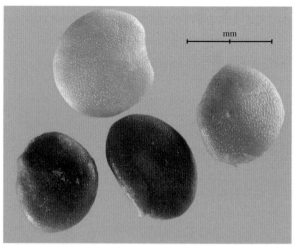

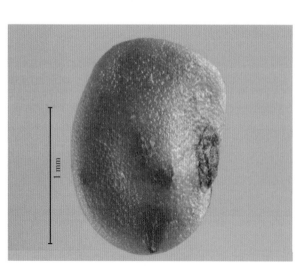

（图片选自www.forestryimages.org，wildflowers.clockworkorrery.com，upload.wikimedia.org，www.missouriplants.com，www.pfaf.org，www.fungoceva.it，ncwings.carolinanature.com，www.inspection.gc.ca，www.agroatlas.ru）

北美刺龙葵（*Solanum carolinense*）

【近似种】银毛龙葵

学　　名：*Solanum elaeagolifolium* Cav.

异　　名：*Solanum dealbatum* Lindl.，*Solanum elaeagnifolium* f. *albiflorum* Cockerell，*Solanum elaeagnifolium* f. *benkei* Standl.，*Solanum elaeagnifolium* var. *angustifolium* Kuntze，*Solanum elaeagnifolium* var. *argyrocroton* Griseb.，*Solanum elaeagnifolium* var. *grandiflorum* Griseb.，*Solanum elaeagnifolium* var. *leprosum*（Ortega）Dunal，*Solanum elaeagnifolium* var. *obtusifolium*（Dunal）Dunal，*Solanum flavidum* Torr.，*Solanum incanum* Pav. ex Dunal，*Solanum leprosum* Ortega，*Solanum obtusifolium* Dunal，*Solanum pyriforme* var. *uniflorum* Dunal，*Solanum roemerianum* Scheele，*Solanum saponaceum* Hook，*Solanum texense* Engelm. & A.Gray，*Solanum uniflorum* Meyen ex Nees

别　　名：银叶颠茄、银叶荨麻、水牛荨麻、苦果、白北美刺龙葵

英文名：silver-leaf nightshade，white horse-nettle，tomato weed，silverleaf nettle，bitterleaf nightshade，bitter apple，silverleaf bitter apple，bull nettle，prairie berry

| 形态特征 |

银毛龙葵为直立多年生灌木状草本。

植物茎直立，分支，覆盖着许多细长的橘色刺。表面有稠密的银白色绒毛。植株高30～100cm。

植物叶互生，长2.5～10cm，宽1～2cm，边缘常呈扇形，叶脉上常具刺。

花紫色，偶尔白色，通常直径为2.5cm左右，也可达4.0cm。具5个联合花瓣形成的花冠；五个黄色花药。

果实为光滑球状浆果，直径为1.0～1.5cm，绿色带暗条纹，成熟时呈黄色带橘色斑点。

种子轻且圆，平滑，暗棕色，直径为2.5～4mm，每个果实里大约有75粒种子。

花果期为11月或12月到翌年4月。

| 生物危害 |

银毛龙葵为我国检疫性有害生物。是麦田和牧场的主要杂草。对小麦及棉花作物产量造成影响。植物各部分均有毒，尤其是成熟果实更甚，动物误食会引起中毒。

| 地理分布 |

银毛龙葵原产南北美洲。现在美国、墨西哥、阿根廷、巴西、智利、印度、南非、澳大利亚均有分布。1909年，第一次记录出现于澳大利亚北墨尔本，现已遍布各大洲。

| 传播途径 |

种子传播，或由其多年生的根进行营养繁殖。种子还可由风、水、机械、鸟类、动物（内部或外部）携带传播。

| 管理措施 |

综合治理（Integrated Pest Management，IPM）或综合控制（Integrated Pest Control，IPC），包括植物卫生措施（Phytosanitary Measures）。

| 管理措施 |

综合治理（Integrated Pest Management，IPM）或综合控制（Integrated Pest Control，IPC），包括植物卫生措施（Phytosanitary Measures）。

（图片选自 keyserver.lucidcentral.org，blog.growingwithscience.com，nathistoc.bio.uci.edu，www.redorbit.com，wnmu.edu，www.wildflower.org，upload.wikimedia.org，static.inaturalist.org，www.actaplantarum.org，www.inspection.gc.ca）

银毛龙葵（*Solanum elaeagolifolium*）

【近似种】刺茄

学　　名：*Solanum torvum* Swartz

异　　名：*Solanum ferrugineum* Jacq.，*Solanum largiflorum* CT White，*Solanum mayanum* Lundell，*Solanum verapazense* Standl. & Steyerm.

别　　名：水茄、金纽扣、山颠茄、观赏茄

英文名：turkey berry，devil's fig，prickly Solanum，terongan，wild tomato

| 形态特征 |

刺茄为灌木植物。

植物茎有皮刺。植株高2～3m，有的可高达5m。

植株小枝、叶下面、叶柄及花序梗均被星状毛。小枝疏生基部宽扁的皮刺，皮刺淡黄色或淡红色，长2.5～10mm。

植物叶互生，羽状浅裂。叶卵形至椭圆形，长6～9cm，宽4～11（13）cm。叶先端尖，基部心形或楔形，两边不相等，边缘5～7浅裂或波状，下面灰绿，密被具柄星状毛；脉有刺或无刺；叶柄长2～4cm，具1～2枚皮刺或无刺。

花白色，排成螺状菊聚伞花序。聚伞式圆锥花序腋外生；花梗长5～10mm，被腺毛及星状毛。花萼裂片卵状长圆形，长约2mm；花冠辐状，白色，直径约1.5cm，筒部隐于萼内，裂片卵状披针形，先端渐尖，外面被星状毛；花丝长约1mm，花药长7mm，顶孔向上。

浆果黄色，球形，直径1～1.5cm，无毛；果梗长约1.5cm上部膨大。

种子盘状，直径1.5～2mm。

全年开花结果。

| 生物危害 |

刺茄为我国检疫性有害生物。具刺杂草，有时危害旱地作物。微毒。

| 地理分布 |

刺茄原产于美洲加勒比地区。现广布于热带地区。1827年，在我国澳门发现。现分布于西藏（墨脱）、云南（东南部、南部及西南部）、贵州、广西、广东、海南、香港、澳门、福建、台湾、江西等地。

| 传播途径 |

种子自然或人为传播。

| 管理措施 |

综合治理（Integrated Pest Management，IPM）或综合控制（Integrated Pest Control，IPC），包括植物卫生措施（Phytosanitary Measures）。

中国外来入侵 物种图鉴

外来入侵植物

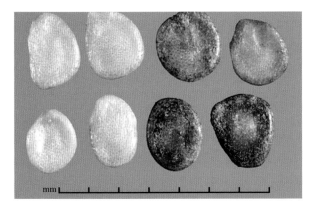

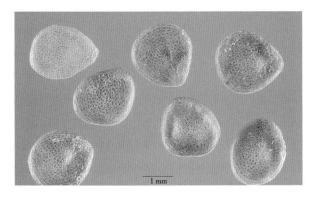

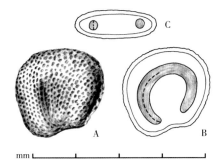

（图片选自www.florablog.it，keyserver.lucidcentral.org，upload.wikimedia.org，publish.plantnet-project.org，delta-intkey.com，luirig.altervista.org，en.wikipedia.org，www.natureloveyou.sg，www.flora.sa.gov.au，plants.usda.gov，idtools.org）

刺茄（*Solanum torvum*）

刺果瓜

学　　名：*Sicyos angulatus* L.

异　　名：*Sicyoides angulata*（Linnaeus）Moench, Methodus，*Sicyos acutus* Rafinesque，Fl. Ludov.

别　　名：刺果藤、棘瓜、单子刺黄瓜、星刺黄瓜

英文名：burcucumber，one seed burcucumber，oneseed burr cucumber，Star-cucumber

| 形态特征 |

刺果瓜为一年生攀援草本植物。

植物茎长5~20m，具纵棱，被开展的硬毛，具3~5分叉的卷须。

植物叶互生，具柄。叶片圆形、卵圆形或宽卵圆形，3~5浅裂，长5~22cm，宽3~30cm，基部具深心形，裂片三角形，两面微糙。

花单性，雌雄同株。雄花排列成总状花序，花序梗长10~20cm。花萼钻形，长约1mm。花冠黄白色，具绿色脉，直径9~14mm，5深裂，裂片先端急尖。雌花直径约6mm，淡绿色，聚成头状，花序梗长1~2cm。

果长卵圆形，长10~15mm，先端渐尖，外面散生柔毛和长刚毛，黄色或土灰色，内含1粒种子。

种子椭圆状卵形，长7~10mm，光滑。

花期为7~10月。果期为8~11月。

| 生物危害 |

刺果瓜通过竞争或占据本地物种生态位来排挤本地物种，它们与本地物种竞争生存空间，直接扼杀当地物种。

还分泌释放化学物质，抑制其他植物生长，减少本地物种的种类和数量，甚至导致物种濒危或灭绝。

由于其入侵农田，常使玉米、大豆等旱地作物减产。

| 地理分布 |

刺果瓜原产于北美洲，后作为观赏植物引入欧洲，因逃逸成为杂草。现已在欧洲、亚洲和大洋洲的多个国家发生。我国于2003年首次在大连发现。2005年首次报道其危害性。现主要分布于北京、辽宁、山东、四川、云南、台湾等地。

| 传播途径 |

人为传播或自然扩散。

管理措施

综合治理（Integrated Pest Management，IPM）或综合控制（Integrated Pest Control，IPC），但不包括植物卫生措施（Phytosanitary Measures）。

管理措施

综合治理（Integrated Pest Management，IPM）或综合控制（Integrated Pest Control，IPC），但不包括植物卫生措施（Phytosanitary Measures）。

中国外来入侵 物种图鉴

外来入侵植物

Sicuos angulatus

（图片选自upload.wikimedia.org，www.weedinfo.ca，objects.liquidweb.service，extension.cropsciences.illinois.edu，www.cabi.org，www.wiseacre-gardens.com，www.minnesotawildflowers.info，www.fnanaturesearch.org，www.all-creatures.org，luirig.altervista.org，www.oardc.ohio-state.edu，commons.wikimedia.org，mississippientomologicalm.useum.org.msstate.edu）

刺果瓜（*Sicyos angulatus*）

【www.efloras.org】*Sicyos angulatus* Linnaeus, Sp. Pl. 2, 1013. 1753.（as angulata）.

One-seed bur or star cucumber, nimble-Kate, sicyos anguleux One-seed bur or star cucumber, nimble-Kate, sicyos anguleux

Stems moderately to densely villous-puberulent, hairs glandular-viscid, mixed with stipitate-glandular hairs.

Leaves: Petiole 1-7（-10）cm; blade orbiculate-angulate to broadly ovate-angulate or shallowly（3-）5-lobed, 4-12 cm × 6-17 cm, terminal lobe deltate-acuminate to ovate-acuminate, basal sinus narrow to broad, margins evenly and minutely green-apiculate-mucronulate, ± ciliate, hairs gland-tipped, surfaces hispidulous-hirsute; proximal pair of lateral veins divergent from edge of basal sinus.

Inflorescences: Staminate 10-21（-34）-flowered, peduncle plus floral axis 30-220 mm; pistillate 8-16-flowered, peduncle 20-50 mm.

Flowers: Staminate: corolla white to greenish white, 4-5 mm, stamens prominently exserted; pistillate: sepals not foliaceous, linear to linear-triangular, 0.5-1 mm, corolla 1-2 mm（essentially without a tube）; stigmas 3-lobed.

Pepos Ovoid-beaked, 9-15 mm, echinate, spinules retrorsely barbellate, also densely arachnoid-villous to minutely villosulous, hairs often gland-tipped.

【en.wikipedia.org】*Sicyos angulatus*

The vine produces long branching annual stems that climb over shrubs and fences or trail across the ground. The stems are hairy, pale green and furrowed. The alternate leaves have three to five palmate lobes and can be 8 inches（20 cm）across. The margin is slightly toothed, the upper surface of the blade is usually hairless and the under side has fine hairs, especially on the veins. The petiole is thick and hairy, and about 5 inches（13 cm）long. The leaf is deeply indented where it is attached to the petiole. Opposite some of the junctions formed by the petiole and stem, grow branched tendrils, and at others there are flower shoots. The flowers are monoecious, with separate male and female blooms. The male flowers are in long-stemmed racemes. Each flower is about 0.3 inch（0.8 cm）wide, with a calyx with five pointed teeth, a whitish, green-veined corolla with five lobes, and a central boss of stamens. The small female flowers are bunched together on a short stalk, each having its ovary enclosed in a spiny, hairy fruit; one seed is produced by each flower. The fruit is about 0.5 inch（1.3 cm）long, green at first but becomes brown with age; it is dispersed by animals which come into contact with its bristly surface.

【www.cabi.org】*Sicyos angulatus*（burcucumber）

The leaves of *S. angulatus* are thin, 5-lobed, up to 25 cm across and borne on stout, pubescent petioles 2.5 to 10 cm long（Mann et al., 1981）. They are alternate, broadly heart-shaped and finely toothed（Anon., 2010）. Stems are hairy and form a creeping vine up to 6 m long, with numerous branched tendrils（Anon., 2010）. New vines can form by growth from axillary buds（Mann et al., 1981）. The root system consists of a shallow branched taproot（Anon., 2010）. S. angulatus is monoecious with 5-petalled, green and white flowers（Mann et al., 1981）. The male flowers appear in a corymbose raceme on a very long peduncle and the female flowers appear in a capitate cluster on a short peduncle（Torrey and Gray, 1969）. The calyx is green, five-toothed, and pubescent. The corollas of both sexes are white with green striations, and consist of five petals fused at the base into an open bowl, and free and spread at the tips. Staminate flowers form on either paniculate or racemose inflorescences. The anthers unite to form a central column. Pistillate flowers are borne on a compact cyme, in a globose cluster of 8-20 flowers. The pistil consists of a superior ovary, a slender style, and 3 stigmas (Medley and Burnham, 2011). The bur-like fruits are small and spiny, 1.0-1.5 cm long, one-seeded, produced in clusters of 3-20, initially green, turning brown, indehiscent, containing a single brown flattened seed (Anon., 2010). Each fruit contains one seed, and 3 to 15 fruits are borne in a cluster (Mann et al., 1981). The seeds are large (15 mm by 10 mm), dark brown to black, compressed, smooth, and covered with a crustaceous pericarp (Britton and Brown, 1947). The pericarp is actually covered with four types of "hairs" consisting of barbed prickles (8 to 10 mm long), long jointed hairs, short pointed conical hairs, and unicellular conical hairs (Barber, 1909).

野燕麦

学　名：*Avena fatua* L.

别　名：燕麦草、乌麦、香麦、铃铛麦

异　名：*Anelytrum avenaceum* Hack，*Avena fatua* var. *glabrata* Peterm.，*Avena fatua* var. *intermedia* Hartman，*Avena fatua* var. *intermedia* Husn.，*Avena fatua* var. *intermedia* Vasc.，*Avena fatua* ssp. *meridionalis* Malzev，*Avena fatua* var. *vilis*（Wallr.）Hausskn.，*Avena fatua* var. *vilis*（Wallr.）Malzev，*Avena intermedia* Lindgr.，*Avena intermedia* T. Lestib.，*Avena lanuginosa* Gilib.，*Avena meridionalis*（Malzev）Roshev.，*Avena patens* St.-Lag.，*Avena pilosa* Scop.，*Avena sativa* var. *fatua*（L.）Fiori，*Avena sativa* var. *sericea* Hook.，*Avena septentrionalis* Malzev，*Avena vilis* Wallr.

英文名：wild oat，Common wild oat

| 形态特征 |

野燕麦为一年生草本植物。

植物秆单生或丛生，直立或基部膝曲，有2~4节。植株高30~150cm。

植物叶片长10~30cm，宽4~12mm，叶鞘光滑或基部被柔毛，叶舌膜质透明，长1~5mm。

圆锥花序呈金字塔状开展，分枝轮生，长10~40cm。小穗长17~25mm，含2~3小花，其柄弯曲下垂。颖披针形，几等长，具9~11脉。外稃质地硬，下半部与小穗轴均有淡棕色或白色硬毛，第一外稃长15~20mm。芒自外稃中部稍下处伸出，长2~4cm，膝曲。

花果期为4~7月。

| 生物危害 |

野燕麦在海拔4 300m以下均可分布，常见于荒野或田间，根系发达，分蘖能力强，为农田恶性杂草，可与农作物争水、争光、争肥，降低作物产量。

种子易混杂于作物中，降低作物品质。

野燕麦能传播小麦条锈病、叶锈病。还是小麦黄矮病毒病和多种害虫的中间寄主和越冬越夏的栖息场所。

| 地理分布 |

野燕麦原产于欧洲南部及地中海沿岸。现在欧、亚、非三洲的温寒地带均有分布，北美也有输入。该种是世界性的恶性农田杂草。

19世纪中叶曾先后在我国香港和福州采到标本。现主要分布于北京、天津、河北、山西、内蒙古、辽宁、吉林、黑龙江、上海、江苏、浙江、安徽、福建、江西、山东、河南、湖北、湖南、广东、广西、海南、重庆、四川、贵州、云南、西藏、陕西、青海、宁夏、新疆、台湾、香港、澳门等地。

| 传播途径 |

种子随人为传播或随鸟类自然传播。

| 管理措施 |

综合治理（Integrated Pest Management，IPM）或综合控制（Integrated Pest Control，IPC），但不包括植物卫生措施（Phytosanitary Measures）。

野燕麦（*Avena fatua*）

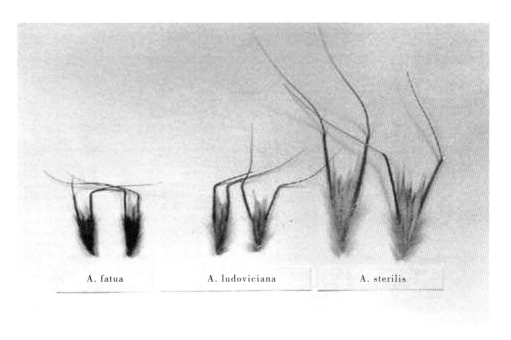

野燕麦（*Avena fatua*）、法国野燕麦（*Avena ludoviciana*）和不实野燕麦（*Avena sterilis*）

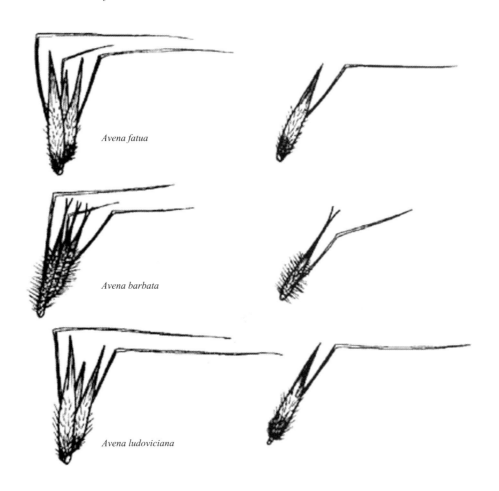

野燕麦、裂稃燕麦（*Avena barbata*）和法国野燕麦

（图片选自www.cropscience.bayer.com，keys.lucidcentral.org，www.conabio.gob.mx，nathistoc.bio.uci.edu，luirig.altervista.org，newfs.s3.amazonaws.com，calphotos.berkeley.edu，plants.usda.gov，www.focusnatura.at，www.greekflora.gr，www.plantwise.org，wswa.org.au，upload.wikimedia.org，idtools.org）

【www.efloras.org】*Avena fatua* Linnaeus, Sp. Pl. 1: 80. 1753.

Annual. Culms erect or geniculate at base, 50-150 cm tall, unbranched, 2-4-noded. Leaf sheaths glabrous or basal sheaths puberulous; leaf blades 10-30 cm, 4-12 mm wide, scabrid or adaxial surface and margins pilose; ligule 1-5 mm. Panicle narrowly to broadly pyramidal, 10-40 cm, nodding; branches scabrid. Spikelets 1.7-2.5 cm, florets 2 or 3, all florets awned; rachilla easily disarticulating below each floret at maturity, each floret with a bearded callus, internodes hirsute or glabrous; glumes lanceolate, subequal, herbaceous, 9-11-veined, apex finely acute; callus hairs up to 4 mm; lemmas 1.5-2 cm, leathery, glabrous to densely hispid in lower half, green and scaberulous above, awned from near middle, apex shortly 2-4-toothed; awn 2-4 cm, geniculate, column twisted, blackish brown. Fl. and fr. Apr–Sep. $2n = 42$.

【www.plantwise.org】/【www.cabi.org】*Avena fatua*（wild oat）

A. fatua is an annual tufted grass with erect culms（Häfliger and Scholz, 1981）. Plant height varies from 25 to 120 cm（Holm et al., 1977; Yamaguchi and Sasaki, 1985）. Blades are dark green, grow up to 40 cm and show a membranous ligule, which is 1 to 6 mm long and often irregularly toothed. Rooney（1990）found upright and prostrate forms; the former are mostly taller than the crop stand. Sheaths are smooth or slightly hairy, especially in younger plants.

The inflorescence of *A. fatua* is a loose, open panicle with 2 to 3-flowered pedicelled spikelets. As a specific trait of Avena species, lemmas have 2 to 3 awns arising from the back, which are mostly dark-coloured, bent and 3 to 4 cm long（Holm et al., 1977）. Each of the 2 to 3 florets has an oval abscission scar at its base, causing them to fall separately.

Grains are 6 to 8 mm long and usually of mass 11 to 18 mg, but grains of mass 25 mg may also be found. Increasing seed mass enhances competitiveness and seed production（Peters, 1985）. Without interference A. fatua can produce more than 20 tillers and 1 500 seeds per plant（Morrow and Gealy, 1982）, but in crop stands only 1 to 5 tillers and 200（50 to 1 000）seeds per plant are reached（Hanf, 1990）.

The height of wild oat lines is related to climate and the number of tillers is correlated with the USA state of origin（Somody et al., 1980a）. Morphological characteristics of *A. fatua* selections can also vary within relatively narrow geographical limits（Somody et al., 1981a）.

In a study by Yang et al.（1999）that looked at the genomic structure of *A. fatua*, a translocation not previously observed in reports on other hexaploid *Avena* species was found. If this translocation is found to be unique to A. fatua, then this information, combined with more traditional morphological data, will add support to the view that A. fatua is genetically distinct from other hexaploid Avena species.

【近似种】法国野燕麦（参见长颖燕麦）

学　　名：*Avena ludoviciana* Durieu

别　　名：野燕麦、不实野燕麦

异　　名：*Avena ludoviciana* var. *psilathera*（Thell.）Parodi，*Avena persica* Steud.，*Avena sterilis* subsp. *ludoviciana*（Durieu.）Gillett & Magne，*Avena sterilis* var. *ludoviciana*（Durieu.）Husn.，*Avena sterilis* var. *psilathera* Thell.，*Avena trichophylla* C. Koch

英文名：wild oat，sterile oat

| 形态特征 |

法国野燕麦为一年生草本植物。

植物叶片线型，轮生，长60cm，宽0.5～1.5cm。叶舌膜状。基部叶具叶鞘。

小穗含2～5小花；外稃背部具扭曲长芒，但第三小花无芒。小穗成熟后整体脱落。带稃颖果长约1.5cm，暗褐色，被毛。芒扭曲，长3～8cm，上部明显弯曲成直角。

| 生物危害 |

法国野燕麦为我国检疫性有害生物。它是极具危险性的田间恶性杂草，其生命力、竞争力及生态可塑性都很强；传播途径多，繁殖系数大，在发生区容易形成单一群落，可降低自然生态系统的稳定性和物种多样性，造成严重的经济损失和景观破坏。常危害小麦等冬季作物。

| 地理分布 |

法国野燕麦起源于欧洲和地中海地区。现美国、墨西哥、巴西、阿根廷、智利、乌拉圭、苏联、德国、葡萄牙、巴基斯坦、印度、斯里兰卡、阿富汗、黎巴嫩、摩洛哥、南非、澳大利亚等国。

| 传播途径 |

法国野燕麦的刺苞挂在农作物及衣物、毛皮上进行自然传播。也可随粮食调运而人为传播。

| 管理措施 |

综合治理（Integrated Pest Management，IPM）或综合控制（Integrated Pest Control，IPC），包括植物卫生措施（Phytosanitary Measures）。

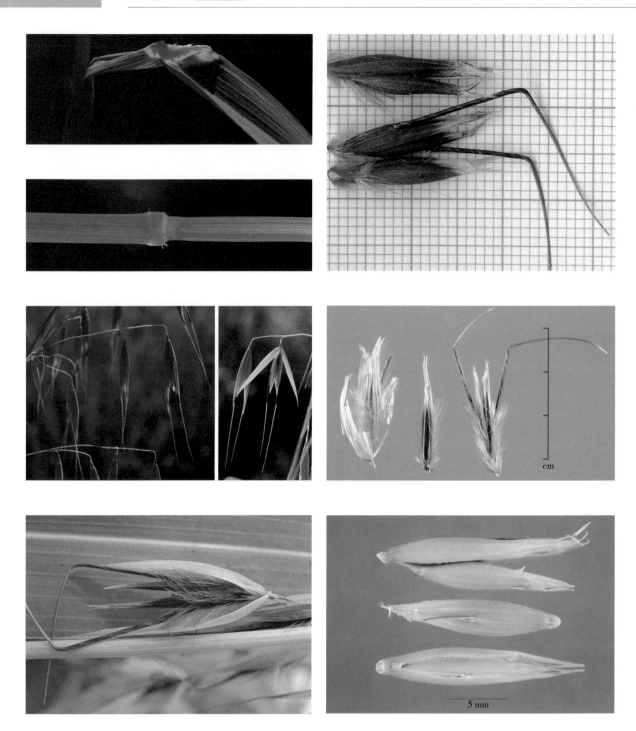

(图片选自idao.cirad.fr,luirig.altervista.org,bugwoodcloud.org)

法国野燕麦(*Avena ludoviciana*)

长颖燕麦（法国野燕麦）

【www.efloras.org】*Avena sterilis* Linnaeus subsp. *ludoviciana*（Durieu）Nyman, Consp. Fl. Eur. 810. 1882.

Avena ludoviciana Durieu, Actes Soc. Linn. Bordeaux 20: 41. 1855.

Annual. Culms solitary or tufted, erect or ascending, 50-120 cm tall, unbranched, 2-4-noded. Leaf sheaths glabrous or basal sheaths puberulous; leaf blades up to 60 cm, 4-13 mm wide, scaberulous, glabrous; ligule 3-4 mm. Panicle loose, open, pyramidal, 13-30 cm, nodding; branches coarsely scabrid. Spikelets 2-3 cm, florets 2 or 3, 2-awned; rachilla disarticulating only below lowest floret, florets falling together at maturity, only lowest floret with a bearded callus, internodes glabrous; glumes narrowly elliptic-oblong, subequal, as long as spikelet, 7-9-veined, apex finely acuminate; callus hairs up to 5 mm; lemmas 1.8-2.5 cm, leathery, hispid, finally brown in lower half, green and scabrid above, awned at about lower 1/3, apex finely 2-fid; awn 3-6 cm, fairly slender, strongly geniculate, column dark brown, pubescent. $2n = 42$.

Arable weed, adventive. Yunnan [native to SW Asia and Europe].

This is a noxious weed of arable land, especially fields of cereals, native to the Mediterranean region and SW Asia, but now widespread in warm-temperate regions of the world. It has been recorded in China only from Yunnan. The typical subspecies, *Avena sterilis* subsp. *sterilis*, is distinguished by its larger, 3-5 cm spikelets with 3-5 florets, 9-11-veined glumes, and stouter, 6-9 cm awns. Both subspecies occur over the whole range of the species（参见*Avena sterilis*）.

【近似种】不实野燕麦

学　　名：*Avena sterilis* L.

别　　名：冬野燕麦

异　　名：*Avena affinis* Bernh. ex Steud.，*Avena algeriensis* Trab.，*Avena byzantina* var. *solida*（Hausskn.）Maire & Weiller，*Avena fatua* var. *ludoviciana*（Durieu）Fiori，*Avena ludoviciana* Durieu，*Avena ludoviciana* var. *ludoviciana* Durieu，*Avena macrocalyx* Sennen，*Avena macrocarpa* Moench，*Avena melillensis* Sennen & Mauricio，*Avena nutans* St.-Lag.，*Avena persica* Steud.，*Avena sativa* var. *ludoviciana*（Durieu）Fiori，*Avena sativa* var. *sterilis*（L.）Fiori，*Avena sativa* ssp. *sterilis*（L.）de Wet，*Avena sensitiva* hort. ex Vilm.，*Avena solida*（Hausskn.）Herter，*Avena sterilis* var. *algeriensis*（Trab.）Trab.，*Avena sterilis* ssp. *ludoviciana*（Durieu）Nyman，*Avena sterilis* var. *ludoviciana*（Durieu）Husn.，*Avena sterilis* ssp. *macrocarpa*（Moench）Briq.，*Avena sterilis* var. *solida*（Hausskn.）Malzev，*Avena sterilis* subsp. *sterilis*，*Avena turonensis* Tourlet

英文名：animated oat，sterile oat，wild oat

| 形态特征 |

不实野燕麦为一年生草本植物。

成年植株的茎红色，丛生，直立，植株高达80～220cm。

植物幼苗叶鞘圆柱形，叶舌平截，长约8mm。叶长约60cm，是叶宽的20～50倍；叶片或多或少被有软毛，舌叶膜状，有锯齿，叶耳缺失。成年植株下部叶片较宽且有软毛（3～5mm），叶缘常有纤毛。

圆锥花序，长约30cm、宽约20cm；花绿色，微带红色，着生于分散状直立或稍有弯曲的圆锥花序上。小穗状花序水平或松垂，30～40mm长，生有3～4朵花。颖片远长于稃片。小穗中各小花不分离。颖果在成熟时随整个小穗脱落。带稃颖果长约2.5cm，密被褐色至暗褐色长毛；外稃背部具膝曲长芒、扭转、暗褐色。

| 生物危害 |

不实野燕麦为我国检疫性有害生物。可为害麦类、豆类及玉米、葡萄、橄榄树等作物，造成其作物产量和品质下降。

| 地理分布 |

不实野燕麦现分布于俄罗斯、日本、缅甸、印度、巴基斯坦、斯里兰卡、阿富汗、阿尔及利亚、阿拉伯半岛、肯尼亚、马耳他、摩洛哥、南非、埃及、埃塞俄比亚、法国、希腊、意大利、葡萄牙、突尼斯、土耳其、英国、美国、澳大利亚、克里特岛、秘鲁、阿根廷、新西兰等国家及地区。

| 传播途径 |

不实野燕麦的果实与种子可自然或人为传播。

| 管理措施 |

综合治理（Integrated Pest Management，IPM）或综合控制（Integrated Pest Control，IPC），包括植物卫生措施（Phytosanitary Measures）。

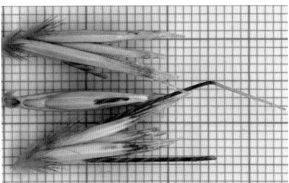

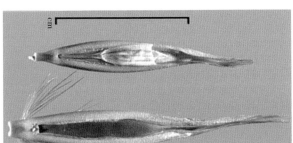

（图片选自keys.lucidcentral.org，www.pfaf.org，www.flowersinisrael.com，www.monde-de-lupa.fr，bugwoodcloud.org，www.actaplantarum.org，idtools.org，www.unavarra.es，www.plantwise.org）

不实野燕麦（*Avena sterilis*）

【近似种】裂稃燕麦

学　　名：*Avena barbata* Pott ex Link

别　　名：细茎野燕麦、细茎燕麦

异　　名：*Avena alba* var. *barbata* Maire & Weiller，*Avena almeriensis* Gand.，*Avena barbata* subvar. *hirsute*（Moench）E. Morren，*Avena deusta* Ball，*Avena fatua* L. var. *barbata*（Pott ex Link）Fiori & Paoletti，*Avena hoppeana* Scheele，*Avena sallentiana* Pau，*Avena sativa* var. *barbata*（Pott ex Link）Fiori，*Avena sesquitertia* hort. ex Steud.，*Avena sterilis* L. subsp. *barbata*（Pott ex Link）Gillet ex Magne，*Avena strigosa* subsp. *barbata* THELL.，*Avena striqosa* subsp. *barbata*（Pott ex Link）Thell.

英文名：slender oat，barbed oat，bearded oat，bearded oats，slender wild oat

| 形态特征 |

植物秆粗壮，直立，光滑无毛，高30~100cm。

植物叶舌膜质，长1~2mm；叶片扁平，宽达5mm，无毛，上面粗糙。

圆锥花序疏松开展，下垂，分枝纤细，长6.2~9.6cm，光滑无毛；小穗卵状长披针形，含2~3小花，长约25mm；小穗轴节间长约3.5mm，被长柔毛，具关节，成熟时各小花之间易断落；颖近相等，长约25mm，具9脉，光滑无毛；外稃长披针形，第一外稃长约23mm，1/2以下全被长约7mm浅褐色长毛，顶端深裂成2枚，裂片渐尖呈芒状，长3~4mm，背部约1/2处伸出1芒，长约40mm，膝曲，芒柱扭转，长约16mm，芒针细直，长约24mm；内稃较外稃短，长约12mm，具2脊，脊上具短纤毛；子房长卵形，被柔毛。

| 生物危害 |

裂稃燕麦为我国检疫性有害生物。可为害麦类、豆类及玉米等作物，造成作物产量和品质下降。

| 地理分布 |

广泛分布于亚洲、欧洲、北美、大洋洲、非洲等区域。中国主要是引种栽培（如南京中山植物园）。

| 传播途径 |

裂稃燕麦果实与种子可自然或人为传播。

| 管理措施 |

综合治理（Integrated Pest Management，IPM）或综合控制（Integrated Pest Control，IPC），包括植物卫生措施（Phytosanitary Measures）。

裂稃燕麦(*Avena barbata*)

(图片选自www.cabi.org、luirig.altervista.org、www.focusnatura.at、www.parcocurone.it、www.sardegnaflora.it、keyserver.lucidcentral.org、www.greekflora.gr、upload.wikimedia.org、idtools.org)

水盾草

学　名：*Cabomba caroliniana* A. Gray
异　名：*Cabomba aquatica* DC.，*Cabomba australis* Speg.，*Cabomba caroliniana* var. *pulcherrima* R. M. Harper，Cabomba *pulchurrima*（R. M. Harper）Fassett
别　名：绿菊花草
英文名：Carolina fanwort，fanwort，cabomba，Carolina water-shield，fish grass，gray fanwort，green cabomba，green grass chrysanthemum，purple cabomba，Washington grass，Washington plant，water shield grass

| 形态特征 |

水盾草为多年生草本植物。

植物茎长可达1.5m，分枝，幼嫩部分有短柔毛。

植物沉水叶对生，叶柄长1~3cm，叶片长2.5~3.8cm，掌状分裂，裂片3~4次，二叉分裂成线形小裂片。浮水叶少数，在花枝顶端互生，叶柄长1~2.5cm，叶片盾状着生，狭椭圆形，长1~1.6cm，宽1.5~2.5mm，边全缘或基部2浅裂。

花单生枝上部，沉水叶或浮水叶腋。花梗长1~1.5cm，被短柔毛。萼片浅绿色，无毛，椭圆形，长7~8mm，宽约3mm。花瓣绿白色，与萼片近等长或稍大，基部具爪，近基部具一对黄色腺体雌蕊6枚，离生，花丝长约2mm，花药长1.5mm，无毛。心皮3枚，离生，雌蕊长3.5mm，被微柔毛。子房1室，通常具3胚珠。

花期为7~10月。

| 生物危害 |

当水盾草入侵水体且大量死亡、腐烂后，会大量耗氧，从而危害渔业生产。国外已有水盾草入侵后对生物多样性造成影响的报道。

| 地理分布 |

水盾草原产于美国至巴西地区。现已引种至加拿大、日本、澳大利亚、东南亚、南亚等多国家或地区。1993年，在浙江首次发现。现主要分布于上海、江苏、浙江等地。

| 传播途径 |

自然扩散或人为传播。

| 管理措施 |

综合治理（Integrated Pest Management，IPM）或综合控制（Integrated Pest Control，IPC），但不包括植物卫生措施（Phytosanitary Measures）。

(图片选自www.aquarium-planten.com，naturalaquariums.com，www.acquariofiliafacile.it，cn.bing.com，nas.er.usgs.gov，luirig.altervista.org，www.ct.gov)

水盾草（*Cabomba caroliniana*）

【www.cabi.org】*Cabomba caroliniana*（Carolina fanwort）

Aquatic，Herbaceous，Perennial，Seed propagated，Vegetatively propagated plant

C. caroliniana is an herbaceous, submersed, rooted aquatic species（ISSG, 2008）that often grows in water from 0.4-1.2 m and up to 6 m deep（Yu et al., 2004; Schooler et al., 2006）. The plant has both submersed and floating leaves. Submersed leaves are oppositely arranged and 1-3.5 cm × 1.5-5.5 cm on petioles up to 4 cm long and finely dissected, having 3-200 terminal segments. Floating leaves are blades 0.6-3 cm × 1-4 mm with margins either notched or entire at base. Flowers are from 6-15 mm in diameter, flowers are white to purplish or yellow. Petals are obtuse or notched, with 3-6 stamens, 2-4 pistils and 3 ovules. Fruits are 4-7 mm, the 1-3 seeds are 1.5-3 mm × 1-1.5 mm long with tubercles in 4 rows（Flora of North America, 1993）. The plant is fully submerged and occasionally produces floating leaves and flowers. The plant is rooted, but can survive in a free-floating state for six to eight weeks（ISSG, 2008）. The plant produces fragile rhizomes, the erect shoots are green to olive green and sometime reddish brown, and are simply upturned extensions of the horizontal rhizomes（Washington State Department of Ecology, 2008）.

【www.eddmaps.org】*Cabomba caroliniana* Gray

C. caroliniana is fully submerged except for occasional floating leaves and emergent flowers（Australian Department of the Environment and Heritage 2003）. The roots grow on the bottom of water bodies and the stems can reach the surface. Parts of the plant can survive free-floating for six to eight weeks. It is a perennial, growing from short rhizomes with fibrous roots. The branched stems can grow up to 10m long and are scattered with white or reddish-brown hairs.

The underwater leaves are divided into fine branches, resulting in a feathery fan-like appearance. These leaves are about 5 cm across and secrete a gelatinous mucous which covers the submerged parts of the plant. The floating leaves, however, are small, diamond-shaped, entire, and borne on the flowering branches. The solitary flowers are less than 2 cm across and range in colour from white to pale yellow and may also include a pink or purplish tinge. The flowers emerge on stalks from the tips of the stems（Australian Department of Environment and Heritage, 2003）.

Submersed leaves: petiole to 4 cm; leaf blade 1-3.5 cm × 1.5-5.5 cm, terminal segments 3-200, linear to slightly spatulate, to 1.8 mm wide. Floating leaves: blade 0.6-3 cm × 1-4 mm, margins entire or notched to sagittate at base. Flowers 6-15 mm diam.; sepals white to purplish [yellow] or with purple-tinged margins, 5-12 mm × 2-7 mm; petals colored as sepals but with proximal, yellow, nectar-bearing auricles, 4-12 mm × 2-5 mm, apex broadly obtuse or notched; stamens 3-6, mostly 6; pistils 2-4, mostly 3, divergent at maturity; ovules 3. Fruits 4-7 mm. Seeds 1-3, 1.5-3 mm × 1-1.5 mm, tubercles in 4 longitudinal rows

The submersed leaves of *Cabomba caroliniana* are similar in form to those of Limnophila（Scrophulariaceae; introduced in southeastern United States）. The latter has whorled leaves in contrast to the opposite leaves of Cabomba.

Part 2

外来入侵
动物

昆　虫

蔗扁蛾

学　名：*Opogona sacchari*（Bojer）

异　名：*Alucita sacchari* Bojer，*Gelechia ligniferalla* Walker，*Gelechia sanctaehelenae* Walker，*Hieroxestis ligniferella*，*Hieroxestis plumipes*，Butler *Hieroxestis sanctaehelenae* Walker，*Hieroxestis subcervinella* Walker，*Laverna plumipes* Butler，*Opogona sanctaehelenae* Walker，*Opogona subcervinella*（Walker），*Tinea subcervinella* Walker

别　名：香蕉蛾、香蕉谷蛾

英文名：banana moth，sugarcane moth

形态特征

成虫　体呈黄灰色，具强金属光泽，腹面色淡。体长7.5～9.0mm，翅展18.0～26.0mm，雄虫略小。头部鳞片大而光滑，头顶色暗且向后平覆，额区则向前弯覆，二者之间由一横条蓬松的竖毛分开，颜面平斜、鳞片小且色淡。下唇须粗长，斜伸微翘；下颚须细长卷折，喙极短小。触角细长纤毛状，长达前翅的2/3，梗节粗长稍弯。胸背鳞片大而平滑，翅平覆。前翅深棕色，披针形，中室端部和后缘各有一黑色斑点，并具多条断续的褐纹，雄蛾多连成较完整的纵条斑；雌蛾前翅基部有一黑色细线，可达翅中部。后翅色淡，披针形，后缘具长缘毛，雄蛾翅基具长毛束。足基节宽大扁平，紧贴体下，后足胫节具长毛，中距靠上。腹部平扁，腹板两侧具褐斑列。停息时，触角前伸；爬行时，形似蜚蠊，速度快，并可做短距离跳跃。

卵　淡黄色。卵圆形，长0.5～0.7mm、宽0.3～0.4mm。

幼虫　乳白色透明。老熟幼虫长30mm，宽3mm。头红棕色，胴部各节背面有4个毛片，矩形，前2后2排成2排，各节侧面亦有4个小毛片。

蛹　棕色。触角、翅芽、后足相互紧贴，与蛹体分离。

生物危害

蔗扁蛾为我国检疫性有害生物。是广谱性贪食性害虫。不仅为害观赏植物，而且为害香蕉、甘蔗、玉米、马铃薯等农作物及温室栽培植物等经济植物，对花卉产业、热带农业和制糖业构成巨大威胁。

| 地理分布 |

蔗扁蛾原产于非洲热带、亚热带地区。现分布于欧洲、南美洲、西印度群岛、美国等国家或地区。

1987年，蔗扁蛾随进口巴西木而传入广州。之后，随着巴西木在中国种植的普及，蔗扁蛾也随之扩散。1997年，在北京发现此虫。目前蔗扁蛾已经扩散到广东、海南、福建、河南、新疆、四川、上海、南京、浙江等省市。

| 传播途径 |

蔗扁蛾主要以幼虫、蛹随巴西木等观赏植物及其繁殖材料、香蕉等水果寄主植物的调运而远距离传播。也可依靠成虫飞翔而进行自然传播，但飞行能力较弱，一般一次只能飞行10m左右。

| 管理措施 |

综合治理（Integrated Pest Management，IPM）或综合控制（Integrated Pest Control，IPC），包括植物卫生措施（Phytosanitary Measures）。

(图片选自mothphotographersgroup.msstate.edu,en.wikipedia.org,cn.bing.com)

蔗扁蛾(*Opogona sacchari*)

【www.cabi.org】 *Opogona sacchari* (banana moth)

Larva The larvae are dirty-white and somewhat transparent (so that the intestines can be seen). They have a bright reddish-brown head with one lateral ocellus at each side and clearly visible, brownish thoracic and abdominal plates. They are 21-26 mm long with a diameter of 3 mm. The presence of older larvae can be detected by characteristic masses of bore-meal and frass at the openings of boreholes.

Pupa The pupae are brown, less than 10 mm long, and are formed in a cocoon, spun at the end of a mine, measuring 15 mm. As maturation approaches, the pupae work themselves partially out of the tissue to allow emergence of the adult. Two bent hooks, characteristic of the species, show at the end of the abdomen on the abandoned protruding pupal skin.

Adult The adult is nocturnal, 11 mm long with a wingspan of 18-25 mm. It is bright yellowish-brown. The forewings may show longitudinal darker brown banding, and in the male a dark-brown spot towards the apex. The hindwings are paler and brighter (Süss, 1974; D'Aguilar and Martinez, 1982). At rest, the long antennae point forwards.

【onlinelibrary.wiley.com】 *Opogona sacchari*

The larvae of *O. sacchari*, dirty-white and somewhat transparent (so that the intestines are visible through body wall), have a bright reddish-brown head with one pair of lateral ocelli on each side (stemmata) and clearly visible brownish thoracic (segments 2-3) and abdominal (1-8) plates. Abdominal segment 9 has one large and 2 small brownish plates and the anal shield is visible on abdominal segment 10. Abdominal prolegs A3-A6 with 43-45 hooks each and 20-22 hooks on anal proleg A10. Pretarsal claw of prothorax prolonged, with 2 axial lobes. Larvae typically measure 26-35 mm in length with a diameter of 3 mm for the last instar. The larvae can be positively identified on the following criteria: one pair of lateral ocelli; characteristic chaetotaxy (Davis & Peña, 1990; see also Süss, 1974) – separation of the spiracle from the pinaculum bearing L2 on the first eight abdominal segments, by the large number of crochets (A3-A6 = 43-45, A10 = 20-22); complete encirclement of abdominal sternite by a band of small, secondary spines. In *O. omoscopa*, only one anterior ocellus (or stemma) has been observed, the first eight abdominal spiracles are united with L2 on a common pinaculum, the crochets are fewer in number, and the abdominal sternite A3-A6 have spines only along the anterior margin (Davis & Peña, 1990).

The pupae are brown and less than 10 mm long and are formed in a cocoon, spun at the end of a feeding chamber/larval tunnel, measuring 15 mm. Two bent hooks, characteristic of the species, are visible at the end of the abdomen on the abandoned protruding skin. The pupa of *O. sacchari* is very similar to that of *O. omoscopa*, but can be distinguished by the raised spiracle on A8 and the larger cremaster spines (Davis, 1978; Davis & Peña, 1990).

The adult is nocturnal, 11 mm long with a wingspan of 18-25 mm, bright yellowish-brown. The forewings may show longitudinal darker brown banding, and in the male a dark-brown spot towards the apex. The hindwings are paler and brighter (Süss, 1974; D'Aguilar & Martinez, 1982) with a frenulum of 5-7 bristles in the female, 1 bristle in the male (Davis & Peña, 1990) and a unique dorsal hindwing hair pencil rising from the base of the wing in the male (Robinson & Tuck, 1997). At rest, the long antennae point forwards. Most distinctive are the relatively large size compared with other *Opogona* spp., the male and female genitalia, and the unique hair pencil.

湿地松粉蚧

学　　名：***Oracella acuta***（Lobdell）
异　　名：*Oracella acuta* Ferris，*Pseudococcus acutus* Lobdell
别　　名：火炬松粉蚧
英文名：loblolly pine mealybug，nantah Oracella，acute mealybug

| 形态特征 |

雌成虫呈梨形。腹部向后尖削，触角7节。

若虫椭圆形至不对称椭圆形，长1.02~1.52mm，足3对。

初孵若虫孵化后聚集在雌成虫的蜡包内，天气适宜时爬出，在松树枝、梢、叶处不停活动，并随气流被动扩散。末龄后期，虫体分泌蜡质物形成白色蜡包，覆盖虫体。

雄成虫分有翅型和无翅型两种。

与当地松粉蚧区别：湿地松粉蚧雌成虫梨形，腹部向后尖削，触角7节；当地松粉蚧雌成虫纺锤形，触角8节。

| 生物危害 |

湿地松粉蚧主要以若虫为害湿地松松梢、嫩枝及球果。受害的松梢轻者抽梢，针叶伸展长度均明显地减少。严重时梢上针叶极短，不能伸展或顶芽枯死、弯曲，形成丛枝。老针叶大量脱落可达70%~80%，其他针叶也因伴发煤污病影响光合作用。球果受害后，发育受限制，变小而弯曲，影响种子质量和产量。由于引入的湿地松、火炬松和加勒比松加速了害虫的扩散，对本地马尾松、南亚松等构成严重威胁。

| 地理分布 |

湿地松粉蚧原产于美国。1988年随湿地松进入中国广东省台山，到1994年已扩散至广东省多个县市。现分布于广东、广西、福建等地。

| 传播途径 |

湿地松粉蚧主要通过松属植物苗木、接穗等繁殖材料嫩枝及新鲜球果的携带而作远距离传播。

| 管理措施 |

综合治理（Integrated Pest Management，IPM）或综合控制（Integrated Pest Control，IPC），但不包括植物卫生措施（Phytosanitary Measures）。

（图片选自www.invasive.org，wiki.bugwood.org，www.forestryimages.org）

湿地松粉蚧（*Oracella acuta*）

【www.cabi.org】 *Oracella acuta* (**loblolly pine mealybug**)

O. acuta is a bisexual species with multiple generations annually. This species is distinguished by the morphology of the adult female. No descriptive information on the morphology of the immature or adult male stages exists at present (2010). The live adult females are pink and covered with a white powdery, waxy secretion. The females are uniquely found enclosed within whitish resinous cells that are open at one end, from which their pygidia protrude. The cells are usually attached to the stem at the base of the needle.

The slide-mounted adult females are 1.6-2.2 mm long and 1-1.5 mm wide (Lobdell, 1930; Kosztarab, 1996). The dorsum is distinguished by four to seven pairs of cerarii on the submargin; each with two to three minute conical setae, two to three slender setae and a few trilocular pores. Oral collar ducts and anal ring bar are absent and tubular ducts are absent or rarely occur with only a few ducts irregularly distributed over the dorsum (Ferris, 1950; Kosztarab, 1996; Miller, 2005). Anal lobes are reduced to small protuberances. Venter possesses seven-segmented antennae with filamentous segments. The well-developed legs are five-segmented with translucent pores on femur and tibia, and the claw is slightly curved and without a denticle. Large, distinctive multilocular pores are located on the posterior borders of the third to eighth abdominal segments (Ferris, 1950; Kosztarab, 1996), with a few scattered in the cephalic region (Lobdell, 1930). The circulus is large and with a divided transverse fold. Sparse trilocular pores are scattered over the abdominal segments.

【www.forestryimages.org】 *Oracella acuta* **Lobdell**

强大小蠹

学　　名：***Dendroctonus valens*** LeConte
异　　名：*Dendroctonus beckeri* Thatcher，*Dendroctonus rhizophagus* Thomas & Bright
别　　名：红脂大小蠹
英文名：red turpentine beetle，red turpentine

| 形态特征 |

雄成虫　体长5.9～8.1mm，平均6.47mm。新羽化的成虫体棕黄色，后变为红褐色，少数黑褐色。额不规则隆起，在复眼上缘下方至口上脊边缘1/3处有一对瘤突，瘤突间凹下。触角柄节长，鞭节5节，锤状部3节，扁平近圆形。前胸背板长为宽的0.73倍，两侧弱弓形，基部2/3近平行。鞘翅长为宽的1.5倍，为前胸背板长度的2.2倍，两侧直伸，基部2/3近平行，后部阔圆形，基缘弓形，生一列约12个中等大小、隆起的重叠齿和几个更小的亚缘齿；鞘翅刻点细小深陷；沟间部宽度约为刻点沟宽度的1.5倍，具大量杂乱的小横齿，平均每个齿约为一条沟间部宽度的1/3，中域后部的齿不超过沟间部宽度的一半。

雌成虫　体长7.5～9.6mm，平均8.28mm。雌虫与雄虫相似，但额中部在复眼上缘有一明显的圆形凸起；前胸背板上的刻窝较大；鞘翅坡面上的粗突和鞘翅中部的锯齿状突均较大。

卵　圆形至长椭圆形，乳白色，有光泽，长0.9～1.1mm，宽0.4～0.5mm。

幼虫　蛴螬形，无足，体白色。老熟幼虫体长平均11.8mm，头宽1.79mm，腹部末端有胴痣，上下各具有一列刺钩，呈棕褐色，每列有刺钩3个，上列刺钩大于下列刺钩。虫体两侧除有气孔外，还有一列肉瘤，肉瘤中心有一根刚毛，呈红褐色。

蛹　平均体长7.82mm。翅芽、足、触角贴于体侧。

| 生物危害 |

红脂大小蠹为我国检疫性有害生物。主要为害已经成材且长势衰弱的大径立木，在新鲜伐桩和伐木上为害尤其严重。但与北美洲发生情况不同的是，在我国发生区，它不仅攻击树势衰弱的树木，也攻击健壮树，导致发生区内寄主大面积死亡。

| 地理分布 |

红脂大小蠹原产于美国、加拿大、墨西哥、危地马拉和洪都拉斯等美洲地区。1998年，首次在我国山西省阳城、沁水发现，可能与20世纪80年代后期山西从美国进口木材有关。现分布于山西、陕西、河北、河南等地。

| 传播途径 |

红脂大小蠹害虫随原木进口而传入我国。成虫飞行能力很强，在早春，其飞行距离超过16km。

可自然扩散或人为传播。

| 管理措施 |

综合治理（Integrated Pest Management，IPM）或综合控制（Integrated Pest Control，IPC），包括植物卫生措施（Phytosanitary Measures）。

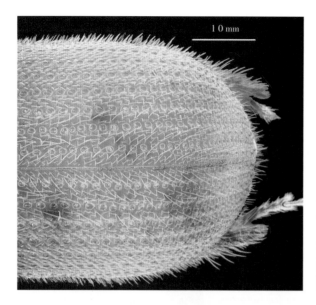
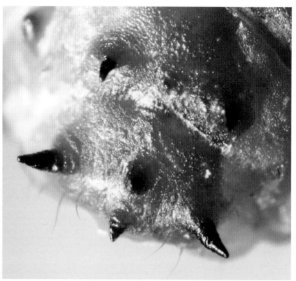

（图片选自baike.baidu.com，bugguide.net，www.cabi.org，www.monarthrum.info）

强大小蠹（*Dendroctonus valens*）

【www.cabi.org】*Dendroctonus valens*（red turpentine beetle）

Egg The eggs are shiny, opaque white, ovoid cylindrical, and about 1 mm long. They are often laid in groups of 10-40 or more, along one side of the gallery away from the area of main beetle activity. The egg masses are covered with compacted frass.

Larva The larva is grub-like, legless and white, except for a brown head capsule and a small brown area at the hind end. A row of small, pale-brown tubercles become evident along each side of the body as the larva grows. The fully-grown larva can be up to 10-12 mm long. Pupa The pupa is white and slightly shorter than the larva. The legs and antennae are held against the body in the pupal or resting stage.

Adult The adult beetles, which are the largest of the Dendroctonus genus, are typically 6-10 mm long and quite stout, 2.1 times as long as wide. At first, the beetle is called a callow adult and is tan, but it rapidly darkens to a reddish-brown. The frons is moderately convex with three elevations, the upper one just below the end of the epicranial suture and the lower two laterally sited on the median frons. The epistomal process is broad, more than 0.55 times as wide as the distance between the eyes. The arms of the epistomal process are oblique about 20° from horizontal and are prominently and roundly elevated. The surface of the epistomal process is longitudinally and broadly concave. The pronotum is 0.73 times as long as wide, 2.2 times as long as the pronotum. Elytral declivity with all interstriae shining, interstriae 1 not elevated, interstriae 2 neither narrower nor more impressed than interstriae 1 and 3, all declivital interstriae coarsely granulated, granulations distributed sometimes confused and sometimes regularly.

美国白蛾

学　　名：***Hyphantria cunea*** Drury

异　　名：*Hyphantria budea*（Hubner）, *Hyphantria candida*（Walker）, *Hyphantria liturata*（Goeze）, *Hyphantria mutans*（Walker）, *Hyphantria pallida*（Packard）, *Hyphantria punctata*（Fitch）, *Hyphantria punctatissima*（Smith）, *Hyphantria suffusa*（Strecker）, *Hyphantria textor*（Harris）

别　　名：秋幕毛虫、秋幕蛾、美国白灯蛾

英文名：mulberry moth, American white moth, blackheaded webworm, fall webworm, redheaded webworm

| 形态特征 |

成虫 体长9~14mm，翅展23~44mm，体白色。复眼黑褐色，下唇须小，端部黑褐色，口器短而纤细。胸部背面密布白毛，多数个体腹部白色无斑点，少数个体腹部白色上有黑点。雄蛾触角双栉状，黑色，长5mm，内侧栉齿较短，约为外侧栉齿的2/3，下唇须外侧黑色，内侧白色，多数前翅散生几个或多个黑褐色斑点。雌蛾触角锯齿状，褐色，前翅多为纯白色，少数个体有斑点。后翅一般为纯白色或近边缘处有小黑点。成虫前足基节及腿节端部为橘黄色，胫节和跗节外侧为黑色，内侧为白色。前中跗节的前爪长而弯，后爪短而直。

老熟幼虫 体长28~35mm，头黑色或红色，具光泽。体黄绿色至灰黑色，背线、气门上线、气门下线浅黄色。背部毛瘤黑色，体侧毛瘤多为橙黄色，毛瘤上着生白色长毛丛。腹足外侧黑色。气门白色，椭圆形，具黑边。根据幼虫的形态，可分为黑头型和红头型两型，其在低龄时就能够明显分辨。三龄以后，从体色、色斑、毛瘤及其上的刚毛颜色上更易区别。

蛹 体长8~15mm，宽3~5mm，暗红褐色。雄蛹瘦小，雌蛹较肥大。蛹外被有黄褐色薄丝质茧，茧上的丝混杂着幼虫的体毛共同形成网状物。腹部各节除节间外，布满凹陷刻点，臀刺8~17根，每根钩刺的末端呈喇叭口状，中凹陷。

| 生物危害 |

美国白蛾为我国检疫性有害生物。其食性杂，繁殖量大，适应性强，传播途径广，扩散快（每年可向外扩散35~50km），也是危害严重的世界性检疫害虫。其幼虫食性很杂，被害植物主要有白腊、臭椿、法桐、山檀、桑树、苹果、海棠、金银木、紫叶李、桃树、榆树、柳树、杨树等200多种植物，严重威胁养蚕业，林果业和城市绿化。据国家林业局植树造林司总工程师吴坚（2006年）介绍，由于美国白蛾侵入中国后，失去了原有天敌的控制，其种群密度迅速增长并蔓延成灾，每年给我国林业造成的损失已经高达660亿元。

地理分布

美国白蛾分布于美国、加拿大、东欧各国及日本、朝鲜等国。1979年，首次在辽宁省发现；1985年，在陕西省西安市武功县发现。目前在北京、天津、河北、辽宁、山东、陕西、河南等7个省市已经出现了美国白蛾发生疫情。

传播途径

美国白蛾可随寄主或木质包装远距离人为传播，也可自然扩散。

管理措施

综合治理（Integrated Pest Management，IPM）或综合控制（Integrated Pest Control，IPC），包括植物卫生措施（Phytosanitary Measures）。

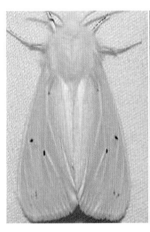

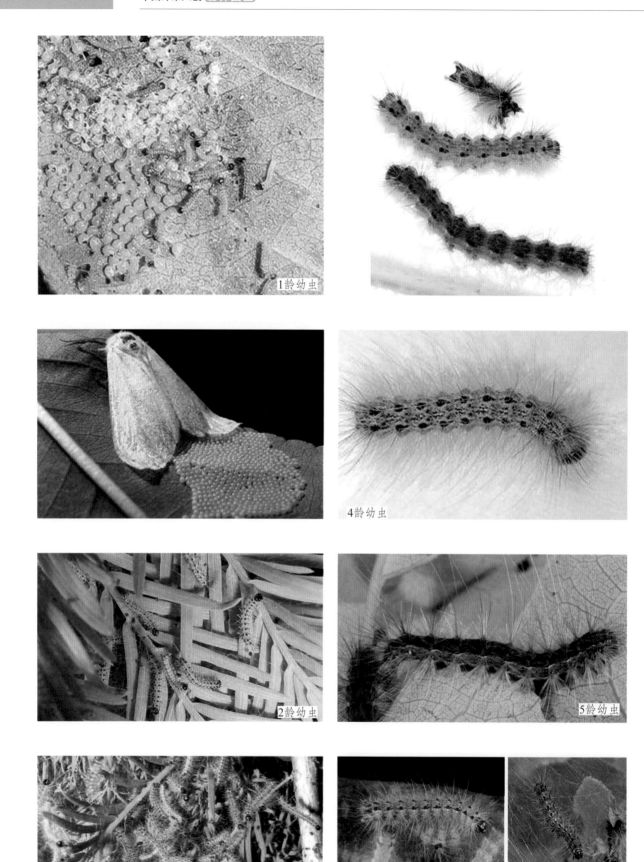

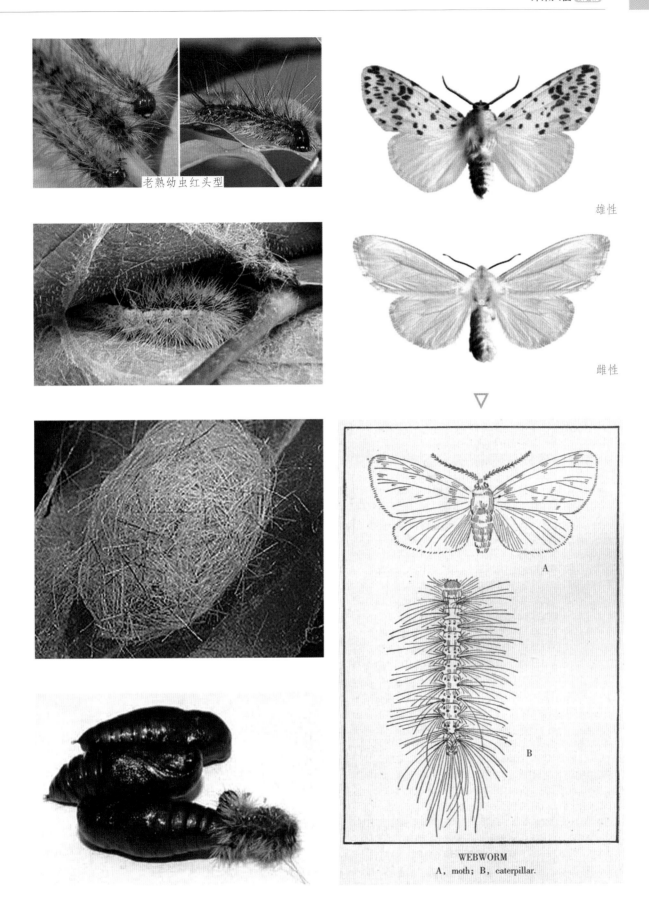

(图片选自entnemdept.ufl.edu，www.discoverlife.org，en.wikipedia.org，mothphotographersgroup.msstate.edu，upload.wikimedia.org）

美国白蛾（*Hyphantria cunea*）

【www.cabi.org】*Hyphantria cunea*（mulberry moth）

Larva Brownish-grey, attains 40 mm when fully developed, and has 12 small warts surmounted by characteristic tufts of hair. There are two forms, those with red heads and those with black heads.

Pupa The pupa has 12 characteristic appendages at the posterior end.

Adult Moth with a wingspan of 25-30 mm; forewings are white or have black spots arranged in a number of rows; hindwings are also white with a small black spot on the leading part.

【entnemdept.ufl.edu】*Hyphantria cunea*

Eggs: The egg mass of *Hyphantria cunea* is almost iridescent green in color.

Larvae Mature larvae are hairy and either have a lime green body with black spots or can have darker color, especially in the later instars. The head capsules in some populations can be either red or black. In other populations, they are entirely black.

Adult The adult fall webworm moth is bright white, with a hairy body. In the southern part of its range, the moth is white with dark wing spots while in the northern part of its range it is nearly always pure white（MPG 2010）. Adult moths have a wingspan of between 1.4-1.7 inches（35-42 mm）. The bases of the front legs are orange or bright yellow.

【en.wikipedia.org】Fall webworm

桉树枝瘿姬小蜂

学　　名：*Leptocybe invasa* Fisher et La Salle
异　　名：无
别　　名：无
英 文 名：blue gum chalcid wasp，Eucalyptus gall wasp

| 形态特征 |

桉树枝瘿姬小蜂个体较小。雌成虫体长为1.1~1.4mm，褐色，略带蓝绿色金属光泽。头扁平，单眼3个，呈三角形排列；复眼近红色；触角分为柄节、梗节、索节和棒节4部分。头部骨化程度较弱，易皱缩。翅透明，翅脉浅棕色。

幼虫　微小，白色，无足。

| 生物危害 |

桉树枝瘿姬小蜂主要为害桉树苗木及幼林，在叶片、主脉、叶柄及当年生枝条上形成虫瘿，危害严重时导致苗木倒伏、落叶、植株矮化、枝梢枯死，严重影响树木生长，以一年生左右幼林受害最重，植株受害率可达100%，受害林分林木产量严重下降。对华南、西南等地区的桉树种植造成极大威胁。

| 地理分布 |

桉树枝瘿姬小蜂原产于澳大利亚。现在法国、意大利、新西兰、葡萄牙、西班牙、希腊、阿尔及利亚、摩洛哥、肯尼亚、叙利亚、坦桑尼亚、乌干达、南非、约旦、伊朗、以色列、泰国、土耳其、越南、印度、柬埔寨、黎巴嫩等有分布。

2007年，首次在我国广西与越南交界处发现，2008年相继在海南和广东发现。现分布于广西、海南、广东等地。

| 传播途径 |

桉树枝瘿姬小蜂行孤雌生殖，繁殖能力强，扩散迅速。自然扩散靠成虫飞行，人为携带繁殖材料则是其远距离传播的主要方式。

| 管理措施 |

综合治理（Integrated Pest Management，IPM）或综合控制（Integrated Pest Control，IPC），但不包括植物卫生措施（Phytosanitary Measures）。

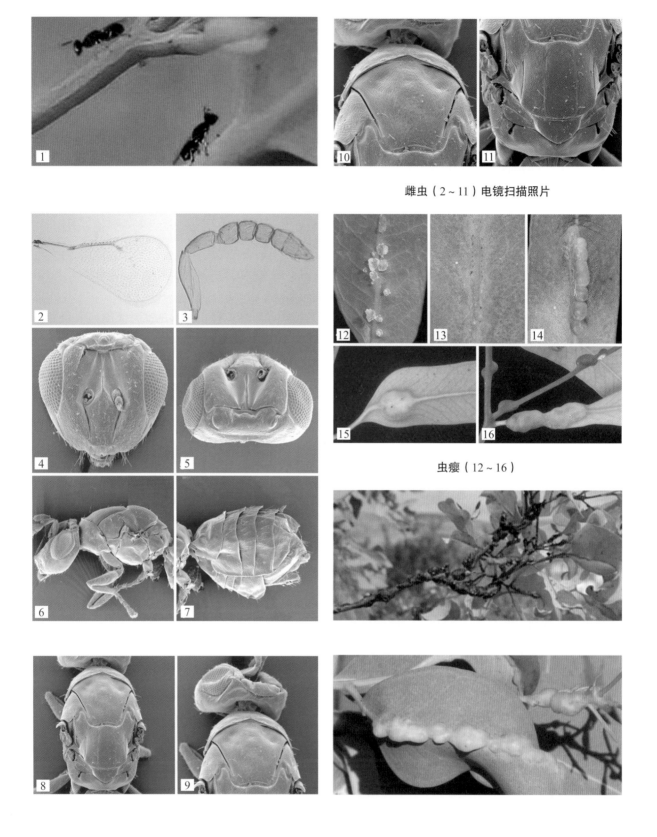

雌虫（2～11）电镜扫描照片

虫瘿（12～16）

（图片选自www.waspweb.org，www.fao.org，upload.wikimedia.org，www.revistagua.cl）

桉树枝瘿姬小蜂（*Leptocybe invasa*）

【www.fao.org】*Leptocybe invasa*

The female chalcid is a small wasp, brown in colour with a slight to distinctive blue to green metallic shine (TPCP, 2005). The average length is 1.2 mm. With the exception of one record describing males in Turkey, only females of this species, which reproduce by parthenogenesis, have been observed. Larvae are minute, white and legless.

【www.waspweb.org】*Leptocybe invasa* Fisher & La Salle

【commons.wikimedia.org】Category:*Leptocybe invasa*

稻水象甲

学　名：*Lissorhoptrus oryzophilus* Kuschel
异　名：*Lissorhoptrus simplex* auctt., nec Say
别　名：稻水象、稻根象
英文名：rice water weevil，American water weevil，lesser water weevil

| 形态特征 |
成虫　体长2.6~3.8mm。喙与前胸背板几等长，稍弯，扁圆筒形。前胸背板宽。鞘翅侧缘平行，比前胸背板宽，肩斜，鞘翅端半部行间上有瘤突。雌虫后足胫节有前锐突和锐突，锐突长而尖，雄虫仅具短粗的两叉形锐突。

卵　圆柱形，两端圆。

幼虫　体白色，头黄褐色。

蛹　长约3mm，白色。

| 生物危害 |
稻水象甲成虫取食水稻叶片。幼虫为害水稻根部。为害秧苗时，可将稻秧根部吃光。

| 地理分布 |
稻水象甲原产于美国东部、古巴等地。1976年进入日本，1988年扩散到朝鲜半岛。1988年首次发现于河北省唐山市。现分布于河北、辽宁、吉林、山东、山西、陕西、浙江、安徽、福建、湖南、云南、台湾等地。

| 传播途径 |
稻水象甲可自然扩散，也可随稻秧、稻谷、稻草及其他寄主植物、交通工具等远距离人为传播。成虫也可借气流飞翔迁移或者随水流传播扩散。

| 管理措施 |
综合治理（Integrated Pest Management，IPM）或综合控制（Integrated Pest Control，IPC），包括植物卫生措施（Phytosanitary Measures）。

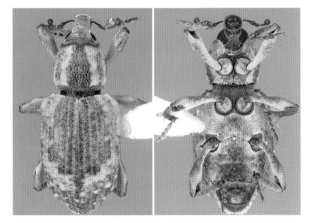

（图片选自www.cabi.org，www.cropscience.bayer.com，bugguide.net）

稻水象甲（*Lissorhoptrus oryzophilus*）

【 www.cabi.org 】 *Lissorhoptrus oryzophilus* (rice water weevil)

Eggs White; cylindrical and elongate, 0.8 mm long by 0.2 mm wide, length three to five times greater than the width; oviposited in submerged leaf sheaths on the lower half of the rice plant.

Larvae White, legless grubs; presence of paired curved dorsal tracheal hooks on the second through seventh abdominal segments, apical segment of the hook is sclerotized and used to pierce root tissue and sequester oxygen, basal segment of the hook is tubular and flexible and separated from the trachea by chitinized rings; four larval instars with head capsule widths varying in size from 0.16 mm to 4.5 mm; body lengths of 1.5 mm (first-instar larvae) to 8 mm (fourth-instar larvae). Larvae were described by Lee and Morimoto (1988).

Pupa White; formed in an oval, water-tight mud cell attached to plant roots; resembles the adult in size and shape.

Adult Dark-brown to black with grey scales; small, oblong (2.8 mm long by 1.2-1.8 mm wide); in sexually dimorphic weevils, the female is more robust than the male and the first two ventral abdominal sternites are flat to convex at the midline of the female, whereas they are broadly concave in the male; females have a large darkened area on the elytra and a deep notch in the seventh tergal segment.

【 en.wikipedia.org 】 *Lissorhoptrus oryzophilus*

Eggs The pearly white eggs are cylindrical (0.8 mm long and 0.14 mm long) of pearly with a very thin corion.

Larvae The larvae are aquatic and live their entire lives in the rhizosphere. They are white and grow up to 1 cm long at 4th instar stage.

Pupae The pupae is a small silk cocoon encased in mud (0.5-0.9 cm long).

Adults The adults are 3.3-3.7 mm long, including the rostrum. The exoskeleton ranges in color from dark beige, brown, or dark-brown. Along the center of the elytra, some rice water weevils have an elongated dark brown to brownish-black mark. The middle pair of legs have hydrophobic hairs that allow it to swim (Hix et al., 2000).

入侵红火蚁

学　名：***Solenopsis invicta*** Buren

异　名：*Solenopsis saevissima* var. *wagneri* Santschi；*Solenopsis wagneri* Santschi

别　名：红火蚁、外引红火蚁、泊来红火蚁（台湾）

英文名：red imported fire ant，fire ant，red imported，RIFA

| 形态特征 |

工蚁　头部近正方形至略呈心形，长1.00～1.47mm，宽0.90～1.42mm。头顶中间略微下凹，无带横纹的纵沟。唇基中齿发达，长约为侧齿的一半，有时不在中间位置；唇基中刚毛明显，着生于中齿端部或近端；唇基侧脊明显，末端突出呈三角尖齿，侧齿间中齿基以外的唇基边缘凹陷。

复眼椭圆形，最大直径为11～14个小眼长，最小直径为8～10个小眼长。触角柄节长，小型工蚁柄节端可伸达或超过头顶。

前胸背板前侧角圆至轻微的角状，罕见突出的肩角。中胸侧板前腹边厚，厚边内侧着生多条与厚边垂直的横向小脊。并胸腹节背面和斜面两侧无脊状突起，仅在背面和其后的斜面之间呈钝圆角状。

后腹柄结略宽于前腹柄结，前腹柄结腹面可能有一些细浅的中纵沟，柄腹突小，平截，后腹柄结后面观长方形，顶部光亮，下面2/3或更大部分着生横纹与刻点。

生殖型雌蚁　有翅型雌蚁体长8～10mm。头及胸部棕褐色，腹部黑褐色。头部细小，触角呈膝状。胸部发达，前胸背板显著隆起。翅2对。

雄蚁　体黑色，体长7～8mm，着生2对翅。头部细小，触角呈丝状。胸部发达，前胸背板显著隆起。

大型工蚁（兵蚁）　体长6～7mm，形态与小型工蚁相似，体橘红色。腹部背板呈深褐色。

卵　圆形，乳白色，直径0.23～0.30mm。

幼虫　共4龄，各龄均乳白色。四龄长度分别为：1龄0.27～0.42mm；2龄0.42mm；3龄0.59～0.76mm。将发育为工蚁的4龄幼虫0.79～1.20mm，而将发育为有性生殖蚁的4龄幼虫体长可达4～5mm。1～2龄体表较光滑，3～4龄体表被有短毛，4龄上颚骨化较深，略呈褐色。

蛹　为裸蛹，乳白色。工蚁蛹体长0.70～0.80mm，有性生殖蚁蛹体长5～7mm。触角、足均外露。

| 生物危害 |

入侵红火蚁为我国检疫性有害生物。其食性杂，觅食能力强，喜食昆虫和其他节肢动物，也猎食无脊椎动物、脊椎动物、植物，还可取食腐肉。

该蚁可取食149种野生花草的种子，为害57种农作物。

入侵红火蚁群体生存、发育需要大量糖分，因此工蚁常取食植物汁液、花蜜或在植物上"放牧"蚜虫、介壳虫。在美国南方，已有13个州超过0.15亿亩的土地被入侵火蚁所占据，对这些受侵害地区造成的经济损失，估计每年达数十亿美元。

入侵红火蚁还袭击人类。一旦被其叮咬，会引起毒性反应，轻者皮肤瘙痒、起泡，重者则会有严重过敏反应，如呼吸困难甚至休克。

| 地理分布 |

入侵红火蚁原产于南美洲，包括巴西、巴拉圭和阿根廷等国家。该虫1930年入侵美国。现在美国南部13个州以及波多黎各、新西兰、澳大利亚等有分布。

2004年7月在我国台湾发现。现已经在台湾、香港、广东、澳门、福建、广西、湖南等地分布。

| 传播途径 |

自然扩散（生殖蚁飞行或随洪水流动扩散）和人为传播（园艺植栽污染、草皮污染、土壤废土移动、堆肥、园艺农耕机具设备、空货柜污染、车辆污染等）。

| 管理措施 |

综合治理（Integrated Pest Management，IPM）或综合控制（Integrated Pest Control，IPC），包括植物卫生措施（Phytosanitary Measures）。

（工蚁）

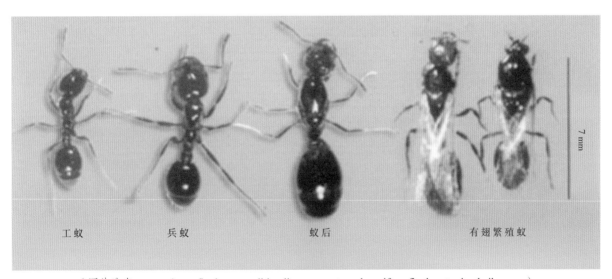

| 工蚁 | 兵蚁 | 蚁后 | 有翅繁殖蚁 |

（图片选自entnemdept.ufl.edu，en.wikipedia.org，entomology.ifas.ufl.edu，tupian.baike.com）

入侵红火蚁（*Solenopsis invicta*）

【www.cabi.org】*Solenopsis invicta*（red imported fire ant）

Worker Ants Worker ants are wingless, dark reddish-brown with black abdomens, and range from 1.5 to 5 mm long. Workers in the genus Solenopsis are polymorphic, meaning they are physically differentiated into more than two different body-forms（Holway, 2002 in ISSG, 2014）. A new colony's first workers, called minims, are smaller than those of later generations. Minors are slightly larger, and medias larger still （Taber, 2000）. The largest is the major worker, which（in later generations）can reach lengths of up to 5 mm. The sting of *S. invicta* ants can be found at the tip of the abdomen under close examination.

Developmental Stages The eggs are spherical and creamy-white and the larvae are legless, cream-coloured and grub-like with a distinct head capsule. The pupae resemble the worker ants and are initially creamy-white turning darker before the adult ants emerge（eclose）. The eggs, larvae and pupae are referred to as a brood.

Sexuals or Winged Reproductives The females are reddish-brown whereas the males are shiny and black with a smaller head. These ants stay in the colony until the conditions exist for their nuptial flight. The queen ants are mated female reproductives, are larger than the worker ants（9 mm）and remove their wings following a nuptial flight.

【entnemdept.ufl.edu】*Solenopsis invicta* Buren

【entomology.ifas.ufl.edu】Red imported fire ant

【www.sms.si.edu】*Solenopsis invicta*

【en.wikipedia.org】Red imported fire ant

苹果蠹蛾

学　名：***Cydia pomonella***（Linnaeus）

异　名：*Carpocapsa pomonana* Treitschke，*Carpocapsa pomonella* Linnaeus，*Enarmonia pomonella* Linnaeus，*Grapholitha pomonella*，*Laspeyresia pomonella* Linnaeus，*Phalaena pomonella* Linnaeus

英文名：codling moth，walnut worm

形态特征

成虫　体长8mm，翅展15～22mm，体灰褐色。前翅臀角处有深褐色椭圆形大斑，内有3条青铜色条纹，其间显出4～5条褐色横纹；翅基部外缘突出略成三角形，杂有较深的斜形波状纹；翅中部颜色最浅，淡褐色，也杂有褐色斜形波状纹。前翅R_{4+5}脉与M_3脉的基部明显，通过中室；R_{2+3}脉的长度约为R_{4+5}脉基部至R_1脉基部间距离的1/3；组成中室前缘端部的一段R_3脉的长度约为连接R_3脉与R_4脉的分横脉（S）的3倍；R_5脉达到外缘；M_1与M_2脉远离；M_2与M_3脉接近平行；Cu_2脉左起自中室后缘2/3处；臀脉（1A）1条，基部分叉很长约占臀脉的1/3。后翅黄褐色，基部较淡；Rs脉与M_1脉基部靠近；M_3脉与Cu_1脉共柄。雌虫翅缰4根，雄虫1根。

雄性外生殖器抱器瓣在中间有明显的颈部。抱器腹在中间凹陷，外侧有1尖刺。抱器端圆形，有许多毛。阳茎短粗，基部稍弯。阳茎针6～8枚，排列成两行。

卵　椭圆形，长1.1～1.2mm，宽0.9～1.0mm，极扁平，中央部分略隆起，初产时像一极薄的蜡滴，半透明。随着胚胎发育，中央部分呈黄色，并显出1圈断续的红色斑点，后则连成整圈，孵化前能透见幼虫。卵壳表面无显著刻纹，放大100倍以上时，则可见不规则的细微皱纹。

幼虫　初龄幼虫体多为淡黄白色。成熟幼虫14～18mm，多为淡红色，背面色深，腹面色浅。头部黄褐色。前胸盾片淡黄色，并有褐色斑点，臀板上有淡褐色斑点。头部眼群毛O_1与A_3的连接不通过单眼1（最多仅相切）。上唇上缘较平直，下缘呈"W"形，但中央缺刻较浅；表面有6对对称排列的毛，其中4对沿上唇下缘分布，另2对位于上唇中区。上颚具齿5个，但仅有3个较发达。前胸气门群4、5、6位于同一毛片上；足群7a、7b；中胸和后胸亚背群1、2毛及气门上群3、3a分别位于同一毛片上，气门群4、5位于同一毛片上，足群仅有7a毛。腹节1～8气门上群3、3a位于同一毛片上，气门群4、5位于同一毛片；腹节9的4、5毛位于同一毛片或与6毛相连。腹节1～6足群7a、7b、7c位于同一毛片。腹节7～8足群7a、7b位于同一毛片；腹节9仅有7a毛且1、3毛位于同一毛片，1、2、3毛（D_1、D_2、SD_1）排成三角形。

蛹　全体黄褐色。复眼黑色。喙不超过前足腿节。雌蛹触角较短，不及中足末端；雄蛹触角较长，接近中足末端。中足基节显露，后足及翅均超过第3腹节而达第4腹节前端。雌蛹生殖孔开口第八节、第九腹节腹面，雄虫开口第九腹节腹面，肛孔均开口第十腹节腹面。雌雄蛹肛孔两侧各有2根钩状毛，加上末端有6根（腹面4根，背面2根）共10根。第1腹节背面无刺；腹节2～7背面的前后缘各有1排刺，前面一排较粗，大小一致，后面一排细小；腹节8～10背面有1排刺，第10腹节上的刺为7~8根。

生物危害

苹果蠹蛾为我国检疫性有害生物。也是世界上仁果类果树的毁灭性蛀果害虫。该虫以幼虫蛀食苹果、梨、杏等的果实，造成大量虫害果，并导致果实成熟前脱落和腐烂，蛀果率普遍在50%以上，为害重的可达70%~100%，严重影响水果生产和销售。1987年，苹果蠹蛾随旅客携带水果传入甘肃，在敦煌市立足，而后迅速扩散。到1992年已遍布敦煌市，导致30多个大中型果园受害，年均损失40多万元。

地理分布

苹果蠹蛾原产于欧亚大陆南部，属古北、新北、新热带、澳洲、非洲区系共有种。现已广泛分布于世界六大洲几乎所有的苹果产区。在我国周边，已经分布于印度、朝鲜、哈萨克斯坦、吉尔吉斯坦、塔吉克斯坦、乌兹别克斯坦、土库曼斯坦、格鲁吉亚。

20世纪50年代前后，经由中亚地区传入新疆；20世纪80年代中期传入甘肃省，之后持续向东扩张。2006年在黑龙江省发现，这一部分可能由俄罗斯远东地区传入。目前分布于新疆全境、甘肃省的中西部、内蒙古西部以及黑龙江南部等局部地区。

传播方式

苹果蠹蛾主要以幼虫随果品、果制品、包装物及运输工具远距离传播。此外，成蛾可在田间飞行，但最大飞行距离仅500m，自身扩散能力较弱。

管理措施

综合治理（Integrated Pest Management，IPM）或综合控制（Integrated Pest Control，IPC），包括植物卫生措施（Phytosanitary Measures）。

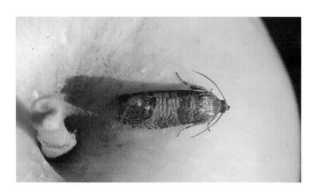
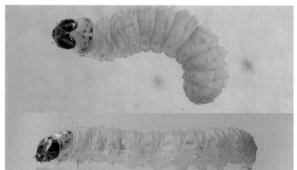

外来入侵**动物**

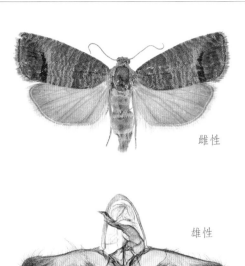
雌性

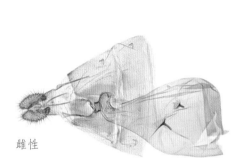
雄性

雌性

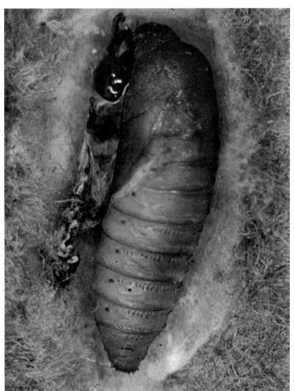

雄性

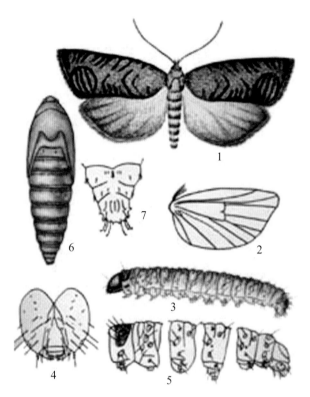

1.成虫；2.后翅脉相；3.幼虫；4.幼虫体节侧面；5.幼虫头部正面观；6.蛹；7.雌蛹腹末部

（图片选自www.hantsmoths.org.uk，www.andermattbiocontrol.com，ukmoths.org.uk，media.eol.org，en.wikipedia.org，hidroponia.mx，www2.nrm.se，idtools.org）

苹果蠹蛾（*Cydia pomonella*）

【idtools.org】*Cydia pomonella*（Linnaeus）

Adult Forewings are gray with silvery striations. FWL: 6.5-11.0 mm. The ocellus is dark purplish brown and is edged with metallic gold or copper scales. Hindwings are grayish brown and males have a fold along the base of the cubital vein that contains a hair pencil with long black sex scales. Male genitalia are characterized by a ventrally projecting spur at the base of the cucullus. Female genitalia are characterized by a short sclerotized ductus bursae.

Adults may appear similar to dark individuals of *Cydia splendana*. *Cydia pomonella* can be separated from *C. splendana* by the metallic scales surrounding the ocellus and the hair pencil on the male hindwing. A genitalic dissection can be used to confirm identity.

【entoweb.okstate.edu】Codling Moth, *Cydia pomonella*

The adult moth has gray to brown front wings with gray crosslines and a characteristic coppery brown spot near the tip of each wing. The hindwings are pale with fringed borders. The wing spread is about 3/4 inch. The eggs are transparent, flattened, and about the size of a pinhead. They develop a red ring and darken just before hatching. Young larvae are white with a dark head but older larvae are pink with a mottled brown head. They are about 3/4 inch long when fully developed.

【www.butterfliesandmoths.org】Codling Moth, *Cydia pomonella*（Linnaeus, 1758）

【www.andermattbiocontrol.com】Codling moth

【en.wikipedia.org】Codling moth

【ukmoths.org.uk】Codling Moth, *Cydia pomonella*（Linnaeus, 1758）

【www.irac-online.org】Codling Moth, *Cydia Pomonella*

【eol.org】*Cydia pomonella*

三叶斑潜蝇

学　　名：***Liriomyza trifolii*** Burgess in Comstock
异　　名：*Agromyza phaseolunata* Frost，*Liriomyza alliivora* Frick，*Liriomyza alliovora* Frick，
　　　　　Liriomyza phaseolunata（Frost），*Oscinis trifolii* Burgess in Comstock
别　　名：三叶草斑潜蝇
英 文 名：American serpentine leafminer，chrysanthemum leaf miner，serpentine leaf miner

| 形态特征 |

雌虫体长2.3mm；雄虫体长1.6mm。头顶和额区黄色，眶全部黄色。额宽为眼宽的2/3倍。头鬃褐色，头顶内、外鬃着生处黄色，具2根等长的上眶鬃及2根较短小的下眶鬃，眼眶毛稀疏且向后倾。触角3节，均黄色，第3节圆形，触角芒淡褐色。中胸背板黑色，无光泽，中背鬃3+1根，第3、第4根稍短小，小中毛不规则排成4列，向后减少至2列。背板两后侧角靠近小盾片处黄色，小盾片黄色，具缘鬃4根。中胸侧板下缘黑色，腹侧片大部分黑色，仅上缘黄色。前翅翅长1.1～1.5mm，前缘脉加粗达中脉M_{1+2}脉末端，亚前缘脉末端成一皱褶并止于前缘脉折断处；中室小，M_{3+4}脉后段为中室长度的3～4倍。平衡棒黄色。各足基节和股节黄色，胫节及跗节呈黑色。腹部可见7节，各节背板黑褐色，第2节背板前缘及中央常呈黄色，3～4背板中央亦常为黄色，形成背板中央不连续的黄色中带纹。腹节腹板黄色，各节中央略呈褐色。

雌虫产卵鞘锥形，黑色；雄虫第7腹节短钝，黑色，外生殖器端阳具中央明显收窄。
卵　米色，稍透明，0.2～0.3mm×0.10～0.15mm。
幼虫　无头蛆，成熟时长达3mm。初龄幼虫刚孵化时无色，后变为浅橙黄色。末龄幼虫为橙黄色。幼虫具后胸气门1对，形如三面锥体。每个后胸气门有3个气门孔，其中有1个气门孔位于三面锥体的顶端。
蛹　呈卵圆形，腹部稍扁平，1.3～2.3mm×0.5～0.75mm。蛹颜色变化大，从浅橙黄色到金棕色。

| 生物危害 |

三叶草斑潜蝇为我国检疫性有害生物。亦为世界性蔬菜、瓜类和观赏植物最重要的害虫之一。在北美，三叶草斑潜蝇是菊花上的一种重要害虫。在美国引起的蔬菜损失相当大，如1980年在旱芹上造成的损失就达900万美元。同时，三叶草斑潜蝇还是植物病毒的媒介昆虫。

| 地理分布 |

三叶草斑潜蝇起源于北美洲。自20世纪60年代以后，随国际贸易传播蔓延到世界其他国家，但主要集中于温带地区，较少分布于热带。

目前三叶斑潜蝇的发生地区有韩国、日本、菲律宾、印度、塞浦路斯、以色列、黎巴嫩、土耳

其、也门、奥地利、比利时、保加利亚、克罗地亚、捷克、丹麦、芬兰、法国、德国、匈牙利、冰岛、意大利、马耳他、荷兰、挪威、波兰、葡萄牙、罗马尼亚、俄罗斯、斯洛伐克、斯洛文尼亚、西班牙、瑞典、瑞士、英国、南斯拉夫、贝宁、科特迪瓦、埃及、埃塞俄比亚、几内亚、肯尼亚、马达加斯加、毛里求斯、马约特岛、尼日利亚、留尼汪、塞内加尔、南非、苏丹、坦桑尼亚、突尼斯、赞比亚、津巴布韦、加拿大、美国巴哈马、巴巴多斯、百慕大、哥斯达黎加、古巴、多米尼加共和国、瓜德罗普、危地马拉、马提尼克岛、特立尼达和多巴哥、巴西、哥伦比亚、法属圭亚那、圭亚那、秘鲁、委内瑞拉、美属萨摩亚、密克罗尼西亚、关岛、北马里亚纳群岛、萨摩亚、汤加。

声明已根除三叶草斑潜蝇的国家有：丹麦、芬兰、匈牙利、挪威和英国。

三叶草斑潜蝇先后在我国台湾（1988年）、菲律宾（1984年）、日本（1990年）、印度（1993年）和韩国（1994年）等地相继出现。

2005年，在广东省中山市潭洲镇蔬菜种植基地发现，并迅速蔓延。现已经分布于台湾、广东、海南、云南、浙江、江苏、上海、福建等地。为全球性入侵物种。

| 传播途径 |

三叶草斑潜蝇可随寄主植物（包括切花）进行远距离扩散。成虫也能作有限的飞行扩散。

| 管理措施 |

综合治理（Integrated Pest Management，IPM）或综合控制（Integrated Pest Control，IPC），包括植物卫生措施（Phytosanitary Measures）。

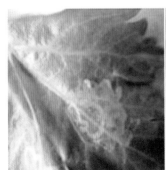

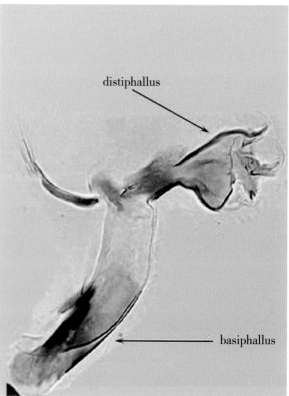

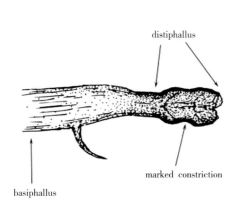
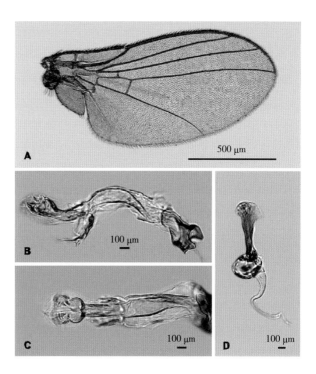

(图片选自tupian.baike.com，baike.sogou.com，image.baidu.com，www.hortoinfo.es，entomology.ifas.ufl.edu，keys.lucidcentral.org，www.scielo.org.mx）

三叶斑潜蝇（*Liriomyza trifolii*）

[www.cabi.org] *Liriomyza trifolii* (American serpentine leafminer)

Egg *L. trifolii* eggs are 0.2-0.3 mm × 0.1-0.15 mm, off white and slightly translucent.

Larva This is a legless maggot with no separate head capsule, transparent when newly hatched but colouring up to a yellow-orange in later instars and is up to 3 mm long. *L. trifolii* larvae and puparia have a pair of posterior spiracles terminating in three cone-like appendages. Spencer (1973) describes distinguishing features of the larvae. Petitt (1990) describes a method of identifying the different instars of the larvae of *L. sativae*, which can be adapted for use with the other *Liriomyza* species, including *L. trifolii*.

Puparium This is oval and slightly flattened ventrally, 1.3-2.3 mm × 0.5-0.75 mm with variable colour, pale yellow-orange, darkening to golden-brown. The puparium has posterior spiracles on a pronounced conical projection, each with three distinct bulbs, two of which are elongate. Pupariation occurs outside the leaf, in the soil beneath the plant.

Menken and Ulenberg (1986) describe a method of distinguishing *L. trifolii* from *L. bryoniae*, *L. huidobrensis*, and *L. sativae* using allozyme variation patterns as revealed by gel electrophoresis.

Adult *L. trifolii* is very small: 1-1.3 mm body length, up to 1.7 mm in female with wings 1.3-1.7 mm. The mesonotum is grey-black with a yellow blotch at the hind-corners. The scutellum is bright yellow; the face, frons and third antennal segment are bright yellow. Male and female *L. trifolii* are generally similar in appearance.

Head The frons, which projects very slightly above the eye, is just less than 1.5 times the width of the eye (viewed from above). There are two equal ors and two ori (the lower one weaker). Orbital setulae are sparse and reclinate. The jowls are deep (almost 0.33 times the height of the eye at the rear); the cheeks form a distinct ring below the eye. The third antennal segment is small, round and noticeably pubescent, but not excessively so (vte and vti are both on a yellow ground).

Mesonotum Acrostical bristles occur irregularly in 3-4 rows at the front, reducing to two rows behind. There is a conspicuous yellow patch at each hind-corner. The pleura are yellow; the meso- and sterno-pleura have variable black markings.

Wing Length 1.3 -1.7 mm, discal cell small. The last section is M_{3+4} from 3-4 times the length of the penultimate one.

Genitalia The shape of the distiphallus is fairly distinctive but could be mis-identified for *L. sativae*. Identification using the male genitalia should only be undertaken by specialists.

Colour The head (including the antenna and face) is bright yellow. The hind margin of the eye is largely yellow, vte and vti always on yellow ground.

The mesopleura is predominantly yellow, with a variable dark area, from a slim grey bar along the base to extensive darkening reaching higher up the front margin than the back margin. The sternopleura is largely filled by a black triangle, but always with bright yellow above.

The femora and coxa are bright yellow, with the tibia and tarsi darker; brownish-yellow on the fore-legs, brownish-black on the hind legs. The abdomen is largely black but the tergites are variably yellow, particularly at the sides. The squamae are yellowish, with a dark margin and fringe.

Although individual specimens may vary considerably in colour, the basic pattern is consistent.

[entomology.ifas.ufl.edu] *Liriomyza trifolii*

松突圆蚧

学　　名：*Hemiberlesia pitysophila* Takagi
英文名：pine needle hemiberlesian scale，pine armoured scale，pine greedy scale，pine-needle scale insect

| 形态特征 |

介壳　雌介壳圆形或椭圆形，隆起，白色或浅灰黄色。介壳直径约2mm；有蜕皮壳2个，大小为1.25mm×0.71mm。1、2龄若虫蜕皮壳为橙黄色，偏于介壳一端近边缘部分，略有环状带纹。

雄介壳长椭圆形，前端稍宽，后端略狭，大小为0.9mm×0.5mm。头端微隆起，淡褐色，有蜕皮壳1个，白色，位于介壳前端中央；尾端扁平，蟹青色。

雌成虫　体宽梨形，淡黄色，长0.7～1.1mm。头胸部最宽，0.5～0.9mm。体侧边第二至第四腹节稍突出，臀板较宽，稍硬化，虫体除臀板外，均为膜质。触角疣状，具刚毛1根。口器发达。胸气门2对。臀叶2对，中臀叶（L1）突出，长略大于宽，顶端圆，每边有2凹刻，外侧大内侧小，基部硬化部分深入臀板中；第二臀叶（L2）斜向内，小而硬化，不二分；在中臀叶和第二臀叶间有1对顶端膨大的硬化棒。臀棘细而短，其长度不超过中臀叶，在中臀叶间有1对，在中臀叶和第二臀叶间有1对，第二臀叶前各3对。肛孔位于臀板基部。背管腺细长，中臀角间1个，中臀叶与第二臀叶间3个，在第七和第八腹节间；在第二臀叶前2纵列：一列4～8个，在第六和第七腹节间；另一列5～7个，在第五和第六腹节间。另外，在后胸到第五腹节的边缘均有管腺分布。第四腹节有亚缘腺。腹面的管腺细小，分布在头胸部和第一至第五腹节的边缘，在前后胸气门间呈横带，在口器附近及后胸气门下方也有分布。无围阴腺。口器前面近体边缘处的背面有一圆形突起。

雄成虫　体橘黄色，细长，长0.80mm，翅展1.10mm。翅1对，膜质，翅面光滑，翅脉2条。翅较短而阔，白色透明且有金属光泽。后翅退化成平衡棒，平衡棒前侧有1个膜质纺锤形小片，端部有1条长刺毛。口器退化，只存遗迹，头紧连前胸不能活动，单眼为半球形的突起。前胸无气门。后胸基腹片上生有2对气门；侧板上面着生平衡棒。触角10节，长约0.20mm，每节有数根毛，柄节粗短，鞭节各节大小形状相似。足3对，发达，足跗节1节，爪1个，冠球毛无明显膨大的末端。腹部由9个可见节构成。

雌雄成虫近似，体长约0.65mm，宽约0.50mm。雄性常比雌性狭，臀板附属物相同。

卵　椭圆形，无雕刻纹，长约0.35mm，宽约0.14mm。卵壳白色透明，表面有细颗粒。

1龄若虫　体卵圆形扁平，淡黄色，长0.25～0.35mm。头胸略宽。眼1对，发达，着生于触角下侧方。若虫触角4节，柄节较粗，基部3节较短，第四节长，其长度约为基部3节之和的3倍。口器发达。触角毛发达，末节上的毛特别长。胸气门2对，胸足3对，发达，转节上有1根长毛。跗冠毛、爪冠毛各1对。腹面沿体缘有1列刚毛。背面从中胸到体末的边缘有管腺分布。臀叶2对，中臀叶较大而突出，外缘有齿刻；第二臀叶小，不二分。中臀叶间有长、短刚毛各1对。

2龄若虫　性分化前近圆形，淡黄色，长约0.35mm，宽约0.34mm，足完全消失，触角退化只留遗迹，腹部末端出现了臀板；性分化后，雄若虫比雌若虫身体稍窄，体长约0.55mm，宽0.35mm，2龄后期雄性若虫体型与2龄初期雄性若虫相似，体形进一步变狭。

雄蛹　预蛹黄色，棒槌状，后端略小，长约0.72mm，宽约0.40mm，前端出现眼点。在前蛹期出现一些成虫器官芽体如触角、复眼、翅、足和交配器。蛹淡黄色，棒槌状，长约0.75mm，宽约0.41mm；触角、足及交配器淡黄色而稍显透明；口器完全消失。

| 生物危害 |

松突圆蚧为我国检疫性有害生物。其主要为害松属植物，如马尾松、湿地松、黑松等，其中以马尾松受害最重。主要为害松树的针叶、嫩梢和球果。害虫在寄主叶鞘内或针叶、嫩梢、球果上吸食汁液，使针叶和嫩梢生长受到抑制，严重影响松树造脂器官的功能和针叶的光合作用，致使被害处变色发黑、缢缩或腐烂，针叶枯黄。受害严重时，针叶脱落，新抽的枝条变短、变黄，甚至导致松树全株枯死。

| 地理分布 |

松突圆蚧原产于日本和我国台湾。20世纪70年代末在广东发现。现分布于我国台湾、香港、澳门、广东、广西、福建和江西等地。

| 传播途径 |

松突圆蚧除了可以借助若虫爬行或借助风力作近距离扩散外，蚧虫的若虫、雌成虫主要随寄主苗木、接穗、鲜球果及原木、枝梢、盆景等调运而作远距离传播。

| 管理措施 |

综合治理（Integrated Pest Management，IPM）或综合控制（Integrated Pest Control，IPC），包括植物卫生措施（Phytosanitary Measures）。

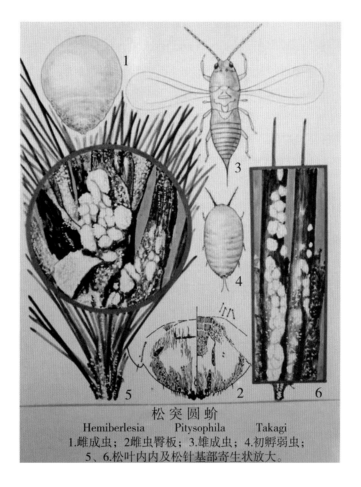

（图片选自image.baidu.com，course.cau-edu.net.cn，www.nmsgsfz.com）

松突圆蚧（*Hemiberlesia pitysophila*）

【 www.cabi.org 】 *Hemiberlesia pitysophila* (pine needle hemiberlesian scale)

In life, scale cover of the adult female of *H. pitysophila* is oval, convex, dirty white to mid-brown, with buff to red-brown submarginal exuviae that may be paler or darker than the secreted scale. The scale cover of the developing male is similar to but smaller than that of the female, and elongate oval in shape.

The absence of perivulvar pores from the pygidium suggests that this species is probably ovoviviparous. In the female, the first-instar larva possess legs and well-developed antennae; the second instar lacks legs and its antennae are reduced to stubs. The second instar is smaller than the adult female and lacks a vulva. In the male, the first two instars resemble those of the female and feed; the third and fourth developmental stages are the non-feeding prepupa and pupa. The adult male probably has well-developed legs and antennae, one pair of wings, long genitalia and no mouthparts.

When studied under a compound light microscope, the slide-mounted adult female is membranous and pyriform. The pygidium has large median lobes, separated by at least 1/3 of the width of a median lobe; second lobes reduced but sclerotized; and third lobes very reduced or completely obsolete. A large anal opening is situated near the posterior margin of the pygidium (less than 2.3 times its length from base of median lobes) . Paraphyses are present, shorter than the lobes, situated only on the margin between the third lobes. Perivulvar pores are present; also marginal plates reaching beyond the tips of the median lobes, each moderately to little fringed, the outermost plates being the smallest and least elaborate in shape.

椰心叶甲

学　名：***Brontispa longissima***（Gestro）
异　名：*Brontispa castanea* Lea，*Brontispa froggatti* Sharp，*Brontispa longissima* var. *javana* Weise，*Brontispa longissima* var. *selebensis* Gestro，*Brontispa reicherti* Uhmann，*Brontispa simmondsi* Maulik，*Oxycephala longipennis* Gestro，*Oxycephala longissima* Gestro
别　名：红胸叶甲、椰长叶甲、椰棕扁叶甲
英文名：coconut hispine beetle，coconut leaf hispid，new hebrides coconut hispid，Two-coloured coconut leaf beetle

|形态特征|

成虫　体扁平狭长，雄虫比雌虫略小。体长8~10mm，宽约2mm。头部红黑色，头顶背面深，有绒毛，柄节长2倍于宽。触角间突超过柄节的1/2，由基部向端部渐尖，不平截，沿角间突向后有浅褐色纵沟。前胸背板黄褐色，略呈方形，长宽相当；具不规则粗刻点；前缘向前稍突出，两侧缘中部略内凹，后缘平直；前侧角圆，向外扩展，后侧角具一小齿；中央具一大黑斑。鞘翅两侧基部平行，后渐宽，中后部最宽，往端部收窄，末端稍平截；中前部有刻点8列，中后部10列，刻点整齐。鞘翅颜色因分布地不同而有所不同，有时全为红黄色（印尼爪哇），有时全为蓝黑色（所罗门群岛）。足红黄色，粗短，跗节4节。雌虫腹部第5节可见腹板为椭圆形，产卵期为不封闭的半圆形小环；雄虫该节腹板为尖椭圆形，生殖器褐色，长约3mm。

卵　椭圆形，两端宽圆。长1.5mm，宽1.0mm。卵壳褐色，表面有细网纹。

幼虫　白色至乳白色。

蛹　体浅黄至深黄色，长约10.0mm，宽2.5mm。头部具1个突起，腹部第2~7节背面具8个小刺突，分别排成两横列，第8腹节刺突2个，靠近基缘。腹末具1对钳状尾突。

|生物危害|

椰心叶甲为我国检疫性有害生物。属毁灭性害虫。为害寄主幼嫩的心叶部位。叶片受害后可出现枯死症状，严重时导致植株死亡。1929年，椰心叶甲传入印度尼西亚苏拉威西省，致使该省一些地区10%~15%的椰树致死，椰农不得不放弃椰树种植。

1975年，椰心叶甲由印尼传入我国台湾，造成台湾地区17万椰子树死亡，目前疫情仍未得到有效控制。最近几年，椰心叶甲在越南大爆发，为害1 000多万株椰树，造成大约有50万株椰子树死亡。

|地理分布|

椰心叶甲原产于印度尼西亚与巴布亚新几内亚。现广泛分布于太平洋群岛及东南亚。越南、印度尼西亚、泰国、新加坡、菲律宾、澳大利亚、巴布亚新几内亚、所罗门群岛、新喀里多尼亚、萨摩亚群岛、法属波利尼西亚、新赫布里底群岛、俾斯麦群岛、社会群岛、塔西提岛、关岛、马来西亚、斐济群岛、瓦努阿图、法属瓦利斯和富图纳群岛、马尔代夫、马达加斯加、毛里求斯、塞舌尔、日本等均有分布。

1975年，椰心叶甲由印尼传入我国台湾。2002年6月，在海南省首次发现。现分布于海南、广东、广西、香港、澳门和台湾等地。为全球性入侵物种。

| 传播途径 |

远距离传播是椰心叶甲各个虫态随寄主（主要是种苗、花卉等）调运而人为传播。成虫可借助飞行或气流进行一定距离的自然扩散（可飞行300~500m）。

| 管理措施 |

综合治理（Integrated Pest Management，IPM）或综合控制（Integrated Pest Control，IPC），包括植物卫生措施（Phytosanitary Measures）。

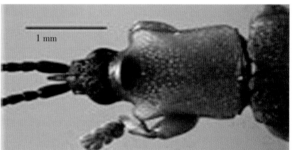

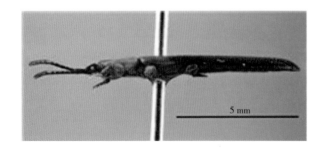

（图片选自www.cabi.org，www.agr.kyushu-u.ac.jp，idtools.org）

椰心叶甲（*Brontispa longissima*）

【www.cabi.org】 *Brontispa longissima* (coconut hispine beetle)

Eggs The brown eggs are elliptical, about 1.5 mm long and 1 mm wide, with each end broadly rounded. The slightly convex upper surface has the chorion with a honeycomb sculpturation. They are fixed to the leaf surface by a cementing substance (Maulik, 1938).

Larvae The larvae are whitish in colour. The first-, second- and final-instar larvae are described in detail by Maulik (1938). The head of the first stage larva is comparatively large compared with the body. The entire cuticle is more densely covered with minute spicules. A seta arises from about the middle of the lateral margin of each thoracic segment, with two setae on each of the abdominal lateral processes. Each process of the tail-shovel bears a large, sharp, curved spine at the inner angle, and a series of five or six setae along the dorsal and ventral margins. The larva is about the same length as the egg, but about 0.75 mm in width.

The second-instar larva resembles the full-grown larva more than the first instar. The lateral abdominal processes are longer, each bearing four setae which are comparatively longer than those of the fully developed larva and situated at different points around the periphery of the apex. There are eight setae on the prothorax (four on each side) and six on the meso- and metathorax (three on each side, two setae on the produced part and one posteriorly). The distinct spine at the inner angle of each prong of the tail-shovel is not prominent as in the first-instar.

The fully developed larva has the body moderately flat, almost parallel-sided, very slightly and gradually narrowed from the prothorax towards the apex, composed of 13 segments (one head, three thoracic and nine abdominal). It is almost 9 mm long and 2.25 mm wide. The anus is situated ventrally on the ninth segment, if the fold at the anal orifice is considered as representing a segment, then the abdomen should be regarded as 10-segmented. The distinct head bears a pair of 2-segmented antennae; a group of five ocelli, three in a line and two in another, situated behind the antennae; a pair of apically bidentate mandibles.

The thoracic segments are broader than they are long, the mesothorax very slightly broader than the prothorax and the metathorax very slightly broader than the mesothorax. The dorsal surface of the prothorax is more strongly sclerotized with a fine median suture and laterally rounded. The meso- and metathorax bears laterally a small knob in the middle, bearing two fine, short setae. There is a pair of well developed 3-segmented legs on each thoracic segment, each terminating in a single claw and fleshy pad-like structure. Each of the first seven abdominal segments is broader than it is long; each of the first three segments is somewhat shorter than the following segments; the eighth and ninth together form the terminal tail-shovel. Each segment bears laterally a moderately long process, ventral to the spiracles. A lateral process is a conical structure projecting horizontally nearly from the middle of the margin, except in the fifth, sixth and seventh segments, in which they appear to arise more from the posterior part. The apical part is subconical and distinctly thinner than the basal part. The entire surface is densely covered with spinules. The apical part bearing three or four setae. The tail-shovel is longer than broad, apically deeply concave, the prongs bent inwards and bluntly-pointed. The upper ridge on each side bearing nine spinules, three smaller on the basal part nearer the large spiracle, four larger on the middle part and two smaller on the bent apical part. There are 4 small spinules on the lower ridge more widely spaced than those on the upper ridge. The upper surface is concave, bearing many irregularly placed transparent areas and the ventral surface is flat. There are 9 pairs of spiracles, one thoracic, between the pro- and mesothorax, situated on a conical structure resembling a lateral process, and one pair dorsally situated on each of the first seven abdominal segments.

The larvae undergo five instars which can be distinguished on the morphometrics (in mm) of the tail-shovel as follows: L1, 0.33, mean 0.13; L2, 0.47, mean 0.20; L3, 0.65, mean 0.29; L4, 0.82, mean 0.37;

L5, 0.94, mean 0.45.

Dorsal habitus views of the first-, second- and final-instar larvae are provided by Maulik（1938）.

Generic and specific keys to separate the larvae of *Brontispa* species are provided by Gressitt（1963）. The key characters are: body not oval with a continuous margin; head visible from above; last abdominal segment with a caudal process; not leaf-mining; meso- and metathoracic segments lacking lateral processes; caudal process with arms widely separated basally, thickened or strongly angulate posteriorly; lateral abdominal processes short and blunt, rarely last two longer; spiracle of last segment elliptical（Generic）; lateral abdominal processes subequal, eighth shorter than greatest width of an arm of caudal process; emargination of caudal process usually broader than long; arms of caudal process parallel-sided externally, at least in central part; emargination of caudal process reaching about half way from apices of arms to spiracles; emargination of caudal process not much broader than long, broadly oval, widest in middle; arm curved and subacute apically; eighth abdominal process shorter than preceding（Specific）.

Pupae The pupae of several *Brontispa* spp. including *B. longissima* are described by Maulik（1938）. The head bears three processes, one median and one on each side. Each lateral process is fleshy, broadest basally, pointed apically, pre-apically bearing small blunt spine. Thorax with meso- and metanotum each bearing a pair of wings. The abdomen is 9-segmented, the eighth and ninth segments are fused. Each of the first six segments bears laterally a pair of spiracles opening dorsally. Each segment bearing spinules arranged as follows: no spinules on first segment; from segments 2 to 7, two groups of four in transverse line, one nearer the basal margin and the other nearer the apical, those of the basal group are comparatively larger and more widely-spaced, those of the apical line are situated in the intervals of the spinules of the basal group; on segment 8 only two in transverse line near the basal margin. Dorso-laterally bearing a longitudinal series of spinules, one on each of the first seven segments, situated a little posterior to and more inward than the corresponding spiracle（on the seventh in a similar position）. Anterior to each spiracle on segments 3 to 6 and in a similar position on 7 is a spine. Laterally on segments 1-7 bearing a group of closely placed spinules bearing moderately long setae, the spinules being less prominent on segments 1-2, or 3. Ventrally, only segments 4-7 bearing spinules. The prongs of the tail-shovel are more slender, somewhat longer than those of the larva, and also lack lateral spines or setae. The last larval exuvium is always retained on the prongs of the tail-shovel. Adults

The adults 8.5-9.5 mm long, 2.00-2.25 mm wide; length of antenna, 2.75 mm. The males are slightly smaller than females（Maulik, 1938）. Their colour varies geographically from reddish-brown in Java, to almost black in the Solomon Islands and Irian Jaya. Some specimens have the elytra brown or black, or have a spindle-shaped black marking on the elytral suture. Considerable overlapping of these forms, which were long regarded as distinct species, occurs（Lever, 1969）. Maulik（1938）examined the male genitalia, especially the median lobes of the different colour varieties, and found no differences. However, the median lobes of distinct species, for example *B. linearis* and *B. longissima*, exhibit obvious structural differences. Maulik（1938）and Gressitt（1963）have provided keys to separate the adults of *Brontispa* species. The relevant key characters for *B. longissima* are: antennae not serrate; central portion of head usually parallel-sided, broader than long; rostrum more than half as long as first antennal segment; pronotum flattish, shiny, with several large impunctate areas; body quite flat and narrow; prothorax laterally distinctly concave; anterior lateral angles of pronotum expanded, expanded portion broadly-rounded, constricted behind, without a minute projection at inner angle.

【www.iucngisd.org】*Brontispa longissima*

红棕象甲

学　名：***Rhynchophorus ferrugineus***（Oliver）

异　名：*Curculio ferrugineus* Olivier，*Cordyle sexmaculatus* Thunberg，*Calandra ferruginea* Fabricius，*Rhynchophorus pascha* v. *papuanus* Kirsch，*Rhynchophorus indostanus* Chevrolat，*Rhynchophorus signaticollis* Chevrolat，*Rhynchophorus pascha* v. *cinctus* Faust，Rhynchophorus *ferrugineus* v. *seminiger* Faust，*Rhynchophorus signaticollis* v. *dimidiatus* Faust

别　名：棕榈象甲、锈色棕象、锈色棕榈象、椰子隐喙象、椰子甲虫、亚洲棕榈象甲、印度红棕象甲

英文名：palm weevil，red palm weevil，Asian palm weevil，sago palm weevil

| 形态特征 |

成虫　体长30~34mm，身体红褐色，前胸具两排黑斑。鞘翅红褐色，有时全部暗黑褐色。身体腹面黑红相间或暗黑褐色上有一不规则红斑。

卵　长椭圆形，乳白或者乳黄色，表面光滑。

幼虫　初孵幼虫为白色，头部黄褐色；老熟幼虫肥胖，纺锤形淡黄白色，头褐色，口器坚硬。

蛹　3~4cm，初孵化时乳白色，后转为黄色、橘黄色，外包裹一层深褐色有光泽的不透明膜，最外面包裹一层取食后的植物纤维作成的茧。

| 生物危害 |

红棕象甲为我国检疫性有害生物。是一种高危检疫性害虫。在东南亚地区严重为害椰子、油棕等棕榈科植物。成虫、幼虫均为害棕榈科植物，后者造成损害更大。受害植株初期表现为树冠周围的叶子变枯黄，后扩展至树冠中心，心叶黄萎。虫口密度高时，树干被蛀空，遇大风容易折断。为害树的生长点时，生长点腐烂，植株死亡。世界许多地区（包括我国）的棕榈科植物，均遭受红棕象甲的严重为害。

| 地理分布 |

红棕象甲原产于亚洲南部及西太平洋美拉尼西亚群岛。目前分布于印度、伊拉克、沙特、阿联酋、阿曼、伊朗、埃及、巴基斯坦、巴林、印尼、马来西亚、菲律宾、泰国、缅甸、越南、柬埔寨、斯里兰卡、所罗门群岛、新喀里、多尼亚、巴布亚、新几内亚、日本、约旦、塞浦路斯、法国、希腊、以色列、意大利、西班牙、土耳其。

1998年，最早在海南文昌县发现红棕象甲严重为害椰子树。近20年来，引进的大量加拿利海枣均遭到严重为害。目前，在我国海南、广东、广西、台湾、云南、西藏、江西、上海、福建、四川、贵州、江苏、浙江等地均有分布。

| 传播途径 |

红棕象甲以成虫飞翔作近距离传播,以各种虫态随侵染的植株调运作远距离传播。

| 管理措施 |

综合治理（Integrated Pest Management，IPM）或综合控制（Integrated Pest Control，IPC），包括植物卫生措施（Phytosanitary Measures）。

Phynchophorus Ferrugincus-Olivier 1790
J.cerdá Martí-Montroy Valencia Mayo 2009

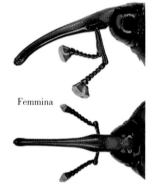

（雄虫）

（雌虫）

（图片选自www.lanzaroteinformation.com，species.wikimedia.org，en.wikipedia.org，upload.wikimedia.org，entbari.org，www.naturamediterraneo.com，www.iberia-natur.com，www.leszoosdanslemonde.com，www.aphis.usda.gov，planthealthportal.defra.gov.uk，onlinelibrary.wiley.com）

红棕象甲（*Rhynchophorus ferrugineus*）

【www.aphis.usda.go】*Rhynchophorus ferrugineus*

Eggs "Whitish-yellow, smooth, very shiny, cylindrical with rounded ends, slightly narrower at the anterior end, averaging 0.98 by 2.96 mm" (EPPO, 2007).

Larvae "Piriforme, apodous, colour, creamy white to ivory, cephalic capsule brown russet-red to brilliant brown-black. Body slightly curved. Last instar is 36 to 47 mm in length by 15 to 19 mm in width" (EPPO, 2007).

Adult male: Length: "19 to 42 mm, width 8 to 16 mm. Body elongate-oval, general colour ferruginous to black, legs lighter coloured than body; elytra dark red to black, shiny or dull, slightly pubescent; black spots on pronotum extremely variable" (EPPO, 2007).

Head: "dull to shiny; smooth to finely punctured; interocular space slightly more than onehalf width of rostrum at base" (EPPO, 2007).

Antennae: "arising laterally from scrobe at base of rostrum; scrobe deep, broad and widely opened ventrally; scape elongate, longer than funicle and club combined or equal to onehalf length of rostrum; funicle with 6 segments; antennal club large usually ferruginous or reddish-brown; broadly triangular with several setae dorsally and ventrally; inner side of spongy area with 8 to 15 setae" (EPPO, 2007).

Pronotum: "with sides gradually curved to apex and abruptly constricted anteriolaterally; slightly pubescent to shiny; posterior margin nearly rounded; colour mostly ferruginous and varying to dark brown and black; underside of pronotum mostly ferrugineus or dark brown, may vary to almost black, very minutely punctured. Scutellum varying from reddish brown to black; somewhat pointed posteriorly, one-quarter to one fith elytral" (EPPO, 2007).

Elytra: "smooth or slightly velvety pubescent, nearly rectangular, with punctuation along the outer edges with 5 deep striae and traces of 4 laterally; length of each elytron two and one-third times its own width" (EPPO, 2007).

Abdomen: "usually ferruginous, but may vary from ferruginous to almost black; first abdominal sternite as long as third and fourth combined but much shorter than second" (EPPO, 2007).

Adult female: "Length 26 to 40 mm, width 10 to 16 mm. Very similar to male in body size, colour, markings on pronotum, except rostral setae absent; snout longer, slender and more cylindrical, setae on front femur absent and on front tibia much shorter" (EPPO, 2007).

The Center for Invasive Species Research produced a video, detailing the "Overview of the Red Palm Weevil" available on the website: http://cisr.ucr.edu/red_palm_weevil.html. The video also provides information on the two color morphs of R. ferrugineus

【onlinelibrary.wiley.com】*Rhynchophorus ferrugineus*

Eggs Creamy white, oblong, shiny; average size 2.62 × 1.12 mm (Menon & Pandalai, 1960). Eggs hatch in 3 days and increase in size before hatching (Reginald, 1973). The brown mouth parts of the larvae can be seen through the shell before hatching.

Larvae Up to 35 mm long; brown head, white body composed of 13 segments; mouthparts well developed and strongly chitinized; average length of fully grown larvae 50 mm, and width (in middle) 20 mm.

Pupae Pupal case 50-95 mm × 25-40 mm; prepupal stage of 3 days and pupal period of 12-20 days; pupae cream, then brown, with shiny surface, greatly furrowed and reticulated; average size 35 × 15 mm.

Adults Reddish brown, about 35 × 10 mm, with long curved rostrum; dark spots on upper side of thorax; head and rostrum comprising about one third of total length. In male, dorsal apical half of rostrum covered by a patch of short brownish hairs; in female, rostrum bare, more slender, curved and a little longer than in male (Menon & Pandalai, 1960). See Booth et al. (1990) for a full description.

Pictures of the pest are available in the EPPO diagnostic protocol for *Rhynchophorus ferrugineus* and *Rhynchophorus palmarum* (OEPP/EPPO, 2007).

【onlinelibrary.wiley.com】*Rhynchophorus ferrugineus* and *Rhynchophorus palmarum*
(EPPO BULLETIN Volume 37, Issue 3, December 2007, Pages: 571–579)

【en.wikipedia.org】*Rhynchophorus ferrugineus*

【www.forestryimages.org】Red palm weevil, *Rhynchophorus ferrugineus* (Olivier, 1790)

【scialert.net】Red Palm Weevil (*Rhynchophorus ferrugineus* Olivier, 1790): Threat of Palms

【planthealthportal.defra.gov.uk】*Rhynchophorus ferrugineus*

【commons.wikimedia.org】*Rhynchophorus ferrugineus*

【species.wikimedia.org】*Rhynchophorus ferrugineus*

【www.wikiwand.com】*Rhynchophorus ferrugineus*

铃木方翅网蝽

学　　名：***Corythucha ciliata*** Say

异　　名：*Corythuca ciliata*，*Tingis ciliata* Say

别　　名：无

英文名：sycamore lace bug

| 形态特征 |

　　成虫　体长3.2～3.7mm，乳白色，头顶和体腹面黑褐色；头兜发达，盔状，头兜突出部分的网格比侧板的略大，从侧面看，头兜的高度较中纵脊稍高。在两翅基部隆起处的后方有褐色斑。头兜、侧背板、中纵脊和前翅表面的网肋上密生小刺，侧背板和前翅外缘的刺列十分明显。前翅显著超过腹部末端，其前缘基部强烈上卷并突然外突，亚基部呈直角状外突，使得前翅近长方形。足和触角浅黄色；腿节不加粗。后胸臭腺孔缘小，且远离侧板外缘。

　　卵　长椭圆形，乳白色，顶部有椭圆形褐色卵盖。

　　若虫　共5龄。1龄若虫体无明显刺突；2龄若虫中胸小盾片具不明显刺突；3龄若虫前翅翅芽初现，中胸小盾片2刺突明显；4龄若虫前翅翅芽伸至第1腹节前缘，前胸背板具2明显刺突；末龄若虫前翅翅芽伸至第4腹节前缘，前胸背板出现头兜和中纵脊，头部具刺突5枚，头兜前缘处有刺突2对，后缘有1对3叉刺突，前胸背板侧缘后端具单刺1枚，中胸小盾片有1对单刺突，腹部背面中央纵列4枚单刺，两侧各具6枚双叉刺突。

| 生物危害 |

　　铃木方翅网蝽刺吸叶片汁液直接为害。通常于悬铃木树冠底层叶片背面吸食汁液，最初造成黄白色斑点和叶片失绿，严重时叶片由叶脉开始干枯至整叶萎黄、青黑及坏死，从而造成树木提前落叶、树木生长中断、树势衰弱至死亡。

　　该害虫可携带悬铃木叶枯病菌（*Gnomonia platani*）和甘薯长喙壳菌（*Ceratocystis fimbriata*）。而这两种病原菌能降低悬铃木树势并导致其死亡。

| 地理分布 |

　　铃木方翅网蝽原产于北美的中东部。1960年从北美传入欧洲，1990年传入南美洲智利，1996年传入韩国，2001年传入日本，2006年传入澳大利亚新南威尔士。现主要分布于美国、加拿大、法国、匈牙利、西班牙、奥地利、瑞士、捷克、保加利亚、希腊、俄罗斯、智利、韩国、日本等国。

　　2006年，我国首次在湖北武汉发现铃木方翅网蝽。现已在上海、浙江、江苏、重庆、四川、湖北、贵阳、河南、山东等地发现。并呈暴发态势。

| 传播途径 |

铃木方翅网蝽借助风力作近距离传播,人为调运感染害虫的苗木或带皮原木是其远距离传播的主要方式。

(图片选自en.wikipedia.org,commons.wikimedia.org,www.cabi.org,entnemdept.ufl.edu)

悬铃木方翅网蝽(*Corythucha ciliata*)

【www.cabi.org】 *Corythucha ciliata* (sycamore lace bug)

Adults are about 3 mm in length and 2 mm in width. The insect is flattened dorso-ventrally and the wing covers and pronotum are dilated laterally forming a broadened lace-like covering of the body. The pronotum is inflated anteriorly into a bulbous "hood" which covers the insect's head. The outer margin of the pronotum and wing covers bear small pointed spines (except for the posterior third); the nervures of the hood, pronotum and wing covers are also armed with a few erect spines. The wing covers each bear a tumid elevation near the anterior inner margin. The body is black, while the hood, pronotal margins and wing covers are whitish except for an irregular brown spot on the tumid elevation of each wing cover.

Egg length is about 0.5 mm, width 0.16-0.18 mm. They are barrel-shaped and rather pointed at the base where it is glued onto the underside of the leaf. The top is not pointed and about 0.1 mm across with a cone-shaped cap resting on a circular base and bearing on the top a number of ridges which converge from the outer margin to the apex. At the apex is sometimes a thread-like filament, usually short. The egg's colour is black, the cap is a dull whitish, though sometimes dark.

Nymphs progress through 5 instars. Some features are common throughout the nymphal stages: All are armed with spines along the margins of the body and head as well as on the back at different points. These spines are of two main types and quite prominent.

"The first nymphal stage can always be recognized by its more slender, less flattened appearance, the absence of long, spine-bearing protuberances, and by each spine having a blunt tip which is usually enlarged and rounded. The minute spinules found covering the body surface in the succeeding instars are absent in this stage. But five facets are present in each eye.

"The second nymphal stage still has only the one spine on the lateral margin of each of the abdominal segments two to nine, but it now arises from an elongated protuberance instead of a conical base and is pointed, while just inside of this is a trumpet-shaped spine. The body is broader and darker in colour. The antennae are still three segmented, but the lateral margins of the pro- and meso-thorax bear two spines on protuberances and a trumpet-shaped spine just inside of these instead of the single blunt spine on a conical base found in instar one. Each eye has six or more facets.

"The third nymphal stage now has four segments in the antennae, but the wing pads are not developed beyond a faint enlargement. The lateral margins of abdominal segments two to nine have three spines, one on a protuberance, one arising from a conical base, and a trumpet-shaped spine just inside of these. Each eye has fifteen or more facets.

"The fourth nymphal stage is easily recognized by the well developed wing pads which are oval in outline and reach to the second abdominal segment. The anterior margin of the pronotum is extended forward to the eyes, or slightly over them, while the posterior margin is produced into a rounded triangular point or apex at the median line.

"The fifth nymphal stage will be known at a glance by the greatly developed wing pads, rather elongated and reaching to the fifth abdominal segment. The anterior margin of the pronotum at the median line is raised and extended a little forward while the posterior margin is quite distinctly triangular. The lateral margins of the abdominal segments one, two and three are without spines."

扶桑绵粉蚧

学　名：***Phenacoccus solenopsis*** Tinsley
异　名：*Phenacoccus cevalliae* Cockerell，*Phenacoccus gossypiphilous* Abbas et al.
别　名：无
英文名：cotton mealybug，solenopsis mealybug

| 形态特征 |

雌成虫活体卵圆形，浅黄色，扁平，表皮柔软，体背被有白色薄蜡粉，体表白色蜡粉较厚实，胸、腹背面的黑色条斑在蜡粉覆盖下呈成对黑色斑点状，其中胸部可见1对，腹部可见3对，腹脐黑色。除去蜡粉后，在前、中胸背面亚中区可见2条黑斑，腹部1~4节背面亚中区有2条黑斑。体缘有蜡突，均短粗，腹部末端4~5对蜡突较长。成虫体长2.5~2.9mm，宽1.60~1.95mm。触角9节，基节粗，其他节较细；单眼发达，突出，位于触角后体缘。常有螺旋形三格腺，25~30个；多格腺仅分布于腹面，五格腺缺。足红色，通常发达，可以爬行。

雄成虫体微小长1.4~1.5mm。触角10节，长约为体长的2/3。足细长，发达。腹部末端具有2对白色长蜡丝。前翅正常发达，平衡棒顶端有1根钩状毛。

| 生物危害 |

扶桑绵粉蚧为我国检疫性有害生物。其寄主植物甚多，已知有57科149属207种，其中以锦葵科、茄科、菊科、豆科为主。以雌成虫和若虫吸食植物汁液为害，主要为害扶桑、棉花等植物的幼嫩部位，包括嫩枝、叶片、花芽和叶柄。受害植株长势衰弱，生长缓慢或停止，呈失水干枯状，造成棉花等植株的花蕾、花、幼铃脱落；可造成扶桑茎叶甚至整个植株扭曲变形，严重时导致死亡。

扶桑绵粉蚧分泌的蜜露还可以诱发煤污病，从而加重对寄主植物危害，影响光合作用，最终导致叶片大量脱落，明显影响植物观赏性与绿化效果。

| 地理分布 |

扶桑绵粉蚧原产于北美洲。目前分布于墨西哥、美国、古巴、牙买加、危地马拉、多米尼加、厄瓜多尔、巴拿马、巴西、智利、阿根廷、尼日利亚、贝宁、喀麦隆、新喀里多尼亚、巴基斯坦、印度、泰国等。

我国于2008年首次在广东发现扶桑绵粉蚧。目前扶桑绵粉蚧已分布于浙江、福建、江西、湖南、广东、广西、海南、四川、云南、海南等地。

| 传播途径 |

扶桑绵粉蚧可以通过空气气流进行短距离扩散，也可借助水、苗床土、人类、家畜和野生动物进行扩散。

管理措施

综合治理（Integrated Pest Management，IPM）或综合控制（Integrated Pest Control，IPC），但不包括植物卫生措施（Phytosanitary Measures）。

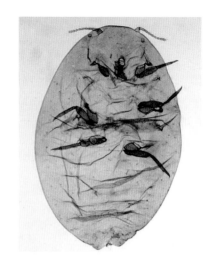

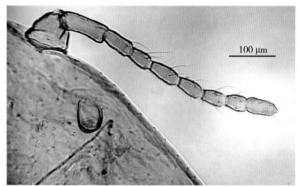

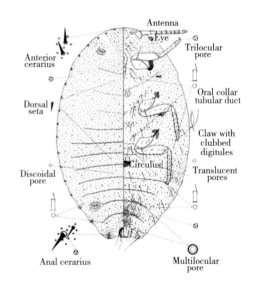

（图片选自pic.baike.soso.com，www.cabi.org，idtools.org）

扶桑绵粉蚧（*Phenacoccus solenopsis*）

【www.cabi.org】 *Phenacoccus solenopsis* (cotton mealybug)

P. solenopsis is a bisexual species with multiple generations annually. Like other mealybugs, this species is distinguished by the morphology of the adult female. Adult females are covered with a powdery, waxy secretion with six pairs of transverse, dark bands that are located across the pro- to meta-thoracic segments. A series of waxy filaments extend from around the margin of the body with the pair of terminal filaments longest. The ovisac is composed of fluffy, loose-textured wax strands (McKenzie, 1967; Kosztarab, 1996) . Adult females range from 2 to 5 mm long and 2 to 4 mm wide.

Slide-mounted females are distinguished ventrally by the presence of nine-segmented antennae, five-segmented legs with translucent pores on meta-femur and meta-tibia, each claw with a minute tooth, two sizes of oral collar tubular ducts, absence of quinquelocular pores, a large circulus, and a series of multilocular pores concentrated around the vulva and submarginal areas of abdominal segments (McKenzie, 1961; 1967; Kosztarab, 1996; Hodgson et al., 2008) . On the dorsum, 18 pairs of cerarii, each with two spinose setae, are located around the marginal area, with evenly distributed trilocular pores, and minute circular pores. Also, oral rim ducts, oral collar tubular ducts, and multilocular pores are absent on the dorsum. Upon hatching, female development consists of first (crawler) , second, and third instars and the adult, whereas males undergo first, second, prepupa, pupa and adult stages of development. Hodgson et al. (2008) provided comprehensive descriptions and illustrations for the immature stages of the solenopsis mealybug. First instars are separated from the other stages by possessing six-segmented antennae, lack of circulus, and quinquelocular pores on the head, thorax and abdomen. Second-instar nymphs are distinguished by having 18 pairs of distinct cerarii around the margin of the body, the lack of quinquelocular pores on the body and the claw with a distinct denticle. The third instar nymph differs by having seven-segmented antennae and a circulus.

刺桐姬小蜂

学　　名：*Quadrastichus erythrinae* Kim
异　　名：无
别　　名：无
英文名：Erythrina gall wasp

| 形态特征 |

雌成虫　体黑褐色，长1.45～1.6mm。具3个红色单眼，略呈三角形排列。触角9节。前胸背板黑褐色，中胸盾片具1个"V"形或起自前缘的倒三角形黑褐色中叶，其余部分黄色。翅透明无色，翅面纤毛黑褐色，翅脉褐色。腹部褐色，背面颜色较腹面深。前、后足基节黄色，中足基节浅白色，腿节棕色。产卵器鞘不突出，藏于腹内。

雄成虫　体色较雌虫浅，长1.0～1.15mm，外生殖器延长，阳茎长而突出，并具1对腹侧突。

| 生物危害 |

刺桐姬小蜂为我国检疫性有害生物。严重为害刺桐属植物。刺桐受害后，叶片常发生卷曲，在叶的正背面和叶柄等处出现畸形，在受害处产生肿大的虫瘿。虫瘿严重发生的叶片和茎干生长迟缓或缺乏活力，可引起大量落叶甚至植株死亡。

| 地理分布 |

刺桐姬小蜂原产于毛里求斯、留尼汪、新加坡、美国（夏威夷）。2005年7月，我国首次在深圳发现。随后在福建省厦门市和海南省三亚市、万宁县也相继发现。现分布于台湾、广东、广西、福建、海南等地。

| 传播途径 |

人为传播或自然扩散。

| 管理措施 |

综合治理（Integrated Pest Management，IPM）或综合控制（Integrated Pest Control，IPC），包括植物卫生措施（Phytosanitary Measures）。

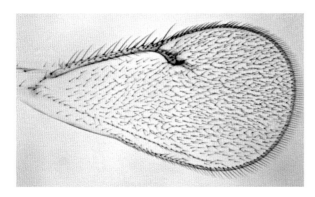
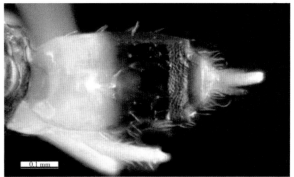

刺桐姬小蜂为害刺桐和象牙红形成的虫瘿

（图片选自pic.sogou.com，www.forestry.gov.cn，commons.wikimedia.org，en.wikipedia.org）

刺桐姬小蜂（*Quadrastichus erythrinae*）

【 www.cabi.org 】/ 【 www.iucngisd.org 】 *Quadrastichus erythrinae* (Erythrina gall wasp)

Female: Length 1.45-1.6 mm. Dark brown with yellow markings. Head yellow, except gena posteriorly brown. Antenna pale brown except scape posteriorly pale. Pronotum dark brown. The mid lobe of mesoscutum with a "V" shaped or inverted triangular dark brown area from anterior margin, the remainder yellow. Scapula yellow. Scutellum, axilla and dorsellum brown to light brown. Propodeum dark brown. Gaster brown. Fore and hind coxae brown. Mid coxa almost pale. Femora mostly brown to light brown. Specimens from Mauritius are generally darker than those from Singapore. Oviposter sheath not protruding, short in dorsal view (Kim Delvare and La Salle, 2004).

Male. Length 1.0-1.15 mm. Pale coloration white to pale yellow as opposed to yellow in female. Head and antenna pale. Pronotum dark brown (but in lateral view, only upper half dark brown; lower half yellow to white). Scutellum and dorsellum pale brown. Axilla pale. Propodeum dark brown. Gaster in anterior half pale; remainder dark brown. Legs all pale. Antenna with 4 funicular segments; without the whorl of setae; F_1 distinctly shorter than the other segments and slightly transverse; about 1.4 wider than long. Ventral plaque extending 0.4-0.5 length of scape and placed in apical half. Gaster shorter than female. Genitalia elongate, with digitus about 0.4 length of the long, exserted aedagus (Kim Delvare and La Salle, 2004).

美洲大蠊

学　　名：*Periplaneta americana* Linnaeus

异　　名：*Blatta americana* L.，*Blatta aurelianensis* Foureroy，*Blatta domicola* Risso，*Blatta heros* Fchscholtz，*Blatta kakkerlac* De Geer，*Blatta orientalis* Sulzer，*Blatta siccifolia* Stoll，*Periplaneta americanacolorata* Rehn.，*Periplaneta stolida* Walker

别　　名：蟑螂、蜚蠊、偷油婆、香娘子、石姜、负盘、滑虫、茶婆虫

英文名：American cockroach，cockroach，ship cockroach

| 形态特征 |

成虫　体大型，雄虫体长27～32mm，雌虫体长28～32mm。赤褐色，头顶及复眼间黑褐色。复眼间距雄虫狭雌虫宽，单眼明显淡黄。前胸背板略作梯形，后缘缓弧形，雄虫约6mm×9.5mm（长×宽），雌虫约7mm×9.4mm；颜色淡黄，中部有一赤褐以至黑褐大斑，其后缘中央向后延伸呈小尾状，其前缘有淡黄"T"形小斑，背板后缘与大斑同色；雄虫在大斑之后、背板后缘中部之前，有左右二浅斜沟，但雌虫不明显。雄虫前翅长26～32mm，雌虫前翅长20～27mm。

雄虫腹部各节背板后侧角为直角，钝圆；雌虫后端数节向后略突出。雄虫肛上板宽大，两侧缘弧形，几成四方形，无色透明，后缘中央有深三角形切口，切口端几达肛上板长度之半；雌虫肛上板略呈三角形，赤褐色，不透明，后缘切口略作小三角形，顶端钝圆，其两侧形成两小叶片，后缘角钝圆。尾须细长多节，比肛上板长一倍，端部细长而尖。雄虫生殖腔中部偏左有一钩刺有时露出腹端，一般从后方可以看出，其尖端呈两个小钩，一大一小，方向互异，由长钩刺向内还有一膜片，上有二乳头；此钩刺及乳头即阳茎叶。雄虫下生殖板较宽短，后缘中央无切口；从腹面看，肛上板的切口两侧叶片从下生殖板的后缘向后露出。雌虫的下生殖板中部隆起，两侧及末端下倾如船底。雄虫腹刺细长，尾须与腹刺均橙黄或赤褐色。

初产卵鞘白色渐变褐至黑色，长约1cm，宽0.5cm，每鞘有卵14～16粒。卵期45～90天化为若虫，若虫约经10次蜕皮化为成虫，后期若虫出现翅芽。成虫寿命约2年。完成一个世代约需两年半。

| 生物危害 |

美洲大蠊的排泄物和蜕落的表皮带有过敏原，可以引发人类皮疹、哮喘等病症。美洲大蠊还能携带多种致病菌，如痢疾杆菌、绿脓杆菌、变形杆菌、沙门氏杆菌、伤寒杆菌，还能携带寄生虫，是家畜及人类许多传染性疾病的重要传播媒介。属于世界性卫生害虫。

| 地理分布 |

美洲大蠊原产于非洲北部，可能是在贩卖黑人时期由非洲带入美洲。美洲大蠊在我国各省市广泛分布。现主要分布在北京、河北、辽宁、黑龙江、上海、江苏、浙江、福建、江西、山东、湖北、广东、广西、海南、四川、贵州、云南、台湾等地。

| 传播途径 |

人为传播或自然扩散。

| 管理措施 |

综合治理（Integrated Pest Management，IPM）或综合控制（Integrated Pest Control，IPC），但不包括植物卫生措施（Phytosanitary Measures）。

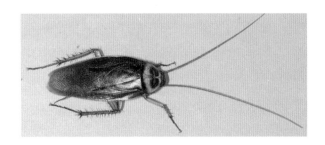

（雌性）

（雄性）

（图片选自en.wikipedia.org，bugguide.net）

美洲大蠊（*Periplaneta americana*）

【animaldiversity.org】*Periplaneta americana*（American cockroach）

Periplaneta americana adults are about 1.125 to 1.375 inches（34-53 mm）long. Their color is a reddish brown except for a submarginal pale brown to yellowish band around the edge of their pronotal shield（Pronotal shield - an expanded version of the top surface plate of the front segment of the thorax）. Both sexes are fully winged. The wings of males extend beyond the tip of the abdomen, while females' do not. They are poor to moderately good fliers.

Early instars of *P. americana* nymphs are uniformly grayish brown dorsally, paler ventrally, and shiny. The cerci（cercus [pl. cerci]- One of a pair of dorsal appendages at the posterior end of the abdomen）are slender, and distinctly tapered from the base with length about 5 times the width. Later instars are reddish brown with lateral and posterior margins of the thorax and lateral areas（sides）of abdominal segments somewhat darker. The cerci are about the same as in the early instars. The widest segments are 2.5 times as wide as long. The antennae are uniformly brown.

The cockroach's walking pattern can be described as follows:

"The cyclic movement of a walking leg consists of two parts, the power stroke（also stance phase or support phase）and the return stroke（also swing phase or recovery phase）. During the power stroke, the leg is on the ground where it can support and propel the body. In a forward-walking animal, this corresponds to a retraction movement of the leg. During the return stroke, the leg is lifted off the ground and swung to the starting position for the next power stroke."（Cruse 1990; Cochran 1980; Smith & Whitman 1992; Bio-Serv 1998）

【en.wikipedia.org】American cockroach

The American cockroach shows a characteristic insect morphology with its body bearing divisions as head, trunk, and abdomen. The trunk, or thorax, is divisible in prothorax, mesothorax and metathorax. Each thoracic segment gives rise to a pair of walking appendages（known as legs）. The organism bears two wings. The forewings, known as tegmina arises from mesothorax and is dark and opaque. The hind wings arise from metathorax and are used in flight, though they rarely do. The abdomen is divisible into ten segments each of which is surrounded by chitinous exoskeleton plates called sclerites, including dorsal tergites, ventral sternites and lateral pleurites.

【bugguide.net】*Periplaneta americana*

德国小蠊

学　　名：**Blattella germanica** Linnaeus

异　　名：*Blattella cunei-vittata* Haitsch，*Blattella niitakana* Bey-Bienko，*Blattella germanica* Linnaeus，*Blattella obliquata* Daldorff，*Blattella stylifera* Chopard，*Blattella transfuga* Ectobia，*Ischnoptera parallela* Tepper，*Phyllodromia bivittata* Saussure，*Periplaneta germanica*（L.），*Phyllodromia cunei-vittata* Hanitsth，*Phyllodromia germanica*（L.），*Phyllodromia germanica* Shiraki，*Phyllodromia magna* Tepper，*Phyllodromia niitakana* Shiraki。

别　　名：德国蟑螂、德国姬蠊

英文名：German cockroach，croton bug，Russian roach，steam fly

| 形态特征 |

成虫　体小型，淡赤褐色。雄虫狭长，体长10~13mm；雌虫较宽短，体长11~14mm。前胸背板具2条内侧平直的纵向黑色条纹。雄虫前翅长9.5~11mm，雌虫前翅长11~13mm。雄虫腹部狭长，第7节背板特化，肛上板半透明；下生殖板左右不对称，形状不正，左后缘有一凹槽，基部两侧各有1侧片即第9节背板侧片，向后延伸部分侧片较宽，端部钝圆，左侧片后端远离凹槽侧缘。雌虫腹部较宽短，基部宽，赤褐色，端部狭，白色，末端钝角，侧缘斜，略向内凹，整体略呈三角形；下生殖板宽大，表面隆起，前侧缘近半圆形，后缘圆弧形，尾须强大多毛。

初产卵鞘乳白色渐变至黄褐色。若虫5~7龄，低龄若虫深褐色，背板纵纹从低龄到高龄逐渐显现。

| 生物危害 |

德国小蠊分泌物可使食物变质，人食后可导致食物中毒。德国小蠊会咬食和破坏食品、纸张、文物、电子设备等，造成损失。还可携带痢疾杆菌、结核杆菌、脊髓灰质炎病毒、乙肝病毒等多种致病病原体，威胁人类健康。该害虫严重破坏人类居住环境，影响生活质量。

| 地理分布 |

德国小蠊原产于南亚，也有学者认为起源于非洲。随着地区间国际贸易而远距离传播，现已遍布全球，在热带、亚热带、温带、寒带均有分布。

在我国，德国小蠊现主要分布于北京、辽宁、黑龙江、上海、福建、广东、广西、四川、贵州、云南、西藏、陕西、新疆等地。

| 传播途径 |

人为传播或自然扩散。

|管理措施|

综合治理（Integrated Pest Management，IPM）或综合控制（Integrated Pest Control，IPC），但不包括植物卫生措施（Phytosanitary Measures）。

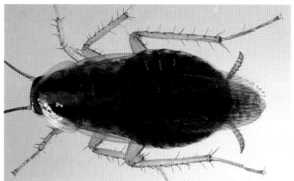
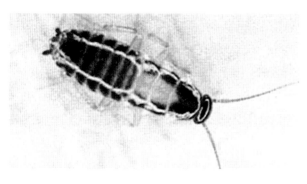

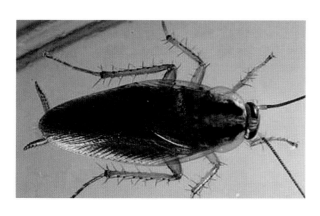

 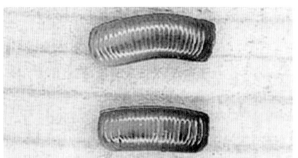

（雌虫及卵鞘）

（图片选自entnemdept.ufl.edu，bugguide.net）

德国小蠊（*Blattella germanica*）

【www.cabi.org】*Blattella germanica*（German cockroach）

An adult *B. germanica* is 10-15 mm long, brown to black, with two distinct parallel bands running the length of the pronotum. The males have thin, slender bodies, the posterior abdomen is tapered, the terminal segments of the abdomen are visible and not covered by tegmina. In contrast, the female has a stout body, the posterior abdomen is rounded and the entire abdomen is just covered by tegmina（Valles, 1996）.

Ramaswamy and Gupta（1981）studied the sensilla of the antennae and the labial and maxillary palps of *B. germanica*. They reported that all these organs contain thick-walled chemoreceptors with fluted shafts and articulated bases. The flaggellar segments of the antennae and the distal segments of the palps contain thin-walled receptors without fluted shafts or articulated bases. The adult male antennae have more thin-walled chemoreceptors than those of the females. At the joints of segments on the palps, scape-head and scape pedicel, hair-plate sensilla can be found. The distal margin of the pedicel, the scape, pedicel and flagellar segments of the antennae and the first segment of the maxillary palps are all sites of campaniform sensilla. Sensilla coeloconica and cold-receptor sensilla are sometimes found on the antennal flagellum.

The nymphs of *B. germanica* are dark-brown to black, with dark parallel bands running the length of the pronotum. The most frequently reported number of moults required to reach adulthood is six, but the number of moults can vary. Development of the nymphs at room temperature takes approximately 60 days（Valles, 1996）. The nymphs eat the moulted skins（Precise Pest Control, 2014）.

【animaldiversity.org】*Blattella germanica*（German cockroach）

German cockroaches are ectothermic organisms. Adults measure 12.7 to 15.88 cm in length（average 13.0 cm）and weigh between 0.1 and 0.12 g（average 0.105 g）. In general, German cockroaches are monomorphic with a flattened, oval shape, spiny legs, and long antennae. They are sexually dimorphic. Males have a thin and slender body, tapered posterior abdomen, visible terminal segments of the abdomen, and do not have tegmina（leathery outer wings）. Females tend to be larger and have a stouter body, rounded posterior abdomen, and tegmina covering the entire abdomen. German cockroaches demonstrate bilateral symmetry at all stages of life.

German cockroaches are light brown in color with two broad, parallel stripes on the dorsal side of the body running lengthwise. Nymphal cockroaches resemble adults in shape; however, they are smaller, darker（dark brown to black）, have only a single stripe down the dorsal side, and have undeveloped wings. Egg capsules are light tan and round（"German Cockroach - Blattella germanica", 2004; Day, August 1996; Jacobs, 2007; Valles, 2008）.

【entnemdept.ifas.ufl.edu】German cockroach

【bugguide.net】*Blattella germanica*

无花果蜡蚧

学　　名：**Ceroplastes rusci**（L.）

异　　名：*Calypticus hydatis*（Costa）Signoret，*Calypticus radiatus* Costa，*Calypticus testudineus* Costa，*Ceroplastes denudatus* Cockerell，*Ceroplastes nerii* Newstead，*Ceroplastes tenuitectus* Green, 1907，*Chermes caricae*（Barnard）Boisduval，*Coccus artemisiae* Rossi，*Coccus caricae* Bernard, Fernald，*Coccus caricae* Fabricius，*Coccus hydatis* Costa，*Coccus rusci* Linnaeus，*Columnea caricae*（Fabricius）Targioni Tozzetti，*Columnea testudinata* Targioni Tozzetti，*Columnea testudiniformis* Targioni Tozzetti，*Lecanium artemisiae*（Rossi）Signoret，*Lecanium radiatum*（Costa）Walker，*Lecanium rusci*（Linnaeus）Walker，*Lecanium testudineum*（Costa）Walker

别　　名：榕龟蜡蚧、拟叶红蜡蚧、锈红蜡蚧、蔷薇蜡蚧

英文名：fig wax scale

| 形态特征 |

成虫　雌成虫前期虫体表皮膜质，略隆起，后期虫体表皮稍硬化，体背部隆起呈半球形。体长2.0～3.5mm，宽1.5～2.5mm，淡褐色。触角6节，第3节有亚分节线。眼在头端两侧突成半球形。足3对，分节正常，胫跗关节硬化，爪下有1小齿，爪冠毛2根，同粗，端部膨大为匙形，跗冠毛2根细长，同形。背面：有8个无腺区，头区1个，背侧各3个，背中区1个。背刺锥状，端钝，均匀分布。背腺多为提篮型二孔腺，也有少量三孔腺，孔腺内有细管，管的末端多叉分支。此外，还有微管腺散布，侧缘数量较多。尾裂浅，肛突短锥形，向体后倾斜。肛板圆滑，没有明显的角，上有背毛3根和长端毛1根。肛板周围的硬化区内有5～13个圆形孔横向排成1或2行。肛筒稍长于肛板。体缘：两眼点间长缘毛6～15根，眼点到前气门刺间每侧有2～4根，两群气门刺间每侧1～8根，后气门刺到体末有10～15根，其中3根尾毛较长。

气门洼浅，气门刺钝圆锥形，大小不一，背面者最大，集成2列（很少有3列），靠背面一列4～5根，靠腹面一列20～23根。腹面：膜质，椭圆形十字腺散布腹面，多数集中在亚缘区。五格腺在气门与体缘间形成约与围气门片同宽的带状气门腺路，每个气门路有33～95个，多格腺在阴门附近及其前腹节成宽带状，第3～5腹节中区有少量分布。杯状管具有细长的端丝，在头区腹面约1～12个，阴门侧1～2个或无。

若虫1、2龄蜡壳长椭圆形，白色，背中有1长椭圆形蜡帽，帽顶有1横沟，体缘有约15个放射状排列的干蜡芒。雌成虫蜡壳白色到淡粉色，稍硬化，周缘蜡层较厚。蜡壳分为9块，背顶1块，其中央有1红褐色小凹，1、2龄干蜡帽位于凹内，侧缘的蜡壳分为8块，近方形，每一侧有3块，前后各有1块；初期每小块蜡壳之间由红色的凹痕分隔开来，每小块中央有内凹的蜡眼，内含白蜡堆积物。后期蜡壳颜色变暗，呈褐色，背顶的蜡壳明显凸起，侧缘小蜡壳变小，分隔小蜡壳的凹痕变得模糊。整壳长1.5～5.0mm，宽1.5～4.0mm，高1.5～3.5mm。

生物危害

无花果蜡蚧为我国检疫性有害生物。其寄主多样，适生区广泛，是许多园林植物和经济果园的重要害虫。该虫除吸食寄主汁液对植物的枝干、嫩梢、叶片和果实造成的直接为害外，还分泌大量的蜜露，诱发煤污病，从而降低寄主植物的生命力，影响园林植物的观赏价值，造成经济果林减产。

地理分布

无花果蜡蚧原产于非洲，最早发现于地中海沿岸地区，现已扩散传播至东洋区、非洲区、新热带区和古北区等动物区系。其中，在热带、亚热带和暖温带分布较广。

2012年，我国首次在广东省茂名市的榕树上和四川省攀枝花市的大叶榕上发现该害虫。现主要分布在广东、四川等地。

传播途径

人为传播或自然扩散。

管理措施

综合治理（Integrated Pest Management，IPM）或综合控制（Integrated Pest Control，IPC），包括植物卫生措施（Phytosanitary Measures）。

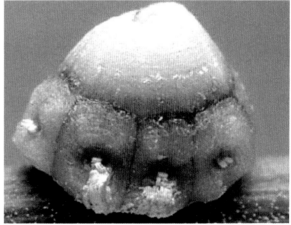

(图片选自www.biodiversidadvirtual.org，shop.agrimag.it，entnemdept.ufl.edu，idtools.org）

无花果蜡蚧（*Ceroplastes rusci*）

【idtools.org】 *Ceroplastes rusci* (Linnaeus)

Stigmatic setae conical with sightly rounded apices; about 25 stigmatic setae laterad of each stigmatic furrow, of 2 intergrading sizes, arranged in 2 or 3 irregular rows; with dorsomedial clear area; tubular ducts without expanded inner filament, present in small numbers near anal plates and on head; dorsal setae cylindrical, apex slightly rounded or acute. Other characters: Marginal setae simple; 1 pair of prevulvar setae (often obscured by anal plates); multilocular pores abundant near vulva, with few on preceding abdominal segments; multilocular pores anterior of anterior spiracle, when present, predominantly with 5 loculi, about same size as pores laterad of anterior spiracle; tibio-tarsal sclerosis present; claw usually with small denticle; claw digitules equal; antennae usually 6-segmented, rarely with 7; area around anal region sclerotized, forming protuberance; anal plates rounded, without distinct angles; each anal plate with 1 subapical seta, 4 apical setae, and 1 subdiscal seta; anal fold with 6 fringe setae; preopercular pores inconspicuous, restricted to area anterior of anal plates; without submarginal tubercles.

Ceroplastes rusci is similar to *C. cirripediformis* Comstock but differs by having apically acute dorsal setae (capitate in *C. cirripediformis*), small denticle on claw (absent from *C. cirripediformis*), and multilocular pores nearly restricted to area around vulva (present on anterior abdominal segments on *C. cirripediformis*).

【entnemdept.ufl.edu】 fig wax scale

枣实蝇

学　　名：***Carpomya vesuviana*** Costa

异　　名：*Carpomyia buchicchii* Rondani，*Carpomyia zizyphae* Agarwal & Kapoor，*Orellia bucchichi* Frauenfeld

别　　名：无

英文名：ber fruit fly

| 形态特征 |

成虫　体黄色，体、翅长2.9~3.1mm。头高大于长，雌雄的头宽相同，淡黄至黄褐色。额表面平坦，两侧近于平行，约与复眼等宽。颜略较额短，侧面观平直，触角沟浅而宽，中间具明显的颜脊。复眼圆形，其高与长大致相等。触角全长较颜短或约与颜等长，第3节的背端尖锐。喙短，呈头状。盾片黄色或红黄色，中间具3个细窄黑褐色条纹，向后终止于横缝略后；两侧各有4个黑色斑点，横缝后亚中部有2个近似椭圆形黑色大斑点，近后缘的中央在两小盾前鬃之间有一褐色圆形大斑点；横缝后另有2个近似叉形的白黄色斑纹。小盾片背面平坦或轻微拱起；白黄色，具5个黑色斑点，其中2个位于端部，基部3个分别与盾片后缘的黑色斑点连接。胸部侧面大部分淡黄至黄褐色，中侧片后缘中间有一黑色小斑点；侧背片部分黑褐色；后小盾片大部分黑色，中间黄色。胸部鬃发达，除肩板鬃和翅侧鬃为黄色至黄褐色外，余均黑色。翅透明，具4个黄色至黄褐色横带，横带的部分边缘带有灰褐色；基带和中带彼此隔离，较短，均不达翅后缘；亚端带较长，伸达翅后缘，带的前端与前端带于R_1和R_{2+3}室内相互连接成倒"V"形。足完全黄色；前股节具1~3根后背鬃和1列后腹鬃；中胫端刺（距）1根。雄虫第5节背板几呈三角形，其宽度不及长度的2倍；第5节腹板后缘向内成"V"形凹陷；雌虫第6节背板略长于第5节背板。

卵　圆形，黄色至黄褐色。

幼虫　蛆形。3龄幼虫体长7.0~9.0mm，宽1.9~2.0mm。口感器具4个口前齿。口脊3条，其缘齿尖锐。口钩具1个弓形大端齿。第1胸节腹面具微刺；第2、3胸节和第1腹节均有微刺环绕；第3~7腹节腹面具条痕；第8腹节具数对大瘤突。前气门具20~23指状突；后气门裂大，长约4~5倍于宽。

蛹　体节11节，初蛹黄白色，后变黄褐色。

| 生物危害 |

枣实蝇为我国检疫性有害生物。其主要以幼虫蛀食果肉进行为害，但不蛀食枣核和种仁。为害时，在果面上可形成斑点和虫孔。内部蛀食后，则形成蛀道。并引起落果和腐烂。被害率可达到60%，造成枣果产量损失20%以上，甚至绝收。

地理分布

枣实蝇原产于印度。现广泛分布于南亚、中亚、东南亚、欧洲东部等国家或地区。2007年，在我国新疆吐鲁番地区的鄯善县、托克逊县、吐鲁番市发现。现主要分布在新疆局部地区。

传播途径

人为传播或自然扩散。

管理措施

综合治理（Integrated Pest Management，IPM）或综合控制（Integrated Pest Control，IPC），包括植物卫生措施（Phytosanitary Measures）。

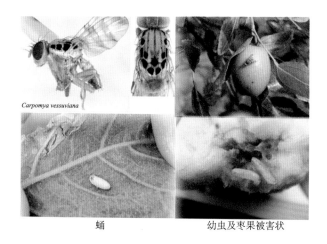

蛹　　　　幼虫及枣果被害状

（图片选自baike.sogou.com，pic.sogou.com，tupian.baike.com，www.researchgate.net）

枣实蝇（*Carpomya vesuviana*）

【delta-intkey.com】*Carpomya vesuviana* Costa

Body. Body plump; predominantly yellow.

Head. Number of frontal bristles three pairs. Number of orbital bristles two pairs. Anterior orbital bristle of male normal, unmodified. Posterior orbital bristles reclinate; acuminate. Ocellar bristles absent or minute, like setulae. Postocellar bristles present. With both inner and outer vertical bristles. Outer vertical, postvertical and postocellar bristles all acuminate. Postocular bristles acuminate; normal. Genal bristle present. Head higher than long. Male and female head width the same. Frontofacial angle about equal to a right angle and angular. Face flat; with distinct antennal grooves and carina; shorter than frons; vertical, or receding; without dark marks. Parafacial spot absent. Frons and parafacial without small silvery markings. Fronto-orbital plate setulose. Frontal stripe setulose (hard to see). Eye round, about as high or slightly higher than long. Antenna considerably shorter than face, or about as long as face. Scape, pedicel, and first flagellomere all relatively short. First flagellomere shorter than face; pointed dorsoapically. Arista longer than first flagellomere; bare or with hairs distinctly shorter than greatest aristal width; hairs both dorsal and ventral. Proboscis short, capitate.

Thorax. Inner scapular bristle present and distinguishable from surrounding vestiture; pale. Outer scapular bristle present and distinguishable from surrounding vestiture; pale. Postpronotal bristle present. Presutural dorsocentral bristle absent. Presutural supra-alar bristle present. Postsutural supra-alar bristle present. Acrostichal bristle present. Postsutural dorsocentral bristle present; bristles aligned with postsutural supra-alar bristles or slightly behind. Intra-alar bristle present, well developed, similar to postalar bristle. Intrapostalar bristles absent. Number of scutellar bristles two pairs. Apical scutellar bristles as long as basals or longer. Anterior notopleural bristle present. Posterior notopleural bristle (s) acuminate. Number of outstanding anepisternal bristles two. Katepisternal bristles present. Anepisternal bristles dark, brown to black. Long, erect setulae on laterotergite absent. Scutal setulae acuminate and pale. Scutellum sparsely setulose. Setulae on scutellum short, decumbent; unicolorous, acuminate. Transverse suture with the lateral branches wide apart. Complete sclerotized postcoxal metathoracic bridge absent or semimembranous. Scutum yellowish, or orange-brown, or black; without a large dark central stripe which broadens basally. Postpronotal lobe entirely pale whitish or yellowish. Posterior half of notopleuron pale whitish or yellowish (except dark on mesal third). Scutum dorsad of notopleuron of the ground color, not whitish or yellowish, or with a pale stripe which extends from the postpronotal lobe dorsad of the anterior notopleuron, not reaching the posterior half of the notopleuron. Dark lyre-like pattern on scutum absent. Discrete shiny black spots on scutum present. Median longitudinal black stripe on scutum present (very thin). Number of pale whitish to yellow postsutural stripes four. Lateral postsutural stripes of scutum extending to intra-alar bristles or beyond (narrow). Area bordering scutoscutellar suture medially with dark brown spot. Discrete pale horizontal stripe along upper anepisternum absent or indistinct (indistinct). Distinct pale vertical anepisternal stripe absent. Katepisternite with pale yellowish or whitish spot present and distinct. Transverse suture without distinct stripe or spot. Katatergite with pale yellowish or whitish spot absent or indistinct. Anatergite with pale yellowish or whitish spot absent or indistinct. Subscutellum yellowish to orange-brown medially, with distinct dark spots laterally. Mediotergite uniformly yellowish to orange-brown (microtrichose, appearing white). Scutum microtrichia in discrete pattern due to density differences, or microtrichia in discrete pattern due to bare areas or completely absent. Dorsum of scutellum flat or slightly convex, not swollen, or convex and swollen. Scutellum normal; with a dark and pale pattern; with two isolated dark spots (& 3 basal); without mark.

Legs. Femora swollen. Fore femur with regular bristles; without ventral spines; with 1 to 3 posterodorsal and 1 posteroventral rows of bristles only. Mid femur and hind femur without spine-like bristles. Middle leg of male without feathering. Femora all entirely of one color; dark mark on fore femur 0% of length of femur; dark mark on middle femur 0% of length of femur; dark mark on hind femur 0% of length of femur.

Wings. Wing partly bare. Cell bc microtrichia absent. Cell c microtrichia covering whole cell. Cell dm entirely microtrichose. Dense microtrichia at end of vein A_1+CuA_2 in male absent. Dominant wing pattern cross-banded. Crossbanded wing patterns Rhagoletis-like. Wing pattern mostly yellowish. Dark longitudinal streaks through basal cells absent. Crossvein r-m covered by a major crossband. Crossvein dm-cu covered by a major crossband which reaches posterior margin of wing. Crossveins r-m and dm-cu not both covered by a single crossband. Cell r_{2+3} apical to r-m with large hyaline area. Anal band absent, or not reaching nearly to wing margin. Cell r_1 and r_{2+3} with distinctly darker spots within dark areas of pattern. Intercalary band absent. Subbasal crossband present. Subbasal and discal crossbands not joined. Marginal hyaline area in cell r_1 present and distinct. Ratio of width of apical band in cell r_{4+5} to length of r-m 1.5. Anterior apical crossband partly to entirely separated from costa by marginal hyaline band or spots. Anterior apical band or costal band extended to vein M. Posterior apical crossband absent. Costal and discal bands not joined on vein R_{4+5}. Discal band transverse, or oblique in anterobasal-posteroapical direction, or absent. Discal and apical crossbands not directly joined. Discal and subapical crossbands not joined. Discal and subapical bands not connected along vein R_{4+5}. Subapical and anterior apical crossbands joined. Outstanding costal spine (s) at subcostal break present. Ratio of length of costal section 3 to costal section 4 0.2. Ratio of pterostigmal length to width 1.5-2. Vein R1 dorsal setation without bare section opposite end of vein Sc. Vein Rs dorsal setation non-setulose. Vein R_{2+3} generally straight. Anteriorly-directed accessory vein emerging from R_{2+3} absent. Vein R_{4+5} dorsal setation absent, or on node only; ventral setation absent or only present on node or close thereafter. Distance between crossvein r-m and costa longer than r-m. R-m crossvein on cell dm at or near middle of cell dm. Cell bm narrow, triangular; ratio of length to width 3; ratio of width to cell cup width 1. Vein M distally straight. Cell dm widens apically gradually from base. Posterodistal corner of cell dm approximately a right angle. Cell cup extension or lobe present, vein CuA_2 abruptly bent; shorter than vein A_1+CuA_2; triangular.

Abdomen. Abdomen ovate or parallel sided. Abdominal tergites separate. Abdomen in lateral view flatter, more flexible. Abdominal tergite 1 broader at apex than at base; without a prominent hump laterally. Pecten of dark bristles on tergite 3 of male absent. Tergal glands on tergite 5 absent. Abdominal tergite 5 normal 6th tergite of female exposed; longer than 5th. Abdominal setulae acuminate and dark. Abdominal microtomentum uniform. Abdominal sternite 5 of male less than 2 × wider than long, not longer than wide. Posterior margin of sternite 5 of male with deep V-shaped posterior concavity. Abdominal tergites 3-5 predominantly yellow to orange brown. Abdominal tergites without medial dark stripe; not brown with medial T-shaped yellow mark; without isolated dark areas on lateral margins of T3-T5; without dark brown transverse bands.

Male terminalia. Epandrium in posterior view with long outer surstyli, which are more than half as long as epandrium; lateral view with outer surstyli distinctly narrower than epandrium, clearly differentiated. Distiphallus present; with extensive medial sclerotization; without stout, curved, basal spine; without basal setulose rod. Sclerite of vesica of distiphallus absent.

Female terminalia. Syntergosternite 7 straight; shorter than preabdomen; base without a laterally projecting flap; conical. Ratio of syntergosternite 7 to abdominal tergite 5 0.3-0.5. Dorsobasal scales of eversible membrane about as large as other scales (?). Aculeus tip gradually tapering, needle-like, with flat cross-section; fused to main part of aculeus, not movable; 0% serrated; with minute serration, visible only with compound microscope. Three sclerotized spermathecae.

椰子织蛾

学　　名：*Opisina arenosella* Walker
异　　名：*Nephantis serinopa* Meyrick，*Ophisina arenosella*，*Opisina serinopa* Meyrick
别　　名：黑头履带虫、椰蛀蛾
英文名：black-headed caterpillar，coconut black-headed caterpillar，palm leaf caterpillar

| 形态特征 |

成虫　翅展8.0~24.0mm。头部灰白色。下唇须乳白色，第2节腹面和内侧密布灰白色长鳞毛，鳞毛端部杂黑色；第3节散布黑褐色鳞片。触角柄节土黄色；鞭节乳白色，杂黑褐色。胸部和翅基片黄色至暗灰色，散布黑色鳞片。前翅狭长，前缘略拱，顶角钝，外缘弧形后斜；R_4、R_5脉共长柄，R_5脉达前缘末端，CuA_1、CuA_2脉均出自中室后角之前，CuA_2脉位于中室后缘2/3处，CuP脉存在；土黄色至灰白色，散布黑色鳞片；前缘基部约1/6黑色，端半部具多条黑色细纵纹；中室中部和翅褶中部各具1枚黑点，均由2~3枚黑色竖鳞形成，中室端部密布暗灰色鳞片，末端具1枚模糊黑点；缘毛与翅同色。后翅Rs与M_1脉、M_3与CuA_1脉共柄，CuP脉存在；灰褐色，缘毛基部1/3灰褐色，端部灰白色。前、中足乳白色，前足转节和腿节腹侧黑色，胫节外侧黑色，跗节具浅褐色环；后足土黄色。腹部2~6节有背刺。

卵　半透明乳黄色，长椭圆形，具有纵横网格。

幼虫　5~8个龄期。雌、雄幼虫大小相似，雄性6~8龄。雄性腹部第9节前缘腹中腺表面有一圆形凹陷，雌虫无此凹陷，这一特征可用于辨别幼虫的性别。成熟幼虫长20~25mm，头和前胸黑褐色，中胸两侧红褐色，中胸背面和后胸及腹部淡绿色，腹部背面及侧面通常有5条褐色纵纹。

蛹　包被在混合寄主碎屑和虫粪的丝质茧中。裸露的蛹褐色至红褐色。腹部第1~4节背面前缘有刺列，其中第2~4节的刺列较为明显。

| 生物危害 |

椰子织蛾的寄主有棕榈科、芭蕉科植物。其为害不同年龄的棕榈科植物，幼虫从植物的下部叶片向上取食，逐渐向其他叶片扩展。幼虫取食叶片时，还在叶背面形成蛀道，蛀道内粪便与其吐丝交织。每个叶片上可能有几头幼虫，同时为害。受侵染严重的植株，叶片会干枯。除顶端少数叶片外，整个树冠均可被侵害。

椰子织蛾的寄主棕榈科植物，不仅具有较高的观赏价值和食用价值，也是最重要的旅游资源之一。同时，棕榈科植物还是十分珍贵的种质资源，保护价值极高。

椰子织蛾具有繁殖能力强、生长周期短、全年可危害的特点，对寄主的危害相当严重。

| 地理分布 |

椰子织蛾原产于南亚，主要分布在印度、斯里兰卡、孟加拉国、巴基斯坦、缅甸、印度尼西亚、泰国和马来西亚等地。

2013年，我国首次在海南省万宁市的棕榈科植物上发现该虫。现主要分布于广东、广西、海南等地。

| 传播途径 |

自然扩散或人为传播。

| 管理措施 |

综合治理（Integrated Pest Management，IPM）或综合控制（Integrated Pest Control，IPC），但不包括植物卫生措施（Phytosanitary Measures）。

（雄蛹和雌蛹）

（图片选自baike.baidu.com，pic.baike.soso.com，baike.sogou.com）

椰子织蛾（*Opisina arenosella*）

【en.wikipedia.org】*Opisina arenosella*

The coconut black-headed caterpillar is identifiable in the larval form as a caterpillar with greenish brown with dark brown head and prothorax, and a reddish mesothorax. There are often brown stripes on the body of the larva.

Post pupation, the caterpillar morphs into a moth which is greyish white in colour. The female is distinguishable from the male in that it has longer antenna, and three faint spots on the forewings, while the males have fringed hairs in the apical and anal margins of the hind wings.

松树蜂

学　　名：*Sirex noctilio* Fabricius
异　　名：*Paururus noctilio*（Fabricius）
别　　名：云杉树蜂、辐射松树蜂
英文名：woodwasp，European woodwasp，horntail，Sirex wasp，steel blue，steel-blue

| 形态特征 |

成虫　体长10～44mm，雌性比雄性成虫体型略大，圆柱形，触角黑色。雌虫头部、胸腹部具蓝色金属光泽，胸足橘黄色，腹部末端呈角突状；雄虫头胸部具蓝色金属光泽，腹部基部和末端呈黑色，中部橘黄色。后足粗大、黑色。

卵　乳白色，呈梭形，长约1.4mm，中部最宽处直径约0.5mm。

幼虫　乳白色，圆筒形，老熟幼虫体长10～20mm，头宽3～5mm。

蛹　离蛹，乳白色，长10～18mm。

| 生物危害 |

松树蜂为我国检疫性有害生物。其主要为害松属、云杉属、冷杉属、落叶松属以及美国松属等种类。松树蜂幼虫的取食和钻蛀能对寄主树木造成严重破坏，其产卵时注入的毒素和共生菌能够严重影响寄主体内的水分平衡、光合产物运输等重要的生理代谢过程，同时降解寄主木质纤维素等大分子物质和破坏树体内部结构，逐步削弱寄主的防御能力从而加速树势的衰弱甚至导致树木死亡。

| 地理分布 |

松树蜂原产于欧亚大陆和北非，后传入澳洲、南美、北美、非洲等地。
2013年，我国首次在黑龙江省杜尔伯特蒙古族自治县发现。现主要分布于内蒙古、吉林、黑龙江等地。

| 传播途径 |

自然扩散或人为传播。

| 管理措施 |

综合治理（Integrated Pest Management，IPM）或综合控制（Integrated Pest Control，IPC），包括植物卫生措施（Phytosanitary Measures）。

外来入侵动物

雌虫背面

（雌虫）

雌虫侧面

雄虫侧面

（雄虫）

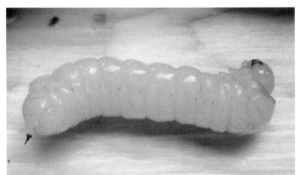

（幼虫）

松树蜂为害火炬松（*Pinus taeda*）

（图片选自www.waspweb.org，entnemdept.ufl.edu，www.invasive.org）

松树蜂（*Sirex noctilio*）

【www.cabi.org】 *Sirex noctilio* (woodwasp)

Eggs The eggs are white, soft, smooth and elongate. They are 1.55 mm long and 0.28 mm wide (Neumann and Minko, 1981).

Larvae The larvae are creamy-white, deeply segmented, usually S-shaped and nearly uniform in diameter. The antennae are one-segmented. The thoracic legs are short and the abdomen has a conspicuous dark brown sclerotic spine (Neumann et al., 1987).

Pupae The pupae are creamy-white and gradually assume the colour of the adults (Neumann et al., 1987).

Adults The adult male is metallic dark blue, except for abdominal segments three to seven. The front and mid-legs are orange-brown, while the hind legs are thickened and black. The wings are amber and 9.3 to 35 mm long. The antennae have 20 segments and are 6.8 mm long (Neumann et al., 1987).

The adult female is metallic dark blue all over, except for the wings and the legs, which are amber. A sheath protects the ovipositor, which projects 2 to 3 mm beyond the abdomen. The body is 12 to 34 mm long. The antennae have 21 segments and are 7.8 mm long (Neumann et al., 1987).

A prominent spine is present in the final abdominal segment in both sexes.

【entnemdept.ufl.edu】 sirex woodwasp

【www.waspweb.org】 *Sirex noctilio* Fabricius

线　虫

松材线虫

学　名：***Bursaphelenchus xylophilus***（Steiner & Buhrer）Nickle

异　名：*Aphelenchoides xylophilus* Steiner & Buhrer，*Bursaphelenchus lignicolus* Mamiya & Kiyohara

英文名：pine wilt nematode，pine wood nematode

| 形态特征 |

雌雄虫都呈蠕虫形，虫体细长，长1mm左右，其基部微增厚。唇区高，缢缩显著。中食道球卵圆形，占体宽的2/3以上，瓣膜清晰。食道腺细长叶状，覆盖于肠背面。排泄孔的开口大致和食道与肠交接处平行，半月体在排泄孔后约2/3体宽处。卵巢单个，前伸；阴门开口于虫体中后部73%处，上覆以宽的阴门盖。后阴子宫囊长，约为阴肛距的3/4。

雌虫尾亚圆锥形，末端宽圆，少数有微小的尾尖突。

雄虫尾似鸟爪，向腹面弯曲，尾端为小的卵状交合伞包裹，退火的交合伞在光学显微镜下不易看见。交合伞在尾端呈铲状，由于边缘向内卷曲，从背面观呈卵形，从侧面观呈尖圆形。雄虫交合刺大，弓形，喙突显著，远端膨大如盘状。

| 生物危害 |

松材线虫为我国检疫性有害生物。松材线虫病又称松枯萎病，是一种毁灭性病害。主要为害松属植物，也危害云杉属、冷杉属、落叶松属和雪松属。通过松墨天牛（*Monochamus alternates* Hope）等媒介昆虫传播到松树体内，从而引发松树病害。被松材线虫感染后的松树，整株干枯死亡，最终腐烂。

实际上，在中国传播松材线虫的媒介昆虫主要是松墨天牛（*Monochamus alternatus* Hope）；在日本，除松墨天牛外，还有小灰长角天牛［*Acanthocinus griseus*（Fabricius）］、褐幽天牛［*Arthopalus rusticus*（Linne）］、黑角伞花天牛（*Corymbia succedanea* Lewis）、短角幽天牛（*Shondylis buprestoides* Linne）、（*Acaloculata fraudatrix* Bates）、显赫墨天牛（*Monochamus nitens* Bates）、双斑泥色天牛（*Uraecha bimaculata* Thomoson）7种；在美国，主要是卡来罗纳墨天牛（*Monochamus carolinensis*）。

松材线虫病在美国、加拿大、墨西哥、日本、韩国等国均有发生，以日本受害最重。20世纪80

年代，松材线虫侵袭中国香港，几乎毁灭了香港广泛分布的马尾松林。1982年在南京中山陵首次被发现，随后相继在安徽、山东、浙江、广东等地形成几个疾病中心，并向四周扩散，使这些省的局部地区发生并流行成灾，导致大批松树枯死。松材线虫病给安徽、浙江两省带来的经济损失就高达5亿～7亿元。

| 地理分布 |

松材线虫原产于北美洲。现在日本、韩国、美国、加拿大、墨西哥、葡萄牙等国均有分布。1982年在南京中山陵首次发现。在短短的十几年内，已经扩散并分布于江苏、浙江、安徽、福建、江西、山东、湖北、湖南、广东、重庆、贵州、云南、台湾、香港等省（区）。

| 传播途径 |

远距离传播主要借助感病苗木、松材、枝桠及其他松木等针叶木制品（包括木质包装等）的调运，而近距离传播则依靠天牛等媒介昆虫的传带。

| 管理措施 |

综合治理（Integrated Pest Management，IPM）或综合控制（Integrated Pest Control，IPC），包括植物卫生措施（Phytosanitary Measures）。

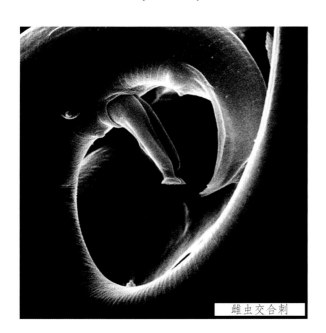

雌虫交合刺

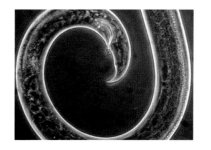

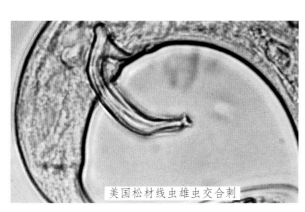

美国松材线虫雄虫交合刺

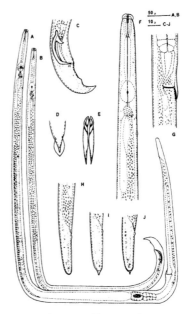

（Mamiya等，1972）

A.雌虫；B.雄虫；C.雄虫尾部；D.雄虫尾部腹面观（示交合伞）；E.交合刺腹面；F.雌虫前端；G.雌虫阴门；H.J.雌虫尾部

媒介昆虫松墨天牛
（*Monochamus alternatus*）

（图片选自www.cabi.org，baike.baidu.com）

松材线虫（*Bursaphelenchus xylophilus*）

【www.cabi.org】 *Bursaphelenchus xylophilus* (pine wilt nematode)

Measurements (after Nickle et al., 1981)

Female lectotypes (n=5): L=0.52 (0.45-0.61) mm; a=42.6 (37-48); b=9.6 (8.3-10.5); c=27.2 (23-31); V=74.7 (73-78) %; stylet=12.8 (12.6-13.0) μm.

Male lectotypes (n=5): L=0.56 (0.52-0.60) mm; a=40.8 (35-45); b=9.4 (8.4-10.5); c=24.4 (21-29); stylet=13.3 (12.6-13.8) μm; spicule=21.2 (18.8-23.0) μm.

Measurements (after Mamiya and Kiyohara, 1972)

Female (n=40): L=0.81 (0.71-1.01) mm; a=40.0 (33-46); b=10.3 (9.4-12.8); c=26.0 (23-32); V=72.7 (67-78) %; stylet=15.9 (14-18) μm.

Male (n = 30): L=0.73 (0.59-0.82) mm; a=42.3 (36-47); b=9.4 (7.6-11.3); c=26.4 (21-31); stylet=14.9 (14-17) μm; spicules=27.0 (25-30) μm.

Description (after Nickle et al., 1981)

Female: Cephalic region high, offset, with six lips. Stylet with small basal swellings. Oesophageal gland lobe slender and about 3-4 body-widths long. Excretory pore located at about the level of the oesophago-intestinal junction, occasionally at the same level as the nerve ring. Hemizonid about 0.67 of a body-width behind the median bulb. Vulva posterior, the anterior lip overhanging to form a flap. Genital tract monoprodelphic, outstretched. Developing oocytes mostly in single file. Post-uterine sac well developed, extending for 0.75 or more of the distance to the anus. Tail subcylindroid with a broadly rounded terminus. Mucron usually absent, but some populations may have a short, 1-2 μm, mucron.

Male: Similar to female in general respects. Spicules large, strongly arcuate so that the prominent transverse bar is almost parallel to the body axis when the spicules are retracted. The apex is bluntly rounded and the rostrum prominent and pointed. The distal tip of each spicule is expanded into a disc-like structure named the cucullus. The tail is arcuate with a pointed, talon-like terminus bearing a small bursa. Seven caudal papillae are present; one pair adanal, a single preanal ventromedian papilla, and two postanal pairs near the tail spike and just anterior to the start of the bursa.

A morphological key to the species of the *xylophilus* group of the genus *Bursaphelenchus* has recently been published by Braasch and Schönfeld (2015).

【en.wikipedia.org】 *Bursaphelenchus xylophilus*

Species of the genus *Bursaphelenchus* are difficult to distinguish because they are similar in morphology. A positive identification can be made with molecular analyses such as restriction fragment length polymorphism (RFLP).

B. xylophilus is distinguished by three characteristics: the spicule is flattened into a disc-shaped cucullus at the tip, the front vulval lip is flap-like, and the tail of the female is rounded.

软体动物

非洲大蜗牛

学　　名：**_Lissachatina fulica_**（Bowdich）

异　　名：_Achatina fulica_（Férussac）, _Helix_（Cochlitoma）_fulica_ Férussac（basionym）,
　　　　　Helix fulica Férussac（original combination）

别　　名：非洲巨蜗牛、露螺、褐云玛瑙螺、东风螺、菜螺、花螺、法国螺

英文名：giant African land snail, African giant snail, kalutara snail

| 形态特征 |

成螺　非洲大蜗牛属于大型贝壳螺，通常体长7~8cm，最大20cm，体重可达32g。贝壳狭窄、锥形，长宽比约为2∶1。壳质稍厚，有光泽，呈长卵圆形。壳高130mm，宽54mm。螺层为7~9个，螺旋部呈圆锥形。体螺层膨大，其高度约为壳高的3/4。壳顶尖，缝合线深。壳面为黄或深黄底色，具焦褐色雾状花纹。其他各螺层具断续棕色条纹。生长线粗而明显，壳内为淡紫色或蓝白色，体螺层上的螺纹不明显；中部各螺层与生长线交错。壳口呈卵圆形，口缘简单、完整。外唇薄而锋利，易碎。内唇贴缩于体螺层上，形成"S"形的蓝白色胼胝部，轴缘外折，无脐孔。足部肌肉发达，背面呈暗棕黑色，遮面呈灰黄色，其黏液无色。

卵　椭圆形，乳白或淡青黄色，外壳石灰质，长4.5~7mm，宽4~5mm。

幼螺　初孵幼螺有2.5个螺层，壳面为黄或深黄底色。具壳，外形呈长卵圆形。螺层有6.5~8个，壳面具焦褐色雾状花纹，壳口呈卵圆形，贝壳可容纳整个足部。生殖系统无附属器官。肾脏长，常为心围膜长的2~3倍。肺静脉无分支。

| 生物危害 |

非洲大蜗牛为我国检疫性有害生物。其为害农作物、林木、果树、蔬菜、花卉（包括木瓜、木霜、仙人掌、面包果、橡胶、可可、茶、柑橘、椰子、菠萝、香蕉、竹芋、番薯、花生、菜豆、落地生根、铁角蕨、谷物等）500多种作物。饥饿时甚至能啃食和消化水泥。还是许多人畜寄生虫[如深奥猫圆线虫（_Aelurostrongylusabstrusus_）、广州管圆线虫（_Angiostrongylus cantonensis_）]和病原菌的中间宿主，尤其是能传播结核病和嗜酸性脑膜炎。

| 地理分布 |

非洲大蜗牛原产于非洲东部的坦桑尼亚（桑给巴尔、奔巴岛），马达加斯加岛一带。但到21世纪已经广泛分布于亚洲、太平洋、印度洋和美洲等地的湿热地区。现分布于日本、越南、老挝、柬埔寨、马来西亚、新加坡、菲律宾、印度尼西亚、印度、斯里兰卡、西班牙、马达加斯加、塞舌尔、毛里求斯、北马里亚纳群岛、加拿大、美国等。

1920年代末至1930年代初，在厦门发现，可能是由一新加坡华人所带的植物而引入。目前在我国福建、广东、广西、云南、海南、台湾等地均有分布。为全球性入侵物种。

| 传播途径 |

非洲大蜗牛主要通过人为方式传播，如通过轮船、火车、汽车、飞机等运输工具，随观赏植物、苗木、板材、集装箱、货物包装箱传播。此外，非洲大蜗牛个体大，外形美观，肉质鲜美，被作为观赏动物或者食材，被引进后饲养繁殖而扩散。

| 管理措施 |

综合治理（Integrated Pest Management，IPM）或综合控制（Integrated Pest Control，IPC），包括植物卫生措施（Phytosanitary Measures）。

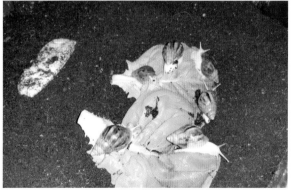

（图片选自www.cabi.org，en.wikipedia.org，landsnails.org，upload.wikimedia.org，www.faunatrhy.cz，www.arkive.org）

非洲大蜗牛（*Achatina fulica*）

【www.cabi.org】 *Lissachatina fulica* (giant African land snail)

L. fulica is distinctive in appearance and is readily identified by its large size and relatively long, narrow, conical shell. Reaching a length of up to 20 cm, the shell is more commonly in the size range 5 to 10 cm. The colour can be variable but is most commonly light brown, with alternating brown and cream bands on young snails and the upper whorls of larger specimens. The coloration becomes lighter towards the tip of the shell, which is almost white. There are from six to nine spirally striate whorls with moderately impressed sutures. The shell aperture is ovate-lunate to round-lunate with a sharp, unreflected outer lip. The mantle is dark brown with rubbery skin. There are two pairs of tentacles on the head: a short lower pair and a large upper pair with round eyes situated at the tip. The mouth has a horned mandible containing some 80 000 teeth (Schotman, 1989) . Eggs are spherical to ellipsoidal in shape (4.5-5.5 mm in diameter) and are yellow to cream in colour.

【idtools.org】 *Lissachatina fulica*

Lissachatina fulica, also known as Achatina fulica is a large snail. The shell of this species is generally narrowly conic with 7-10 whorls and may attain a length of 200 mm (averaging 50-100 mm) and a width of 120 mm when fully mature. The color pattern of the shell will vary widely depending on the diet of the animal but will most often consist of alternating bands of brown and tan. The body of the animal is brown-gray in color and it may be able to extend up to 300 mm in length.

【en.wikipedia.org】 *Lissachatina fulica*

福寿螺

学　名：***Pomacea canaliculata***（Lamarck）

异　名：*Ampullaria australis* d'Orbigny，*Ampullaria canaliculata* Lamarck，*Ampullaria dorbignyana* Philippi，*Ampullaria gigas* Spix，*Ampullaria gualtieri* d'Orbigny，*Ampullaria insularum* d'Orbigny，*Ampullaria levior* Sowerby，*Ampullarium canaliculatus* Lamarck，*Ampullarium insularum* d'Orbigny，*Ampullarium* sp.，*Ampullarius canaliculata* Lamarck，*Ampullarius canaliculatus* Lamarck，*Ampullarius insularum* Hamada & Matsumoto，*Ampullarius insularus* Chang，*Pila canaliculata* Lamarck，*Pila canaliculata* Lamarck，*Pila* sp.，*Pomacea canaliculata chaquensis* Hylton Scott，*Pomacea canaliculate* Lamarck，*Pomacea cuprina* Reeve，*Pomacea gigas* Spix，*Pomacea insularis* d'Orbigny，*Pomacea insularum* d'Orbigny，*Pomacea lineata* Spix

别　名：大瓶螺、苹果螺、雪螺、南美螺

英文名：golden apple snail，apple snail，Argentinian apple snail，channeled apple snail，channeled applesnail，golden miracle snail，golden mystery snail，golden snail，jumbo snail，South American applesnail

| 形态特征 |

成螺　福寿螺属于巨型田螺，个体大，每只重100～150g，最大个体可达250g以上。身体由头部、足部、内脏囊、外套膜和贝壳5个部分构成。头部腹面为肉状足，足面宽而厚实。头部具触角2对，前触角短；后触角长，后触角的基部外侧各有一只眼睛。螺体左边具1条粗大的肺吸管及一个薄膜状的肺囊。肺囊充气后能使螺体浮于水面上，遇到干扰就会排出气体迅速下沉。螺壳螺旋状，颜色随环境及螺龄不同而异，有光泽和若干条细纵纹。爬行时头部和腹足伸出。贝壳短而圆，大且薄，壳右旋，具4～5个螺层，多呈黄褐色或深褐色。成螺贝壳厚，壳高7cm，贝壳的缝合线处下陷呈浅沟，壳脐深而宽。福寿螺雌雄同体，异体交配。

卵　圆形，直径2mm。初产卵粉红色至鲜红色，卵的表面有一层不明显的白色粉状物。卵块椭圆形，大小不一，卵粒排列整齐，卵层不易脱落，鲜红色，小卵块仅数十粒，大的可达千粒。卵于夜间产在水面以上干燥物体或植株的表面，如茎秆、沟壁、墙壁、田埂、杂草等上，初孵化幼螺落入水中，吞食浮游生物等。

幼螺　发育3～4个月后性成熟，除产卵或遇有不良环境条件时迁移外，一生均栖于淡水中。遇干旱则紧闭壳盖，静止不动，长达3～4个月或更长。

| 生物危害 |

福寿螺对水稻生产造成损失，还威胁入侵地的水生贝类、水生植物（水仙花、荷花等），破坏自然食物链结构。

还是卷棘口吸虫 [*Echinostoma revolutum*（Frohlich）]、广州管圆线虫（*Angiostrongylus cantonensis*）的中间宿主。人食用生的或加热不彻底的福寿螺后即可被感染，引发人的嗜酸性粒细胞增多性脑膜炎和脑膜脑炎，引起头痛、头晕、发热、颈部僵硬、面神经瘫痪等症状，严重者会出现瘫痪、嗜睡、昏迷甚至死亡。

| 地理分布 |

福寿螺原产于亚马逊河流域。20世纪70年代引入中国台湾。1981年由巴西籍中国人引入广东。目前已分布于广东、广西、江西、福建、海南、台湾、浙江、湖南、云南、上海、江苏等省市。

| 传播途径 |

福寿螺可以人为携带传播或自然扩散。

| 管理措施 |

综合治理（Integrated Pest Management，IPM）或综合控制（Integrated Pest Control，IPC），但不包括植物卫生措施（Phytosanitary Measures）。

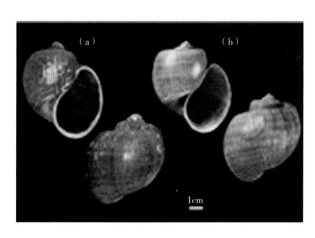

（图片选自en.wikipedia.org，applesnail.net，pic.baike.soso.com，pic.sogou.com，www.cabi.org，upload.wikimedia.org）

福寿螺（*Pomacea canaliculata*）

【www.cabi.org】 *Pomacea canaliculata* (golden apple snail)

Adults The adult shell is thin, smooth and ~35-60 mm in height. It coils dextrally – that is, when viewed with the apex uppermost the aperture is on the right side of the shell. Fully grown females are larger than males. The colour is yellow-brown to greenish-brown or dark chestnut, sometimes with dark brown spiral bands of variable number and thickness. The whorls are rounded and the suture between the whorls is deeply channelled. The shell spire is generally low. The aperture is generally ovoid to kidney-shaped, and the inside lip of the shell is unpigmented.

The operculum (the trap-door like structure attached to the upper part of the animal's foot and used to close the shell aperture when the animal withdraws into the shell) is also brown; it is horny (corneous) in texture and flexible, and is uniformly concave in females, but concave at the centre and becoming convex toward the margins in males.

The foot is oval with a squarish anterior edge. The tentacles are long and tapering, highly extensible and with large but short eye stalks at their outer bases. The snout is short, squarish and with lateral, anterior tips elaborated into long tapering labial palps. The neck is modified on the left into a long, extensible siphon. The mantle cavity is deep and broad, occupying a third to half of the body whorl. In males the penis sheath is visible just behind the mantle edge above the right tentacle. The lung occupies most of the left side of the mantle and the gill is situated in the mantle roof, anterior to the lung and just posterior to the base of the siphon.

Eggs and Hatchlings The eggs are spherical, calcareous, deep pink-red to lighter orange-pink, becoming paler as the calcium hardens, and eventually whitish pink just before hatching. They are laid above water on emergent vegetation and other firm substrates (e.g. bridge pilings, rocks). The height of deposition above water varies from a few centimetres to ~2 metres. The number of eggs per clutch averages ~260, ranging from as few as 12 to as many as ~1 000 (Tamburi and Martín, 2011). Individual egg diameter is ~3.00 mm. One-day-old hatchlings are ~2.6 mm wide and 2.8 mm in height. When they hatch, they drop from where the eggs were laid into the water below.

【applesnail.net】 *Pomacea (pomacea) canaliculata* (Lamarck, 1819)

Shell: The shell of this apple snail species is globose and relatively heavy (especially in older snails). The 5 to 6 whorls are separated by a deep, indented suture (hence the name 'canaliculata' or 'channeled'). The shell opening (aperture) is large and oval to round. Males are known to have a rounder aperture than females. The umbilicus is large and deep. The overal shell shape is similar to that of Pomacea lineata, except the deeper sutures and more globose shape in canaliculata.

The size of these snails varies from 40 to 60 mm wide and 45 to 75 mm high depending on the conditions. The colour varies completely yellow and green (cultivated forms) to brown with or without dark spiral bands (wild form). The shell growth of this species occurs mainly in spring and summer, while it stagnates in fall and winter.

Operculum: The operculum is moderately thick and corneous. The structure is concentric with the nucleus near the centre of the shell. The colour varies light (in young snails) to dark brown. The operculum can be retracted in the aperture (shell opening).

Body: The colour of the body varies from yellow (cultivated), brown to nearly black, with yellow spots on the siphon, but not as much on the mouth as in Pomacea diffusa. When at rest, the tentacles are curled under the shell.

【www.animalspot.net】 *Pomacea canaliculata*

【近似种】田螺（*Cipangopaludina* spp.）

| 形态特征 |

田螺硬壳外观呈圆锥形、高达5~6cm，壳顶略尖，螺层6~7层，壳口卵圆形，褐色，上有同心环状排列生长纹，体柔软头部呈圆柱形。前端突出有吻，腹面有口及一对触角，很敏感灵巧，能伸能缩。足底紧贴着的膜片，称为厣，如圆盖子，当遇不测或需要休息时，田螺便把身体收缩在贝壳里，用厣将贝壳盖严。

福寿螺与田螺相似，但形状、颜色、大小有区别。福寿螺的外壳颜色一般比田螺浅，呈黄褐色，田螺则为青褐色；田螺的椎尾长而尖，福寿螺椎尾则短而平；田螺的螺盖形状比较圆，福寿螺螺盖则偏扁。

| 管理措施 |

综合治理（Integrated Pest Management，IPM）或综合控制（Integrated Pest Control，IPC），但不包括植物卫生措施（Phytosanitary Measures）。

中华圆田螺（*Cipangopaludina cahayensis*）

中国圆田螺（*Cipangopaludina chinensis* Gray）

（图片选自baike.sogou.com）

两栖类

牛 蛙

学　名：***Rana catesbeiana*** Shaw

异　名：*Rana pipiens* Daudin，*Rana taurina* Cuvier，*Rana mugiens* Merrem，*Rana scapularis* Harlan，Rana *conspersa* LeConte，Rana（Rana）*catesbeiana* Boulenger，*Rana nantaiwuensis* Hsü，*Rana mugicus* Angel，*Rana catesbyana* Smith，Rana（Rana）*catesbeiana* Dubois，Rana（*Aquarana*）*catesbeiana* Dubois，Rana（Novirana，Aquarana）*catesbeiana* Hillis & Wilcox，*Rana catesbeianus* Lithobates，Rana（*Aquarana*）*catesbeianus* Dubois

别　名：菜蛙

英文名：American bullfrog，bullfrog，common bullfrog，North American bullfrog

| 形态特征 |

成蛙　牛蛙体大粗壮，雌蛙体长达20cm，雄蛙18cm，最大个体重达2kg以上。皮肤粗糙，肤色随着生活环境而多变，通常背部及四肢为绿褐色，背部带有暗褐色斑纹；头部及口缘鲜绿色；腹面白色；咽喉下面的颜色随雌雄而异，雌性多为白色、灰色或暗灰色，雄性为金黄色。头部宽扁，口端位，吻端尖圆面钝。眼球外突，分上下两部分，下眼皮上有一个可折皱的瞬膜，可将眼闭合。背部略粗糙，有细微的肤棱。四肢粗壮，前肢短，无蹼。雄性个体第一趾内侧有一明显的灰色瘤状突起。后肢较长大，趾间有蹼。因其鸣声很大，远闻如牛叫而得名。

| 生物危害 |

牛蛙对本地两栖类动物造成威胁，甚至影响到生物多样性，如对昆明滇池本地鱼类造成的严重影响，就是典型案例。

| 地理分布 |

牛蛙原产于北美洲落基山脉以东，北到加拿大，南到佛罗里达州北部。我国于1959年从古巴、日本引进大陆养殖生产，全国各地均产。现主要集中在湖南、江西、新疆、四川、湖北等地。

| 传播途径 |

牛蛙可以人为携带传播或自然扩散。

| 管理措施 |

综合治理（Integrated Pest Management，IPM）或综合控制（Integrated Pest Control，IPC），但不包括植物卫生措施（Phytosanitary Measures）。

（幼蛙）

（图片选自upload.wikimedia.org、en.wikipedia.org、baike.baidu.com、www.fao.org）

牛蛙（*Rana catesbeiana*）

【www.cabi.org】 *Rana catesbeiana* (American bullfrog)

R. catesbeiana is not the largest frog species in the world but it is one of the top ten (and the largest true frog in North America) with a maximum body length slightly in excess of 200 mm (typical length 90-152 mm) and body weight up to 0.5 kg. Like most frogs, it undergoes a drastic metamorphosis during its life cycle, passing from a young aquatic life phase with branchial respiration, predominantly plankton feeding, iliophagous or herbivorous, to reach adult life as an animal with pulmonary and skin respiration and a carnivorous feeding habit (Teixeira et al., 2001). Bullfrog tadpoles are also very large by frog standards (80-150 mm) and can take from 12 to 48 months to reach metamorphosis. A bullfrog tadpole's body can be as large as a golf ball with a relatively long, high-finned and muscular tail. The colouration of the tadpole stage is brown to light olive with small black spots scattered across the head and upper body. At metamorphosis the tadpoles resorb their gills and finned tails while transforming into juvenile miniatures of adult bullfrogs but without secondary sexual characters. The colour of adults varies from olive, green or brownish on the dorsum with vague spots or blotches; the head is lighter green, and the legs blotched or banded; the eardrums are conspicuous. The hind feet are fully webbed. The skin is mostly smooth. There are no dorsolateral folds; a short fold extends from the eye over and past the eardrum to the forearm.

Bullfrogs become sexually dimorphic as they mature. Males develop yellow skin pigments on the chin and throat, and the ear covering (tympanic membrane) enlarges to several times the diameter of the eye. On the other hand, as females mature they tend to retain the superficial morphology and colouration of the juvenile stage, e.g. they lack yellow pigmentation on the chin and throat and the tympanic membrane remains about the same diameter as the eye.

Only adult male bullfrogs produce the advertisement call. They do this by trapping air between the lungs and the vocal pouches. The trapped air is forced back and forth over the larynx which generates the sound of the male's call (Gans, 1974). The vocal pouches are located on either side of the throat, just below the jaw hinge, and were thought until recently to be the primary source of amplification and broadcast of sound. However, Purgue (1997) has shown that more sound is emitted by the male's enlarged tympanic membrane than by the vocal pouches. A calling male has longitudinal folds of stretched throat skin that are aligned with the bones of the lower jaw beneath the angle of the mouth. The advertisement call of male bullfrogs communicates a variety of signals. It identifies the location of adult males in reproductive condition. It also attracts other males to form a lek, or calling aggregation (Emlen, 1976), and communicates territorial rights and neighbour-stranger discrimination (Boatright-Horowitz et al., 2000; Bee and Gerhardt, 2001; Bee, 2003).

The eggs of the American bullfrog are very small and appear black or dorsally black with a slightly lighter undersurface. Each egg is surrounded by a jelly capsule with additional jelly that creates a loose cohesion to the entire mass of eggs. A female bullfrog in her first year may produce a single egg mass of only a few hundred eggs, but as she increases in age her egg production increases as well. By her third year she will be capable of producing 20-30 000 (FAO, 2005). At least in some cases, there are two separate clutches produced at consecutive spawning events in a single season. A bullfrog egg mass can be anywhere from 20 cm to over 1 metre across and sits at the surface in order to facilitate oxygen diffusion. The eggs will hatch in 3 to 5 days (Bury and Whelan, 1984).

甲壳类

克氏原螯虾

学　　名：**_Procambarus clarkii_**（Girard）

异　　名：_Procambarus clarki_，_Scapulicambarus clarkii_

别　　名：美国螯虾、小龙虾

英文名：red swamp crayfish，crawdaddy，crawfish，red swamp，crayfish，red swamp，Louisiana crawfish

| 形态特征 |

成虾　克氏原螯虾由头胸和腹部共20节组成，除尾节无附肢外，共有附肢19对，体表具有坚硬的甲壳。

头胸部愈合成为头胸部。头部分为5节，胸部分为8节；头胸部呈圆筒形，前端有一额角，呈三角形。额角表面中部内陷，两侧隆脊，尖端呈锐刺状。头胸甲中部有一弧形颈沟，两侧具粗糙颗粒。

腹部共有7节，其后端有一扁平的尾节与第六腹节的附肢共同组成尾扇。胸足5对，第一对呈螯状、粗大；第二、第三对钳状，后两对爪状。腹足六对，雌性第一对腹足退化，雄性前两对腹足演变成钙质交接器。

性未成熟时，个体呈淡褐色、黄褐色、红褐色等，有时还见蓝色；性成熟后，个体呈暗红色或深红色。常见个体全长4.0~12cm，最大个体为16cm。体重雄性最大为101.7g，雌性最大为120g。

| 生物危害 |

克氏原螯虾破坏当地的原有生态系统。由于其繁殖快，将与原来的鱼类、泥鳅等争夺生存空间，并以幼鱼为食。还会啃食水稻等作物，影响粮食产量。

| 地理分布 |

克氏原螯虾原产于中、南美洲和墨西哥东北部地区。1918年引入日本。我国于1929年经日本引入。1960年代开始养殖，20世纪80~90年代大规模扩散。现分布于我国20多个省市，南起海南岛，北到黑龙江，西至新疆，东达上海，其中尤以华东、华南地区最为密集。

传播方式

人为传播或自然扩散。

| 管理措施 |

综合治理（Integrated Pest Management，IPM）或综合控制（Integrated Pest Control，IPC），但不包括植物卫生措施（Phytosanitary Measures）。

（图片选自neobiota.naturschutzinformationen-nrw.de，www.planetinverts.com，files.akva-tera.webnode.cz，en.wikipedia.org，www.aquarienkrebse.de，my-fish.org，wirbelworld.de，www.fao.org）

克氏原螯虾（*Procambarus clarkii*）

【www.cabi.org】*Procambarus clarkii*（red swamp crayfish）

P. clarkii is a decapod crustacean having two distinct body divisions（Huner and Barr, 1991）. There is the anterior cephalothorax with paired appendages including antennae, mouth parts, and walking legs and the major body organs and the posterior abdomen containing the major abductor muscles for rapid backward escape from real or perceived threats, and paired appendages associated with sperm transfer in males and incubation of eggs and developing embryos in females and the tail 'fan' in both sexes.

The intensity of coloration is dependent on the habitat. Colours are darkest in clear, acid-stained waters and lightest in opaque, muddy waters. The common name for P. clarkii is red swamp crawfish. This common name derives from the red coloration associated with the lateral body surfaces and the appendages. Prior to reaching maturity, the dominant coloration of P. clarkii is greenish-brown with intensity dictated by water clarity. However, red pigment can generally be detected on appendages, especially where walking legs join the body.

P. clarkii exhibits distinct secondary sexual characteristics once the species has reached maturity. Males have elongated, inflated chelae, distinct ischial hooks at the bases of the third and fourth pairs of walking legs and the tips of the gonopods, the first pair of abdominal appendages cornify and assume a very distinct morphology that is species specific in cambarid crawfishes. Chelae of females inflate somewhat as well and the sperm receptacle, located between the walking legs, cornifies and develops a species-specific morphology that is receptive only to the species-specific terminal ends of the male gonopods.

Size at maturation ranges from 5-10 g to as much as 50-60 g with total lengths of 5.5-6.5 cm and 10.5-11.5 cm. The common commercial size is 10-30 g or 7.5-10.5 cm. Therefore, size is not a satisfactory criterion for ascertaining the maturation status or age of this species, as well as any other freshwater crawfish for that matter.

【www.planetinverts.com】*Procambarus clarkii*

【en.wikipedia.org】*Procambarus clarkii*

【animaldiversity.org】*Procambarus clarkii*

爬行类

巴西红耳龟

学　　名：***Trachemys cripta elegans***（Wied.）

异　　名：*Chrysemys elegans* Boulenger，*Chrysemys palustris elegans* Lindholm，*Chrysemys scripta* var. *elegans* Boulenger，*Clemmys elegans* Strauch，*Emys elegans* Wied，*Emys holbrookii* Gray，*Emys sanguinolenta* Gray，*Pseudemys elegans* Cop，*Pseudemys scripta elegans* Cagle，*Pseudemys troostii elegans* Stejneger & Barbour，*Trachemys elegans* Agassiz，*Trachemys holbrookii* Gray，*Trachemys lineata* Gray，*Trachemys scripta elegans* Iverson，*Trachemys scripta elagans* Fong，Parham & Fu（ex errore）

别　　名：巴西龟、红耳龟、可爱龟、秀丽锦龟、红耳彩龟、草草龟、强生龟、麻将龟、七彩龟、红耳滑板龟、密西西比红耳龟

英文名：red-eared slider

| 形态特征 |

成龟　巴西红耳龟的头部宽大，吻钝，头颈处具有黄绿相镶的纵条纹，眼后有一对红色条纹，即耳朵，因此而得名。其背甲扁平，为翠绿色或苹果色，背部中央有条显著的脊棱。盾片上具有黄、绿相间的环状条纹。腹板淡黄色，具有左右对称的不规则黑色圆形、椭圆形和棒形色斑。四肢淡绿色，有灰褐色纵条纹，指、趾间具蹼。

成年雄性个体足的前端具有伸长并弯曲的爪，且位于长而粗的尾部的肛门可显露在臀盾之外。

| 生物危害 |

巴西红耳龟已被国际自然保护联盟收录为世界最严重的100种入侵物种之一。巴西红耳龟排挤本地物种，对入侵地的本土龟造成严重威胁。

巴西红耳龟还是沙门氏杆菌传播的罪魁祸首。在美国，每年有100万~300万的人感染此病菌，其中有14%的病例是由龟类传染所引起。

| 地理分布 |

巴西红耳龟原产于美国中南部，沿密西西比河至墨西哥湾周围地区。巴西红耳龟已经在除南极洲之外的所有大洲上发现有野生个体的存活，并且已在欧洲、非洲、澳洲、亚洲和美国原产地以外

的美洲等世界范围内成功入侵。目前已经分布于日本、智利、南非、意大利、韩国、泰国、澳大利亚、德国、法国、以色列、英国、西班牙、巴哈马、百慕大、巴西、哥斯达黎加、密克罗尼西亚共和国、新西兰、波兰、越南、孟加拉国、印度、丹麦、加拿大等国家。

20世纪80年代，巴西红耳龟经香港引入广东，继而迅速流向全国。由于宠物丢弃，养殖逃逸，错误放生等，导致其在野外普遍发生，致使我国成为世界上巴西红耳龟种群数量最多的国家。尤其以人口较为集中的城市周边水域为主分布。目前野外分布最北端为辽宁沈阳市，南端为海南五指山市，最西端为云南高黎贡山自然保护区，包括河北、河南、陕西、辽宁、四川、湖北、湖南、江西、安徽、山东、山西、江苏、浙江、福建、海南、广东、广西、上海18个省（区、市），均有巴西红耳龟分布。

| 传播途径 |

人为传播或自然扩散。

| 管理措施 |

综合治理（Integrated Pest Management，IPM）或综合控制（Integrated Pest Control，IPC），但不包括植物卫生措施（Phytosanitary Measures）。

雌性爪

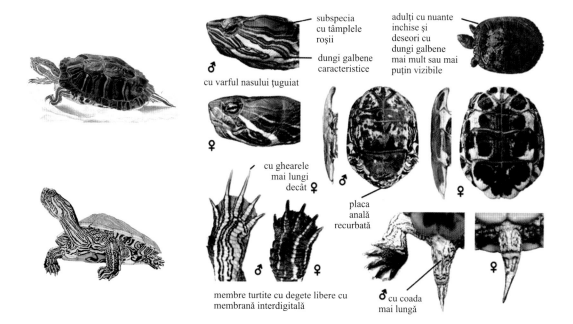

(图片选自commons.wikimedia.org，en.wikipedia.org，upload.wikimedia.org，www.tartarugando.it，thumbs.dreamstime.com)

巴西红耳龟（*Trachemys cripta elegans*）

【en.wikipedia.org】Red-eared slider

The red-eared slider belongs to the order Testudines, which contains about 250 turtle species. It is a subspecies of *Trachemys scripta*. They were previously classified under the name *Chrysemys scripta elegans*.

The species *Trachemys scripta* contains three subspecies: *T. s. elegans* (red-eared slider), *T. s. scripta* (yellow-bellied slider), and *T. s. troostii* (Cumberland slider).

The carapace of this species can reach more than 40 cm (16 inches) in length, but the average length ranges from 15 to 20 cm (6 to 8 inches). The females of the species are usually larger than the males. They typically live between 20 and 30 years, although some individuals have lived for more than 40 years. Their life expectancy is shorter when they are kept in captivity. The quality of their living environment has a strong influence on their lifespans and well being.

The shell is divided into two sections: the upper or dorsal carapace, and the lower, ventral carapace or plastron. The upper carapace consists of the vertebral scutes, which form the central, elevated portion; pleural scutes that are located around the vertebral scutes; and then the marginal scutes around the edge of the carapace. The rear marginal scutes are notched. The scutes are bony keratinous elements. The carapace is oval and flattened (especially in the male) and has a weak keel that is more pronounced in the young. The color of the carapace changes depending on the age of the turtle. The carapace usually has a dark green background with light and dark, highly variable markings. In young or recently hatched turtles, it is leaf green and gets slightly darker as a turtle gets older, until it is a very dark green, and then turns a shade between brown and olive green. The plastron is always a light yellow with dark, paired, irregular markings in the centre of most scutes. The plastron is highly variable in pattern. The head, legs, and tail are green with fine, irregular, yellow lines. The whole shell is covered in these stripes and markings that aid in camouflaging an individual.

Turtles also have a complete skeletal system, with partially webbed feet that help them to swim and that can be withdrawn inside the carapace along with the head and tail. The red stripe on each side of the head distinguishes the red-eared slider from all other North American species and gives this species its name, as the stripe is located behind the eyes where their (external) ears would be. These stripes may lose their color over time.Some individuals can also have a small mark of the same color on the top of their heads. The red-eared slider does not have a visible outer ear or an external auditory canal; instead, it relies on a middle ear entirely covered by a cartilaginous tympanic disc.

The main internal organs of these reptiles are the lungs, heart, stomach, liver, intestines, and the urinary bladder. A cloaca serves both excretory and reproductive functions.

【www.cabi.org】*Trachemys scripta elegans* (red-eared slider)

Adult: The red-eared slider (*Trachemys scripta elegans*) is a medium-sized freshwater turtle (carapace length: 125 to 289 mm; Somma & Fuller 2009; 150 - 350 mm; Obst 1983) It is characterised by prominent yellow to red patches (typically red) on each side of the head (Scalera 2006). Carapace and skin are olive to brown with yellow stripes or spots; males are usually smaller than females and have a long, thick tail (Scalera 2006).

Eggs: The eggs are ovoid in shape, 31 to 43 millimetres long, 19 to 26 millimetres wide and weigh 6.1 to 15.4 grams (Bringsøe 2006).

【www.arkive.org】Yellow-bellied slider turtle (*Trachemys scripta*)

鱼 类

豹纹脂身鲇

学　名：*Pterygoplichthys pardalis*（Castelnau）
异　名：*Acipenser plecostomus*，*Hypostomus plecostomus*，*Hypostomus pardalis*，*Hypostomus punctatus*，*Liposarcus jeanesianus*，*Liposarcus multiradiatus*，*Liposarcus pardalis*，*Liposarcus varius*，*Plecostomus plecostomus*
别　名：清道夫、琵琶鼠、垃圾鱼、豹纹翼甲鲶
英文名：Common Pleco，Albino Pleco，Amazon Sailfin Pleco，Chocolate Pleco，Janitor Fish，Plecostomus，Pleco，Common Plecostomus，Algea Eater，Amazon Sailfin Catfish

| 形态特征 |

成鱼　体长25～40cm，呈半圆筒形，侧宽，背鳍宽大，尾鳍呈浅叉形。体暗竭色、全身灰黑色带有黑白相间的花纹，布满黑色斑点，表面粗糙有盾鳞。头部和腹部扁平，左右两边腹鳍相连形成圆扇形吸盘，须1对。雌性背体宽，倒刺软而柔滑，体色较淡不发黑，胸鳍短而圆；雄性背体狭，倒刺硬而粗糙，体色较深发黑，胸鳍长而尖且具追星。从腹面看，很像一个小琵琶，故又称为琵琶鱼。

| 生物危害 |

成年豹纹脂身鲇食量巨大，除了青苔等藻类，它还会以其他鱼类的鱼卵为食，一天可以吃掉3 000～5 000粒鱼卵，也会吞食鱼苗。由于是外来入侵物种，目前在国内还没有天敌，因而在江河中很容易大量繁殖，从而威胁本地鱼类生存，破坏水生生物生态系统。

| 地理分布 |

豹纹脂身鲇原产于南美洲，广泛分布于亚马逊河流域。1980年，豹纹脂身鲇作为观赏鱼，引入我国。现广布于广东、湖北、台湾、广西、陕西、四川、重庆、江苏、江西、海南、安徽、上海、浙江、福建、云南、吉林等地。

| 传播途径 |

人为传播或自然扩散。

| 管理措施 |

综合治理（Integrated Pest Management，IPM）或综合控制（Integrated Pest Control，IPC），但不包括植物卫生措施（Phytosanitary Measures）。

（图片选自www.planetcatfish.com，www.fishbase.org，nas.er.usgs.gov，rybicky.net，www.ncfishes.com，www.pisces.at，upload.wikimedia.org）

豹纹脂身鲇（*Pterygoplichthys pardalis*）

【www.planetcatfish.com】*Pterygoplichthys pardalis* (Castlenau, 1855)

Pterygoplichthys can be identified by the number of rays in the dorsal fin. More than 10 indicates that it's a *Pterygoplichthys*. Most other plecos have 8 or fewer rays (in particular the larger *Hypostomus* species that are most likely to be confused with *Pterygoplichthys*).

This species is most likely to be confused with *P. disjunctivus*. The pattern on the abdomen is the most useful external key: *P. pardalis* has a spotted pattern, where *P. disjunctivus* has a vermiculated pattern.

Comparison of the genital papilla in mature fish shows the differences in the sexes to the trained eye. In males this is a small yet thick stump which noticeably protrudes from the fish's undercarriage. In females it is less obvious and is recessed or lies flat with the body.

It is important to note that the term "common pleco" is applied to several nondescript loricariid species, especially of the genera *Pterygoplichthys* and *Hypostomus*.

【nas.er.usgs.gov】*Pterygoplichthys pardalis* (Castelnau, 1855)

Weber (1991, 1992) assigned sailfin catfishes to three genera and used the name *Liposarcus pardalis* for this species. Armbruster (1997), after a detailed systematic review, placed the genus *Liposarcus* into the synonymy of *Pterygoplichthys*. Weber (1992) provided a key and distinguishing characteristics and photographs of specimens; Armbruster and Page (2006) present a revised key to species in the genus *Pterygoplichthys* (except *P. ambrosettii*).

Pterygoplichthys and other *suckermouth* armored catfishes (family Loricariidae) can be distinguished from native North American catfishes (Ictaluridae) by the presence of flexible bony plates (absent in *ictalurids*) and a ventral suctorial mouth (terminal in *ictalurids*). *Pterygoplichthys* is often confused with *Hypostomus*: these genera can be distinguished by the number of dorsal fin rays (7-8 in *Hypostomus* vs. 9-14 in *Pterygoplichthys*).

Size: generally to 50 cm TL.

腹锯鲑脂鲤

学　　名：***Pygocentrus nattereri*** Kner
异　　名：*Pygocentrus altus* Gill，*Serrasalmo ternetzi* Steindachner，*Serrasalmus nattereri*（non Günther），*Serrasalmus nattereri*（Kner）
别　　名：食人鲳、食人鱼、水虎鱼
英文名：red-bellied piranha，red pirahna

| 形态特征 |

成鱼　体呈卵圆形，体型侧扁，颈部短。体长10～30cm。体色多变，取决于年龄和生活环境状况。背部从蓝灰色到棕灰色；体侧淡棕色到微橄榄色，并散布银色具金属光泽的小点；身体下部，包括胸鳍和腹鳍，呈淡红色到血红色；背鳍和尾鳍外缘黑色，内侧微白；臀鳍红色具黑缘。背鳍16～18cm，臀鳍28～32cm，尾鳍顶端微凹呈叉形。头大，吻端钝；腭强健，下颚发达，有锐利的牙齿，牙齿为三角形，呈锯齿状排列，作剪刀状咬合。

| 生物危害 |

腹锯鲑脂鲤，又称食人鲳。其性情十分凶猛残暴，常常能将比自己体积大几倍甚至几十倍的动物在数秒内吞噬，尤其是，成群的食人鲳还攻击人类，常将误入水中的人咬伤乃至将其吃掉。还破坏现有鱼类区系，降低生物多样性，对渔业造成重大损失。

| 地理分布 |

腹锯鲑脂鲤原产于南美洲亚马逊河流域。目前分布于巴西、阿根廷、玻利维亚、哥伦比亚、巴拉圭、秘鲁、美国、孟加拉国等国。20世纪80年代初，红腹锯鲑脂鲤被作为观赏鱼引入我国。目前分布于广东、广西、浙江、四川、湖南、江西、北京、天津、辽宁、吉林、福建、海南、台湾等地。

| 传播途径 |

人为传播或自然扩散。

| 管理措施 |

综合治理（Integrated Pest Management，IPM）或综合控制（Integrated Pest Control，IPC），但不包括植物卫生措施（Phytosanitary Measures）。

（图片选自www.zoodia.it，www.arbioitalia.org，seriouslyfish.com，commons.wikimedia.org，www.biopix.dk，www.arkive.org，www.akvarieforum.dk）

红腹锯鲑脂鲤（*Pygocentrus nattereri*）

【en.wikipedia.org】Red-bellied piranha

The red-bellied piranha has a popular reputation as a ferocious predator, despite being primarily a scavenger. As their name suggests, red-bellied piranhas have a reddish tinge to the belly when fully grown, although juveniles are a silver color with darker spots. The largest measured individuals were around 50 centimetres (20 inches) in length and weighed around 3.9 kilograms. The rest of the body is often grey with silver-flecked scales. Sometimes, blackish spots appear behind the gills and the anal fin is usually black at the base. The pectoral and pelvic fins may vary from red to orange. Females can be distinguished from males by the slightly deeper red color of their bellies.

The red-bellied piranha is typically found in white water rivers, such as the Amazon River Basin, and in some streams and lakes. Sometimes, they may inhabit flooded forests such as those found in the Brazilian Amazon. They live in shoals but do not do group hunting behavior, although they may occasionally enter into feeding frenzies. In the case of a feeding frenzy, schools of piranha will converge on one large prey individual, and eat it within minutes. These attacks are usually extremely rare and are due to provocation or starvation. Breeding occurs over a two-month period during the rainy season, but that can vary by area. Females will lay around 5 000 eggs on newly submerged vegetation in nests that are built by the males.

【animal-world.com】Red-bellied Piranha

The Red-bellied Piranha can get up to 13 inches (33 cm) in length in the wild, though in captivity they are generally smaller. A lifespan of 10 years is normal, but a few have lived for over 20 years.

These fish have powerful bodies that are high, thick, and laterally compressed. Like all piranhas, they have a keel-like edge that runs along the upper body from head to dorsal fin and along the belly on the lower body. Members of the genus Pygocentrus are all recognizable by the convex shape of their head and massive, bulldog-like lower jaw. With a large, powerful tail and a streamlined body covered with tiny scales, they are very fast and agile swimmers. They also have a small adipose fin between the tail and dorsal fin, a characteristic of all Characins.

The Red Belly Piranha is gorgeous in its adult coloring. Body colors can be variable, but mostly the back is a steel gray and the rest of the body is a silvery gold with a bright orangish-red or red colored throat, belly and anal fin. It has large black spots on the sides, though they often fade with age, and it sparkles with many shiny scales. In its juvenile form it is more silver colored with dark spots.

The adult Red-bellied Piranha has gorgeous coloring. Colors can be variable, but usually the back is a steel gray and the rest of the body is a silvery-gold with a bright orangish-red or red throat, belly, and anal fin. It has large black spots on its sides, though they often fade with age, and it sparkles with many shiny scales. In its juvenile form it is more silver colored with dark spots.

Some individuals have such intense gold-speckling that they are sometimes called Gold-dust Piranha. There are also two similar species sometimes available in the aquarium hobby. These include the San Francisco Piranha Pygocentrus piraya, which is a yellow-bellied species. Another is the Black Spot Piranha Pygocentrus cariba, which still has the red on its throat and belly, but has a silvery colored body accented with a strong black spot just behind the gill.

尼罗罗非鱼

学　　名：***Oreochromis niloticus***（L.）

异　　名：*Chromis guentheri* Steindachner，*Chromis nilotica*（Linnaeus），*Chromis niloticus*（Linnaeus），*Oreochromis nilotica* Linnaeus，*Oreochromis niloticus niloticus*（Linnaeus），*Perca nilotica* Linnaeus，*Sarotherodon niloticus*（Linnaeus），*Tilapia calciati* Gianferrari，*Tilapia nilotica* Linnaeus，*Tilapia nilotica nilotica*（Linnaeus），*Tilapia nilotious*（Linnaeus）

别　　名：罗非鱼、吴郭鱼、非鲫

英文名：Nile tilapia，cichlid，edward tilapia，mango fish，mozambique tilapia，nilotica，tilapia，Nile

| 形态特征 |

成鱼　体长卵圆形，侧扁，尾柄较短。头略大，背缘稍凹。吻钝尖，吻长大于眼径。口端位。上、下颌几乎等长，上颌骨为眶前骨所遮盖；上、下颌齿细小，3行。眼中等大，侧上位；眼间隔平滑，显著大于眼径。鼻孔细小。前鳃盖骨边缘无锯齿，鳃盖骨无棘；鳃耙细小，基部较宽，末端尖锐。下咽骨密布细小齿群。体侧有9~10条黑色的横带，成鱼较不明显。背鳍发达，起点于鳃盖后缘相对，终止于尾柄前端，硬棘16~17，软条12~13，背鳍鳍条部有若干条由大斑块组成的斜向带纹，鳍棘部的鳍膜上有与鳍棘平行的灰黑色斑条，长短不一。臀鳍末端超过尾柄，硬棘3，软条9~11；胸鳍较长，可达到或超过腹鳍末端，无硬刺，软条14~15；腹鳍胸位，硬刺1，软条15；臀鳍鳍条部上半部色泽灰暗，较下部为甚；尾鳍6~8条近于垂直的黑色条纹，尾鳍末端钝圆形。雄鱼的背鳍和尾鳍边缘有1条狭窄的灰白色带纹。

| 生物危害 |

尼罗罗非鱼适生能力强，杂食性，生长迅速，繁殖力强，有强烈的领域和护幼行为。通过竞争、捕食等作用对本地鱼类威胁很大。在我国广东、广西、海南等省区的一些河流中，已形成地方优势种群，造成本地鱼类减少或消失，物种多样性趋于单调。

| 地理分布 |

尼罗罗非鱼原产于尼罗河流域。目前有塞内加尔、冈比亚、尼日尔、乍得等100多个国家和地区有养殖记录。

1978年我国长江水产研究所首次从尼罗河引进后，尼罗罗非鱼迅速在全国各地推广养殖，成为罗非鱼养殖的主要品种。广东、广西、海南、福建、台湾等地已经形成能够越冬的自然群体。全国养殖范围除上海、青海、宁夏三省（区、市）无罗非鱼生产记录外，其他地区（包括台湾）均有罗非鱼养殖。

|传播途径|

人为传播或自然扩散。

|管理措施|

综合治理（Integrated Pest Management，IPM）或综合控制（Integrated Pest Control，IPC），但不包括植物卫生措施（Phytosanitary Measures）。

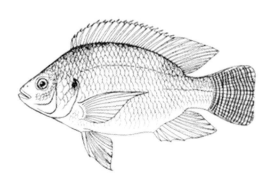

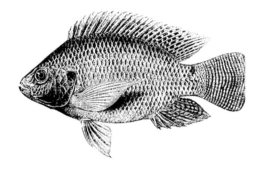

（图片选自www.cabi.org，www.fao.org）

尼罗罗非鱼（*Oreochromis niloticus*）

【 www.fishbase.org 】 *Oreochromis niloticus* (Linnaeus, 1758)

Dorsal spines (total) : 15 - 18; Dorsal soft rays (total) : 11-13; Anal spines: 3; Anal soft rays: 9 - 11; Vertebrae: 30 - 32. Diagnosis: jaws of mature male not greatly enlarged (length of lower jaw 29%-37% of head length) ; genital papilla of breeding male not tassellated. Most distinguishing characteristic is the presence of regular vertical stripes throughout depth of caudal fin.

【 zipcodezoo.com 】 *Oreochromis niloticus niloticus*

The Nile tilapia has distinctive, regular, vertical stripes extending as far down the body as the bottom edge of the caudal fin, with variable coloration. Adults reach up to 60 cm (24 inches) in length and up to 4.3 kg. It lives for up to 9 years. It tolerates brackish water and survives temperatures between 8℃ and 42℃ (46 °F and 108 °F) . It is an omnivore, feeding on plankton as well as on higher plants. Introduced tilapia can easily become an invasive species (see Tilapia as exotic species) . It is a species of high economic value and is widely introduced outside its natural range; probably next to the Mozambique tilapia (*O. mossambicus*) , it is the most commonly cultured cichlid.In recent research done in Kenya, this fish has been shown to feed on mosquito larvae, making it a possible tool in the fight against malaria in Africa.

【 www.iucngisd.org 】 Nile tilapia

Compressed body; caudal peduncle depth equal to length. Cycloid scales. Lacks knobby feature on dorsal surface of snout. Sexual dimorphism not displayed in upper jaw length. First gill arch has 27 to 33 gillrakers. Interrupted lateral line.Spinous and soft ray parts of dorsal fin continuous; dorsal fin with 16 - 17 spines and 11 to 15 soft rays. Anal fin with 3 spines and 10-11 rays. Caudal fin truncated. Colour in spawning season: pectoral, dorsal and caudal fins becoming reddish; caudal fin with numerous black bars (FAO, 2006; FAO, 2007) .

食蚊鱼

学　名：***Gambusia affinis*** Baird et Girard

异　名：*Gambusia affinis affinis* Baird and Girard，*Gambusia affinis katruelis* Biard and Girard，*Gambusia gracilis* Girard，*Gambusia holbrooki* Girard，*Gambusia humilis* Gunher，*Gambusia patruelis* Biard and Girard，*Haplochilus melanops* Cope，*Heterandria affini* Biard and Girard，*Heterandria patruelis* Biard and Girard，*Zygonectes brachypterus* Cope，*Zygonectes gracilis* Girard，*Zygonectes inurus* Jordan and Gilbert，*Zygonectes patruelis* Biard and Girard

别　名：柳条鱼、大肚鱼、山坑鱼、大眼叮当、白头婆、大肚捆

英文名：western mosquitofish，mosquito fish，mosquitofish

|形态特征|

成鱼　食蚊鱼体形细小，形似柳条，故又称柳条鱼。头背平而宽，前方略尖。背部略隆起，呈微弧形。腹部圆。两个背鳍连在一起，起点在胸鳍基部上方，较腹鳍的起点稍靠后，鳍棘部和鳍条部有一个较深的缺刻，背鳍前部的鳍棘上缘为黑色，后部鳍条中央有一条纵行黑色条纹；胸鳍尖长，末端可达腹鳍上方，比腹鳍长，基部黄色，边缘黑色，胸鳍腋部上方一个暗斑；腹鳍末端可达肛门；臀鳍在起点的垂线之前，长有2条棘；尾鳍不分叉，呈楔形。背鳍条7～9，胸鳍条13～14，腹鳍条6，臀鳍条9～10。雄鱼第3～5臀鳍条很长，演化成交接器（生殖鳍）。鱼体背部呈现橄榄褐色，体侧大部分呈半透明灰色，并常有光亮的淡蓝色，腹部为银白色。在背鳍和尾鳍部经常分布一些不很明显的小黑点。雌鱼怀胎时在臀鳍上部有一明显黑斑，称为胎斑。

|生物危害|

食蚊鱼通过捕食浮游动物、土著鱼类的鱼卵或鱼苗、两栖类的卵或幼体，造成当地部分土著水生动物物种濒危或灭绝，进而改变入侵地水生物种群落结构，影响水生生态系统功能。因此，食蚊鱼已经被纳入由国际自然保护联盟公布的全球最严重的100外来入侵种名录中。

|地理分布|

食蚊鱼主要产于美国南方、中美洲和西印度群岛。为控制蚊子和疟疾的发生，食蚊鱼分别于1913年和1924年被引入到我国台湾和大陆。20世纪中后期，为提高渔业产量，我国云南省众多高原湖泊大量引种外来鱼类，食蚊鱼也随着其他鱼类引种被无意引入。目前，食蚊鱼在长江以南（包括台湾）的各地小水体中均有分布，同时在云南高原湖泊中亦广泛分布。

|传播途径|

人为传播或自然扩散。

| 管理措施 |

综合治理（Integrated Pest Management，IPM）或综合控制（Integrated Pest Control，IPC），但不包括植物卫生措施（Phytosanitary Measures）。

（图片选自www.hunter.cuny.edu，www.akvaryum.biz，ninnescahlife.wichita.edu，www.fishbase.org，tropicalfishandaquariums.com，vecteurs.cqeee.org，natural-history.main.jp，www.wpclipart.com）

食蚊鱼（*Gambusia affinis*）

【www.cabi.org】*Gambusia affinis*（western mosquitofish）

G. affinis is a small, stout, robust, dull grey to brown with a terminal and upward pointing mouth adapted for feeding at the surface of the water. It has a small, rounded dorsal fin that originates behind the anal fin. Dorsal fin rays are 7-9 in number, and anal rays are 9-10（FishBase, 2004）. Origin of seventh dorsal fin is opposite anal ray. There are 8 horizontal scale rows between back and abdomen. The first few rays of the anal fin are greatly elongated in adult males. Mature females are larger than males. The maximum total length reported for male and female is 4.0 cm（Billard, 1997）and 7.0 cm respectively（FishBase, 2004）.

【en.wikipedia.org】Mosquitofish

Mosquitofish are small and of a dull grey coloring, with a large abdomen, and have rounded dorsal and caudal fins and an upturned mouth. Sexual dimorphism is pronounced; mature females reach a maximum overall length of 7 cm（2.8 inches）, while males reach only 4 cm（1.6 inches）. Sexual dimorphism is also seen in the physiological structures of the body. The anal fins on adult females resemble the dorsal fins, while the anal fins of adult males are pointed. This pointed fin, referred to as a gonopodium, is used to deposit milt inside the female. Adult female mosquitofish can be identified by a gravid spot they possess on the posterior of their abdomens. Other species considered similar to *G. affinis* include *Poecilia latipinna*, *Poecilia reticulata*, and *Xiphophorus maculatus*; it is commonly misidentified as the eastern mosquitofish.

【nas.er.usgs.gov】*Gambusia affinis*（Baird and Girard, 1853）

Mosquitofish is a small, live-bearing fish, is dull grey or brown in color with no bars of bands on the sides, and has a rounded tail. Its body is short, its head flattened, and its mouth pointed upward for surface feeding. Distinguishing characteristics were provided by Rauchenberger（1989）and Page and Burr（1991）（although the latter authors treated the two forms as subspecies）. *Gambusia affinis* and *G. holbrooki* were long considered subspecies of *G. affinis*, and were only recently recognized as separate species（Wooten et al., 1988; Rauchenberger 1989; Robins et al., 1991）. Complicating matters of identification, most introductions occurred before the recent taxonomic change; furthermore, the origins of introduced stocks were usually unknown or unreported. In addition, both forms were widely available and thought to have been dispersed widely by humans. As a consequence, it often is not possible to determine if many of the earlier records represent introductions of *G. affinis* or of *G. holbrooki*.

【www.iucngisd.org】*Gambusia affinis*

A stout little fish, the back a little arched in front of the dorsal fin and the belly deep in front of the anal. The head is large with a flattened upper surface, the mouth small, upturned and protrusible, and not reaching as far back as the front of the eyes. The eyes are very large relative to the body. The single, soft-rayed dorsal fin is short-based, high and rounded, while the caudal peduncle is long, deep and compressed, and the caudal fin is rounded. The head and trunk are covered with large scales and there is no lateral line. The back is a greenish olive to brownish, the sides grey with a bluish sheen, and the belly a silvery white. A well-defined black spot on the upper rear abdomen is surrounded by a golden patch above and behind the vent. In mature females there is also a black patch above and somewhat forward of the vent. The ventral surface of the head is a steely blue with a diagonal chin stripe below the eyes. The eyes are greyish to olive, the dorsal fin has small black spots, and the caudal fin has several indistinct cross rows of small black spots. The anal, pelvic and pectoral fins are a translucent pale amber. (McDowall, 1990). Males grow to 40mm in length, while females reach 70mm long (FishBase, 2003).

【eol.org】*Gambusia affinis*

The mosquitofish, *Gambusia affinis*, is a small, robust-bellied fish. The relatively large head is flattened on the upper surface, and the small mouth is superior (upturned) and protrusible. The eyes are large relative to the body. Dorsal and caudal fins are rounded and no lateral line is visible (IGGS 2006). The body is usually greenish olive to brown above, grey-blue on the sides, and silvery-white below. The body has a characteristic diamond or net pattern on the body formed by dark pigment at the scale margins. Small black dots are also usually present on the body and tail. A small dark bar below the eye also aids in identification and a black peritoneum (abdominal cavity lining) can often be observed through the belly in living specimens. Melanistic (black or nearly black) individuals are common in some populations. The dorsal fin is single and has only soft rays. Individuals typically have 7-9 soft dorsal rays, 8-10 anal fins, and 29-32 lateral line scales (Hoese and Moore, 1977, Robins et al., 1986, IGGS 2006). The species is sexually dimorphic, with adult males being considerably smaller than females and also possessing a gonopodium-an elongated anal fin that functions as an intromittent organ for sperm transfer during mating. Mature females have a distinct gravid spot located on the posterior abdomen above the rear of the anal fin (Hoese and Moore 1977, GLAVCD undated).

【ninnescahlife.wichita.edu】Western Mosquitofish

Adult Diagnosis: Adults have bodies that are light olive or a dull grey/brown. They have slightly speckled fins without bars or bands and rounded tails. Mosquito fish are characterized by short bodies and flat heads with terminal mouths pointing upward. Females have a rounded, small anal fin and grow to about 6-7centimeters. Males grow to about 4 centimeters and have an elongated intromittent organ (gonopodium).

Juvenile Diagnosis: Juveniles resemble adults in miniature.

【txstate.fishesoftexas.org】*Gambusia affinis*

Maximum size: Female, 53 mm SL (Hughes 1985b); male, 28 mm SL (Hughes 1985).

Coloration: Top of head dusky; bluish black spot below eye; with a conspicuous dark, elongate spot between anterior and posterior nares; back olivaceous with scattered black spots and a few scattered fine black spots; abdomen silvery to whitish, with an anteriorly directed dark triangle above anal origin; post anal region with a dark streak; black pigment surrounding anus and urogenital opening most conspicuous in females with eggs in ovaries; this pigment variable at other times (Peden 1973; Sublette et al., 1990). Fins dusky with fine melanophores; dorsal fin with 1 or 2 irregular rows; caudal fin with a partial, vertical row of small, dark spots. Scales edged in black. Young yellowish (Sublette et al., 1990). No dark band on sides; median fins without large black spots near their bases (Hubbs et al., 1991).

Counts: 6 (rarely 7) dorsal fin rays (Hubbs et al., 1991); 28-31 lateral line scales; 9-10 anal rays; 11-14 pectoral rays (Ross, 2001).

Body shape: Terete. Back nearly straight in profile. Female markedly larger and heavier bodied than the male (Sublette et al., 1990).

Mouth position: Terminal (Goldstein and Simon, 1999).

External morphology: Pectoral fins much larger than pelvic fins. Pelvic fins small, ovate. Anal ovate in female; modified into gonopodium in male. Caudal truncate to slightly rounded (Sublette et al., 1990). Distal end of the fourth fin ray of gonopodium in male parallel or curved in only a weak arch; spines at tip of third anal fin ray of male gonopodium (first enlarged ray) one to three times longer than wide; origin of dorsal fin well behind origin of anal fin (Hubbs et al., 1991).

Internal morphology: Teeth barely moveable (Hubbs et al., 1991), in broad villiform bands (Sublette et al., 1990). Peritonium black (Sublette et al., 1990); intestinal canal short with few convolutions (Hubbs et al., 1991).

【近似种】东方食蚊鱼（Eastern mosquitofish）

学　　名：*Gambusia holbrooki* Girard, 1859

异　　名：*Gambusia affinis holbrocki*（Girard, 1859），*Gambusia affinis holbrooki*（Girard, 1859），*Gambusia holbrookii* Girard, 1859，*Gambusia patruelis holbrooki*（Girard, 1859），*Heterandria holbrooki*（Girard, 1859），*Heterandria uninotata*（non Poey, 1860），*Schizophallus holbrooki*（Girard, 1859），Zygonectes atrilatus Jordan & Brayton, 1878

英文名：eastern mosquitofish，mosquitofish，mosquito-fish，gambies，gambusia，bore-drain Fish，plague Minnow，starling's Perch，top Minnow

【fishesofaustralia.net.au】Eastern Gambusia, *Gambusia holbrooki* Girard 1859

Dorsal-fin rays 6-8（usually 7）; Anal-fin rays 9-11（usually 10）; Vertebrae 31-33; Gill rakers 13-15 Body stout with a deep rounded belly（particularly in females, and more so when carrying a brood of young）; upper surface flattened, especially the head; mouth small, upturned and protrusible, lower jaw a little longer than upper; eyes large. Large scales cover head and trunk（28-32, usually 30-31 along side）; no lateral line. Dorsal fin single, sort-based, soft-rayed, positioned well back on trunk, fin high and rounded; anal fin small and rounded in females, elongate and modified into a gonopodium in males; caudal fin rounded.

Females to 6 cm SL; males to 3.5 cm SL.

Generally greenish olive to brownish on back, sides grey with a bluish sheen, belly silvery; some dark speckling dorsally and on caudal fin; females with a large black blotch surrounded by a golden patch just above vent.

【www.iucngisd.org】*Gambusia holbrooki*

Greenish olive to brown on the back, the sides are grey with a bluish sheen with a belly silvery-white. Females have a distinct black blotch surrounded by a golden patch occurring just above the vent. Males have a highly modified anal fin, the third, fourth and fifth rays of which are elongated and thickened to form a "gonopodium" which is used to inseminate the female. Females are also larger than males with maximum standard lenghts of 60mm and 35mm respectively.

【www.cabi.org】*Gambusia holbrooki*（eastern mosquitofish）

The male *G. holbrooki* is about 35 mm standard length whereas the female is larger（up to 60 mm）with a deeper body, the anal fin unmodified and when pregnant, a gravid spot is visible just above the vent（Lloyd, 1987）. The fish are mostly translucent grey with a bluish sheen on their sides with a silver belly（Lloyd, 1987）. The fins are colourless, with transverse rows of black spots. Some male mosquitofish have irregular black blotching, though some largely melanistic male individuals exist but are uncommon in their native range（Sterba, 1962）and are absent from Australia（Lloyd, 1987）. On the male, the anal fin is modified to form a long, thin intromitent organ, the gonopodium, used for sperm transfer（Lloyd, 1990c）. The body is slightly compressed with a large and flattened head. The eyes are large, and the mouth is small and terminal（Lloyd, 1987）.

(图片选自www.fishbase.org，fishesofaustralia.net.au，www2.fcps.edu，nas.er.usgs.gov，www.dpi.nsw.gov.au，jonahsaquarium.com，www.ittiofauna.org）

东方食蚊鱼（*Gambusia holbrooki*）

主要参考文献

（荷）查尔斯·埃尔顿. 2014. 张润志，任立等译. 动植物入侵生态学[M]. 北京：中国环境科学出版社.

（英）大卫·爱博思. 2012. 鄢建，王洪兵，张艺兵编译. 植物卫生与检疫原理[M]. 北京：中国农业科学技术出版社.

陈仲梅，齐桂臣. 1999. 拉英汉农业害虫名称[M]. 北京：科学出版社.

环境保护部自然生态保护司. 2012. 中国自然环境入侵生物[M]. 北京：中国环境科学出版社.

李志红，杨汉春，沈佐锐. 2004. 动植物检疫概论[M]. 北京：中国农业大学出版社.

刘维志. 2004. 植物线虫志[M]. 北京：中国农业出版社.

农业部全国植保总站. 1995. 瓜菜斑潜蝇[M]. 北京：中国农业出版社.

万方浩，刘全儒. 2012. 生物入侵:中国外来入侵植物图谱[M]. 北京：科学出版社.

萧刚柔. 1997. 拉英汉昆虫 蜱螨 蜘蛛 线虫名称[M]. 北京：中国林业出版社.

徐海根，强胜. 2004. 中国外来入侵物种编目[M]. 北京：中国环境科学出版社.

徐海根，吴军，陈洁君. 2011. 外来物种环境风险评估与控制研究[M]. 北京：科学出版社.

徐正浩，陈再廖，等. 2011. 浙江入侵生物及防治[M]. 杭州：浙江大学出版社.

鄢建，张艺兵，王洪兵. 2010. 贸易性食品农产品植物卫生国际协定应用概论[M]. 北京：中国农业科学技术出版社.

鄢建，张艺兵. 2009. 植物保护国际公约和植物检疫措施国际标准应用概论[M]. 北京：中国农业科学技术出版社.

中华人民共和国动植物检疫局，农业部植物检疫实验所. 1997. 中国进境植物检疫有害生物选编[M]. 北京：中国农业出版社.

周建安，鄢建. 2013. 植物卫生措施国际标准[M]. 北京：中国农业科学技术出版社.

Pimentel, David, Rodolfo Zuniga, Doug Morrison. 2005. Update on the environmental and economic costs associated with alien-invasive species in the United States[J]. Ecological Economics, 52(3): 273-288

Costanza R, et al. 1997. "The Value of the World's Ecosystem Services and Natural Capital" [J]. Nature, 387: 253-260

主要网站

a.share.photo.xuite.net

abiris.snv.jussieu.fr

acorral.es

actaplantarum.org

agron-www.agron.iastate.edu

alberts.ac.in

alienplantsbelgium.be

als.arizona.edu

applesnail.net

aquaworldbg.com

articles.extension.org

asb.com.ar

asperupgaard.dk

baike.baidu.com

baike.sogou.com

bie.ala.org.au

biocache.ala.org.au

biocontrolfornature.ucr.edu

biodiversite.ville-larochesuryon.fr

biopix.com

blog.growingwithscience.com

botanika.wendys.cz

botanix.org

bucket2.glanacion.com

bugguide.net

bugwoodcloud.org

cabezaprieta.org

calphotos.berkeley.edu

canope.ac-besancon.fr

cdn.xl.thumbs.canstockphoto.com

cdn2.arkive.org

chalk.richmond.edu

chestofbooks.com

climbers.lsa.umich.edu

clopla.butbn.cas.cz

cn.bing.com

cnas-re.uog.edu

commons.hortipedia.com

course.cau-edu.net.cn

courses.missouristate.edu

cropgenebank.sgrp.cgiar.org

data.kew.org

data.kew.org，database.prota.org

de.academic.ru

delange.org

delawarewildflowers.org

delta-intkey.com

dryades.units.it

e.share.photo.xuite.net

edis.ifas.ufl.edu

en.hortipedia.com

en.wikipedia.org

entnemdept.ufl.edu

enviroliteracy.org

eol.org

erec.ifas.ufl.edu

etc.usf.edu

eunis.eea.europa.eu

extension.umass.edu

facultystaff.richmond.edu

familist.ro

farm7.staticflickr.com

farm9.staticflickr.com
files.akva-tera.webnode.cz
files.stadsplantenbreda.webnode.com
fioridisicilia.altervista.org
fishesofaustralia.net.au
flora.huh.harvard.edu
flora.nhm-wien.ac.at
florademurcia.es
flora-emslandia.de
flore-bis.lecolebuissonniere.eu
florida.plantatlas.usf.edu
flowers.la.coocan.jp
flowers2.la.coocan.jp
frps.eflora.cn
gardenbreizh.org
gd.eppo.int
gobotany.newenglandwild.org
hasbrouck.asu.edu
healthyhomegardening.com
hidroponia.mx
http2.mlstatic.com
hyg.ycit.cn
i1.treknature.com
i39.servimg.com
i83.servimg.com
idao.cirad.fr
idtools.org
image.slidesharecdn.com
img.botanicayjardines.com
img.over-blog-kiwi.com
img0.ph.126.net
invasoras.pt
jonahsaquarium.com
kaede.nara-edu.ac.jp
keyserver.lucidcentral.org
kplant.biodiv.tw
landsnails.org
lh3.googleusercontent.com
library.taiwanschoolnet.org
lookformedical.com

lowres-picturecabinet.com
luirig.altervista.org
media.eol.org
meltonwiggins.com
mikawanoyasou.org
mississippientomologicalmuseum.org.msstate.edu
mothphotographersgroup.msstate.edu
my-fish.org
nas.er.usgs.gov
nathistoc.bio.uci.edu
naturalaquariums.com
natural-history.main.jp
ncwings.carolinanature.com
neobiota.naturschutzinformationen-nrw.de
newfs.s3.amazonaws.com
newyork.plantatlas.usf.edu
ninnescahlife.wichita.edu
npsot.orgwww.osaka-c.ed.jp
npuir-3d.npust.edu.tw
nyc.books.plantsofsuburbia.com
oak.ppws.vt.edu
objects.liquidweb.service
oformi-akvarium.ru
ograsradgivaren.slu.se
oilcropandmore.info
okeechobee.ifas.ufl.edu
onlinelibrary.wiley.com
ontariowildflowers.com
pestid.msu.edu
pests.agridata.cn
pflanzenbestimmung.info
piantemagiche.it
pic.baike.soso.com
pic.sogou.com
pics.davesgarden.com
plantbook.org
plantevaernonline.dlbr.dk
plantgenera.org
planthealthportal.defra.gov.uk
plantjdx.com

plantnet.rbgsyd.nsw.gov.au

plants.ces.ncsu.edu

plants.ifas.ufl.edu

plants.jstor.org

plants.usda.gov

plants-of-styria.uni-graz.at

plpnemweb.ucdavis.edu

publish.plantnet-project.org

pukubook.jp

rockbugdesign.com

rybicky.net

s3.amazonaws.com

s3-eu-west-1.amazonaws.com

sd.ifeng.com

seriouslyfish.com

shop.agrimag.it

species.wikimedia.org

sr.yuanlin.com

src.sfasu.edu

static.jardipedia.com

stewartia.net，www.fpcn.net

streamwebs.org

stuartxchange.com

susanleachsnyder.com

taliae.actaplantarum.org

thumbs.dreamstime.com

tolweb.org

tribes.eresmas.net

tropical.theferns.info

tropicalfishandaquariums.com

tupian.baike.com

ukmoths.org.uk

unkraeuter.info

upload.wikimedia.org

uses.plantnet-project.org

vecteurs.cqeee.org

vro.agriculture.vic.gov.au

warehouse1.indicia.org.uk

waste.ideal.es

weedecology.css.cornell.edu

weeds.brisbane.qld.gov.au

weeds.dpi.nsw.gov.au

wenku.baidu.com

Wikipedia

wikis.evergreen.edu

wildebloemen.info

wilde-planten.nl

wildflowers.clockwork-orrery.com

wirbelworld.de

wswa.org.au

ww.natureloveyou.sg

ww.wildutah.us

www.aanbodpagina.nl

www.about-garden.com

www.acquariofiliafacile.it

www.actaplantarum.org

www.agr.kyushu-u.ac.jp

www.agroatlas.ru

www.agrokurier.de

www.akvarieforum.dk

www.akvaryum.biz

www.alabamaplants.co

www.all-creatures.org

www.alpinegardensociety.net

www.amazon.com

www.andermattbiocontrol.com

www.aphis.usda.gov

www.aphotoflora.com

www.aqsiq.gov.cn

www.aquarienkrebse.de

www.aquarienpflanzen-shop.de

www.aquariumlife.net

www.aquarium-planten.com

www.arbioitalia.org

www.arkive.org

www.asergeev.com

www.asianplant.net

www.backyardnature.ne

www.baike.com

www.barbechoquimico.com

www.baumschule-horstmann.de

www.biodiversidadvirtual.org

www.bioone.org

www.biopix.dk

www.bivocational.org

www.blumeninschwaben.de

www.botanic.jp

www.botanickafotogalerie.cz

www.botanische-spaziergaenge.at

www.botany.hawaii.edu

www.british-wild-flowers.co.uk

www.businessinsider.com

www.cabdirect.org

www.cabi.org

www.calflora.org

www.carolinanature.com

www.ces.csiro.au

www.china.com.cn

www.conabio.gob.mx

www.cpa.msu.edu

www.cropscience.bayer.com

www.ct.gov

www.ctahr.hawaii.edu

www.darwinfoundation.org

www.discoverlife.org

www.dorsetnature.co.uk

www.dpi.nsw.gov.au

www.eddmaps.org

www.eeaoc.org.ar

www.efferus.no

www.eflora.cn

www.environment.gov.au

www.eppo.int

www.ewrs.org

www.exot-nutz-zier.d

www.fao.org

www.faunatrhy.cz

www.featurepics.com

www.feedipedia.org

www.fireflyforest.com

www.fishbase.org

www.flora.sa.gov.au

www.florablog.it

www.floradecanarias.com

www.floraiwww.syngenta.fr

www.floravascular.com

www.floresefolhagens.com.br

www.florida.plantatlas.usf.edu

www.flowersinisrael.com

www.flowerspictures.org

www.fmcagricola.com.br

www.fnanaturesearch.org

www.focusnatura.at

www.foodmate.net

www.forestry.gov.cn

www.forestryimages.org

www.fpcn.net

www.freenatureimages.eu

www.freshfromflorida.com

www.friendsofqueensparkbushland.org.au

www.fs.fed.us

www.fungoceva.it

www.fws.gov

www.gbif.org

www.geo.arizona.edu

www.greekflora.gr

www.habitas.org.uk

www.he.xinhua.org

www.hear.org

www.hindawi.com

www.hunter.cuny.edu

www.iewf.org

www.inspection.gc.ca

www.invasive.org

www.invasive-species.org

www.invasivespeciesinfo.gov

www.ipmdss.dk

www.issg.org

www.itis.gov

www.ittiofauna.org

www.iucngisd.org

www.iucnredlist.org

www.jcu.edu.au

www.jeffpippen.com

www.jungleseeds.co.uk

www.kalake.ee

www.kansasnativeplants.com

www.kuleuven-kulak.be

www.lesjardinsaquatiques.fr

www.lubera.com

www.mafengwo.cn

www.maltawildplants.com

www.mapaq.gouv.qc.ca

www.mda.state.mn.us

www.minnesotawildflowers.info

www.missouribotanicalgarden.org

www.missouriplants.com

www.moa.gov.cn

www.monarthrum.info

www.nahuby.sk

www.namethatplant.net

www.naturalmedicinefacts.info

www.naturamediterraneo.com

www.natureloveyou.sg

www.nature-museum.net

www.ncbi.nlm.nih.gov

www.ncfishes.com

www.newyork.plantatlas.usf.edu

www.nmsgsfz.com

www.nonnativespecies.org

www.nwvisualplantid.com

www.nzenzeflowerspauwels.be

www.nzpcn.org.nz

www.oardc.ohio-state.edu

www.onlyfoods.net

www.opsu.edu

www.ortobotanico.unina.it

www.pacificbulbsociety.org

www.painelflorestal.com.br

www.parcocurone.it

www.pariscotejardin.fr

www.pesticide.ro

www.pestnet.org

www.pfaf.org

www.photomazza.com

www.pioneercatchment.org.au

www.pisces.at

www.planetcatfish.com

www.plant.csdb.cn

www.plantarium.ru

www.plantnames.unimelb.edu.au

www.plantright.org

www.plantwise.org

www.plant-world-seeds.com

www.ptrpest.com

www.qjure.com

www.redorbit.com

www.researchgate.net

www.revistagua.cl

www.ricesci.cn

www.sardegnaflora.it

www.sarinalandcare.org.au

www.sbs.utexas.edu

www.scielo.org.mx

www.sevenoaksnativenursery.com

www.skolvision.se

www.smmflowers.org

www.sms.si.edu

www.sogou.com

www.soinc.org

www.soortenbank.nl

www.southeastweeds.org.au

www.tartarugando.it

www.theplantlist.org

www.thepoisongarden.co.uk

www.thismia.com

www.thoughtco.com

www.tisanes-indigenes.re

www.tropicalforages.info

www.tropical-plants-flowers-and-decor.com

www.tuinadvies.be
www.uapress.arizona.edu
www.ufrgs.br
www.unavarra.es
www.undkraut.de
www.uniprot.org
www.vaplantatlas.org
www.vaxteko.nu
www.viarural.com.ar
www.vilmorin-tree-seeds.com
www.virboga.de
www.visoflora.com
www.walterreeves.com
www.waspweb.org
www.waza.org
www.weedimages.org
www.weedinfo.ca
www.westafricanplants.senckenberg.de

www.wiseacre-gardens.com
www.wm-sec.com
www.wnmu.edu
www.worldseedsupply.com
www.wpclipart.com
www.yeehua.net
www.yumpu.com
www.zgjhjy.com
www.zhb.gov.cn
www.zhiwutong.com
www.zoodia.it
www2.dijon.inra.fr
www2.nrm.se
www7a.biglobe.ne.jp
xtension.cropsciences.illinois.edu
xueshu.baidu.com
zhidao.baidu.com
zhiwutong.com

附　录

Ⅰ　世界最严重的100种入侵物种

(100 of the World's Worst Invasive Species)

Species 种名	Type 类别	Common names 常用名
Caulerpa taxifolia 杉叶蕨藻	Alga 藻类	Killer Alga (e)
Undaria pinnatifida 裙带菜	alga	Wakame, apron-ribbon vegetable, Asian kelp, Japanese kelp
Bufo marinus = Rhinella marina 海蟾蜍	Amphibian 两栖类	Cane toad, bufo toad, bullfrog, giant American toad, giant neotropical toad, giant toad, marine toad, Suriname toad
Eleutherodactylus coqui 金线雨蛙	amphibian	Caribbean tree frog, common coqui, coqui, Puerto Rican treefrog
Lithobates catesbeianus 北美牛蛙	amphibian	Bullfrog, North American bullfrog, American bullfrog
Eichhornia crassipes 凤眼莲	Aquatic plant 水生植物类	Water hyacinth, floating water hyacinth, water orchid
Acridotheres tristis 家八哥	Bird 鸟类	Common myna, Calcutta myna, house myna, Indian myna, martin triste, talking myna
Pycnonotus cafer 黑喉红臀鹎	bird	Red-vented bulbul
Sturnus vulgaris 紫翅椋鸟	bird	Common starling, English starling, European starling
Mnemiopsis leidyi 淡海栉水母	Comb jelly 栉水母类	American comb jelly, comb jelly, comb jellyfish, sea gooseberry, sea walnut, Venus' girdle, warty comb jelly
Carcinus maenas 普通滨蟹	Crustacean 甲壳类	European green crab, European shore crab, green crab, shore crab
Cercopagis pengoi 多刺水甲	crustacean	Fishhook waterflea
Eriocheir sinensis 中华绒螯蟹	crustacean	Chinese mitten crab, big sluice crab, Chinese freshwater edible crab, Chinese river crab, Shanghai hairy crab
Clarias batrachus 蟾胡鲶	Fish 鱼类	Walking catfish, clarias catfish, climbing perch, freshwater catfish, Thailand catfish
Cyprinus carpio 鲤鱼	fish	Common carp, European carp, fancy carp, feral carp, German carp, grass carp, Japanese domesticated carp, king carp, koi, koi carp, leather carp, mirror carp, Oriental carp, scale carp, wild carp

（续表）

Species 种名	Type 类别	Common names 常用名
Gambusia affinis 食蚊鱼	fish	Mosquitofish, guayacon mosquito, live-bearing tooth-carp, mosquito fish, pez mosquito, western mosquitofish
Lates niloticus 尼罗河鲈鱼	fish	Nile perch, African snook, Victoria perch
Micropterus salmoides 加州鲈鱼	fish	Largemouth bass, American black bass, bass, black bass, green bass, green trout, largemouth black bass, northern largemouth bass, stormundet black bass
Oncorhynchus mykiss 虹鳟	fish	Rainbow trout, Baja California rainbow trout, coast angel trout, coast rainbow trout, coast range trout, hardhead, Kamchatka steelhead, Kamchatka trout, Kamloops, Kamloops trout, lord-fish, redband, redband trout, salmon trout, silver trout, steelhead, steelhead trout, summer salmon
Oreochromis mossambicus 罗非鱼	fish	Tilapia, common tilapia, Java tilapia, kurper bream, Mozambique cichlid, Mozambique mouth-breeder, Mozambique mouthbrooder, Mozambique tilapia
Salmo trutta 褐鳟	fish	Brown trout, blacktail, brook trout, galway sea trout, herling, orange fin, orkney sea trout, salmon trout, sea trout, trout, whiting, whitling
Platydemus manokwari 新几内亚扁虫	Flatworm 扁虫类	New Guinea flatworm, flatworm, snail-eating flatworm
Salvinia molesta 人厌槐叶苹	Floating aquatic fern 漂浮水生蕨类	Giant salvinia
Batrachochytrium dendrobatidis 蛙壶菌	Fungus 真菌类	Bd, chytrid frog fungi, chytridiomycosis, frog chytrid fungus
Cryphonectria parasitica (=*Endothia parasitica*) 栗疫病菌	fungus	Chestnut blight
Ophiostoma ulmi sensu lato 榆蛇喙壳	fungus	Dutch elm disease
Phytophthora cinnamomi 樟疫霉	fungus	Cinnamon fungus, green fruit rot, heart rot, phytophthora root rot, seedling blight, stem canker, wildflower dieback
Arundo donax 芦竹	Grass 牧草类	Giant cane, arundo grass, bamboo reed, cane, cow cane, donax cane, giant reed, reedgrass, river cane, Spanish cane, Spanish reed, wild cane
Imperata cylindrical 白茅	grass	Blady grass, cogon grass, Japanese bloodgrass, speargrass
Spartina anglica 大米草	grass	Common cord grass, rice grass, Townsend's grass, Rice cord grass
Chromolaena odorata 飞机草	Herb 草本植物	Siam weed, bitter bush, Christmas bush, devil weed, camfhur grass, common floss flower, jack in the bush, triffid
Euphorbia esula 乳浆大戟	herb	Leafy spurge, green spurge, spurge, wolf's milk
Hedychium gardnerianum 金姜花	herb	Ginger lily, kahila garland-lily, kahili ginger, wild ginger
Lythrum salicaria 千屈菜	herb	Purple loosestrife, purple lythrum, rainbow weed, spiked loosestrife
Sphagneticola trilobata 三裂叶蟛蜞菊	herb	Creeping ox-eye, Singapore daisy, trailing daisy
Aedes albopictus 白纹伊蚊	Insect 昆虫类	Asian tiger mosquito, forest day mosquito, tiger mosquito

（续表）

(续表)

Species 种名	Type 类别	Common names 常用名
Anopheles quadrimaculatus 四斑按蚊	insect	Common malaria mosquito
Anoplolepis gracilipes 长足捷蚁	insect	Yellow crazy ant, crazy ant, gramang ant, long-legged ant, Maldive ant
Anoplophora glabripennis 光肩星天牛	insect	Asian long-horned beetle, starry sky beetle
Bemisia tabaci 烟粉虱	insect	Cotton whitefly, sweet potato whitefly, sweetpotato whitefly
Cinara cupressi 柏蚜	insect	Cypress aphid
Coptotermes formosanus 台湾乳白蚁	insect	Formosa termite, Formosan subterranean termite
Linepithema humile 阿根廷蚁	insect	Argentine ant
Lymantria dispar 舞毒蛾	insect	Asian gypsy moth, gypsy moth
Pheidole megacephala 褐大头蚁	insect	Big-headed ant, brown house-ant, coastal brown-ant, lion ant
Solenopsis invicta 入侵红火蚁	insect	Red imported fire ant, RIFA, Red Fire ant
Trogoderma granarium 谷斑皮蠹	insect	Khapra beetle
Vespula vulgaris 普通黄胡蜂	insect	Common wasp, common yellowjacket, European Wasp
Wasmannia auropunctata 小火蚁	insect	Cocoa tree-ant, little fire ant, little introduced fire ant, little red fire ant, small fire ant, West Indian stinging ant
Capra hircus 家山羊	Mammal 哺乳动物类	Goat
Cervus elaphus 白臀鹿	mammal	Red deer, deer, elk,[n 1] European red deer
Felis catus 家猫	mammal	Cat, domestic cat, feral cat, house cat
Herpestes javanicus 印度小猫鼬	mammal	Small Asian mongoose, small Indian mongoose
Macaca fascicularis 食蟹猴	mammal	Crab-eating macaque, long-tailed macaque, lion-tailed macaque.
Mus musculus 小家鼠	mammal	Field mouse, house mouse, wood mouse
Mustela ermine 白鼬	mammal	Stoat, ermine, short-tailed weasel
Myocastor coypus 河狸鼠	mammal	Bewerrot, biberratte, coipù, coypu, nutria, ragondin
Oryctolagus cuniculus 欧洲兔	mammal	Rabbit, European rabbit
Rattus rattus 黑家鼠	mammal	Black rat, blue rat, bush rat, European house rat, roof rat, ship rat

(续表)

(续表)

Species 种名	Type 类别	Common names 常用名
Sciurus carolinensis 灰松鼠	mammal	Gray squirrel, grey squirrel, Eastern gray squirrel
Sus scrofa 野猪	mammal	Wild boar, razorback
Trichosurus vulpecula 刷尾负鼠	mammal	Brushtail possum
Vulpes vulpes 赤狐	mammal	Red fox, black or cross fox
Achatina fulica 非洲大蜗牛	Mollusk 软体动物类	Giant African land snail, giant African snail
Dreissena polymorpha 欧亚斑马贻贝	mollusc	Eurasian zebra mussel, wandering mussel, zebra mussel
Euglandina rosea 玫瑰蜗牛	mollusc	Cannibal snail, rosy wolf snail
Mytilus galloprovincialis 地中海贻贝	mollusc	Bay mussel, blue mussel, Mediterranean mussel
Pomacea canaliculata 福寿螺	mollusc	Apple snail, channeled apple snail, golden apple snail, golden kuhol, miracle snail
Potamocorbula amurensis 黑龙江河蓝蛤	mollusc	Amur river clam, Amur river corbula, Asian bivalve, Asian clam, brackish-water corbula, Chinese clam, marine clam
Aphanomyces astaci 螯虾瘟	Protozoan 原生动物类	Crayfish plague
Plasmodium relictum 残片疟原虫	protozoan	Avian malaria
Boiga irregularis 褐色树蛇	Reptile 爬行动物类	Brown catsnake, brown tree snake, brown treesnake
Trachemys scripta elegans 红耳龟	reptile	Red-eared slider, red-eared slider terrapin, slider turtle
Asterias amurensis 多棘海盘车	Sea star 海星类	Northern Pacific seastar, flatbottom seastar, Japanese seastar, Japanese starfish, North Pacific seastar, purple-orange seastar
Acacia mearnsii 黑荆	Shrub 灌木类	Black wattle, Australian acacia
Clidemia hirta 伏地野牡丹	shrub	Koster's curse, faux vatouk, soap bush, soapbush
Fallopia japonica = *Polygonum cuspidatum* 虎杖	shrub	Crimson beauty, donkey rhubarb, fleeceflower, German sausage, Japanese bamboo, Japanese fleece flower, Japanese knotweed, Japanese polygonum, kontiki bamboo, Mexican-bamboo, peashooter plant, reynoutria fleece flower, sally rhubarb
Hiptage benghalensis 风筝果	shrub	Hiptage
Lantana camara 马缨丹	shrub	Angel lips, big sage, blacksage, flowered sage, largeleaf lantana, prickly lantana, Spanish flag, West Indian Lantana, white sage, wild sage
Ligustrum robustum 粗壮女贞	shrub	Bora-bora, Ceylon privét, Sri Lankan privet, tree privet
Mimosa pigra 大含羞草	shrub	Bashful plant, catclaw, catclaw mimosa, giant sensitive plant, giant trembling plant, mimosa

(续表)

(续表)

Species 种名	Type 类别	Common names 常用名
Morella faya 火杨梅	shrub	Candleberry myrtle, fayatree, fire tree, firebush
Opuntia stricta 仙人掌	shrub	Common prickly pear, Araluen pear, Australian pest pear, common pest pear, erect prickly pear, gayndah pear, sour prickly pear, spiny pest pear
Psidium cattleianum 草莓番石榴	shrub	Cattley guava, cherry guava, Chinese guava, purple strawberry guava, strawberry guava
Rubus ellipticus 椭圆悬钩子	shrub	Asian wild raspberry, broad-leafed bramble, Ceylon blackberry, golden evergreen raspberry, Molucca berry, Molucca bramble, Molucca raspberry, robust blackberry, wild blackberry, wild raspberry, yellow Himalayan raspberry
Tamarix ramosissima 多枝怪柳	shrub	Salt cedar
Ulex europaeus 荆豆	shrub	Gorse, Irish furze
Ardisia elliptica 东方紫金牛	Tree 树	Shoebutton Ardisia
Cecropia peltata 号角树	tree	Faux-ricin, pumpwood, trumpet tree, snakewood
Cinchona pubescens 金鸡纳树	tree	Red cinchona
Leucaena leucocephala 银合欢	tree	White leadtree, false koa, faux mimosa, faux-acacia, horse/wild tamarind, jumbie bean, lead tree, wild mimosa, wild tamarind
Melaleuca quinquenervia 白油树	tree	Paperbark teatree, bottle brush tree, broadleaf paperbark tree, broadleaf teatree, broad-leaved paperbark tree, five-veined paperbark tree, paper bark tree, punk tree, white bottlebrush tree
Miconia calvescens 野牡丹	tree	Bush currant, purple plague, velvet tree
Pinus pinaster 海岸松	tree	Cluster pine, maritime pine
Prosopis glandulosa 腺牧豆树	tree	Honey mesquite, mesquite, Texas mesquite
Schinus terebinthifolius 巴西胡椒木	tree	Brazilian holly, Brazilian pepper, Brazilian pepper tree, Christmas berry, Florida holly, Mexican pepper
Spathodea campanulata 火焰树	tree	African tulip tree, fireball, flame of the forest, fountain tree, Indian cedar, Santo Domingo mahogany
Pueraria montana var. *lobata* 野葛	Vine 藤本植物类	Japanese arrowroot, kudzu, kudzu vine
Mikania micrantha 薇甘菊	vine, climber 藤本植物、攀缘植物类	American rope, Chinese creeper, mile-a-minute weed
Banana bunchy top virus（BBTV）香蕉束顶病毒	Virus 病毒类	Banana bunchy top disease, BBTD, abaca bunchy top virus, BBTV, bunchy top, bunchy top virus

（资料来自IUCN/ISSG，2014；Wikipedia，2017）

II 世界十大外来入侵物种

亚洲鲤鱼（鲤鱼、青鱼、草鱼、鲢鱼、鳙鱼等鲤科8种淡水鱼的总称）

乌鳢（*Ophiocephalus argus* Cantor）

葛根（*Pueraria lobata*（Willd.）Ohwi）

野兔（兔属、粗毛兔属与岩兔属中的4个物种）

八哥（*Acridotheres cristatellus* Linnaeus）

缅甸蟒蛇（*Python molurus* Linnaeus）（见*Python molurus bivittatus* Linnaeus）

蔗蟾蜍（*Rhinella marina*（Linnaeus））

东灰松鼠（*Sciurus carolinensis* Gmelin）

杂交蜜蜂（*Apis* Linnaeus）Hybrids

老鼠（Muroidea Illiger）

III 中华人民共和国进境植物检疫性有害生物名录

(更新至2017年6月,441种)

昆 虫

Acanthocinus carinulatus(Gebler)
白带长角天牛

Acanthoscelides obtectus(Say)
菜豆象

Acleris variana(Fernald)
黑头长翅卷蛾

Agrilus spp.(non-Chinese)
窄吉丁(非中国种)

Aleurodicus dispersus Russell
螺旋粉虱

Anastrepha Schiner
按实蝇属

Anthonomus grandis Boheman
墨西哥棉铃象

Anthonomus quadrigibbus Say
苹果花象

Aonidiella comperei McKenzie
香蕉肾盾蚧

Apate monachus Fabricius
咖啡黑长蠹

Aphanostigma piri(Cholodkovsky)
梨矮蚜

Arhopalus syriacus Reitter
辐射松幽天牛

Bactrocera Macquart
果实蝇属

Baris granulipennis(Tournier)
西瓜船象

Batocera spp.(non-Chinese)
白条天牛(非中国种)

Brontispa longissima(Gestro)
椰心叶甲

Bruchidius incarnates(Boheman)
埃及豌豆象

Bruchophagus roddi Gussak
苜蓿籽蜂

Bruchus spp.(non-Chinese)
豆象(非中国种)

Cacoecimorpha pronubana(Hübner)
荷兰石竹卷蛾

Callosobruchus spp.(*maculatus*(F.) and non-Chinese)
瘤背豆象(四纹豆象和非中国种)

Carpomya incompleta(Becker)
欧非枣实蝇

Carpomya vesuviana Costa
枣实蝇

Carulaspis juniperi(Bouchè)
松唐盾蚧

Caulophilus oryzae(Gyllenhal)
阔鼻谷象

Ceratitis Macleay
小条实蝇属

Ceroplastes rusci(L.)
无花果蜡蚧

Chionaspis pinifoliae(Fitch)
松针盾蚧

Choristoneura fumiferana(Clemens)
云杉色卷蛾

Conotrachelus Schoenherr
鳄梨象属

Contarinia sorghicola(Coquillett)
高粱瘿蚊

Coptotermes spp.(non-Chinese)
乳白蚁(非中国种)

Craponius inaequalis(Say)
葡萄象

Crossotarsus spp.(non-Chinese)
异胫长小蠹(非中国种)

Cryptophlebia leucotreta（Meyrick）
苹果异形小卷蛾

Cryptorrhynchus lapathi L.
杨干象

Cryptotermes brevis（Walker）
麻头砂白蚁

Ctenopseustis obliquana（Walker）
斜纹卷蛾

Curculio elephas（Gyllenhal）
欧洲栗象

Cydia janthinana（Duponchel）
山楂小卷蛾

Cydia packardi（Zeller）
樱小卷蛾

Cydia pomonella（L.）
苹果蠹蛾

Cydia prunivora（Walsh）
杏小卷蛾

Cydia pyrivora（Danilevskii）
梨小卷蛾

Dacus spp.（non-Chinese）
寡鬃实蝇（非中国种）

Dasineura mali（Kieffer）
苹果瘿蚊

Dendroctonus spp.（*valens* LeConte and non-Chinese）
大小蠹（红脂大小蠹和非中国种）

Deudorix isocrates Fabricius
石榴小灰蝶

Diabrotica Chevrolat
根萤叶甲属

Diaphania nitidalis（Stoll）
黄瓜绢野螟

Diaprepes abbreviata（L.）
蔗根象

Diatraea saccharalis（Fabricius）
小蔗螟

Dryocoetes confusus Swaine
混点毛小蠹

Dysmicoccus grassi Leonari
香蕉灰粉蚧

Dysmicoccus neobrevipes Beardsley
新菠萝灰粉蚧

Ectomyelois ceratoniae（Zeller）
石榴螟

Epidiaspis leperii（Signoret）
桃白圆盾蚧

Eriosoma lanigerum（Hausmann）
苹果绵蚜

Eulecanium gigantea（Shinji）
枣大球蚧

Eurytoma amygdali Enderlein
扁桃仁蜂

Eurytoma schreineri Schreiner
李仁蜂

Gonipterus scutellatus Gyllenhal
桉象

Helicoverpa zea（Boddie）
谷实夜蛾

Hemerocampa leucostigma（Smith）
合毒蛾

Hemiberlesia pitysophila Takagi
松突圆蚧

Heterobostrychus aequalis（Waterhouse）
双钩异翅长蠹

Hoplocampa flava（L.）
李叶蜂

Hoplocampa testudinea（Klug）
苹叶蜂

Hoplocerambyx spinicornis（Newman）
刺角沟额天牛

Hylobius pales（Herbst）
苍白树皮象

Hylotrupes bajulus（L.）
家天牛

Hylurgopinus rufipes（Eichhoff）
美洲榆小蠹

Hylurgus ligniperda Fabricius
长林小蠹

Hyphantria cunea（Drury）
美国白蛾

Hypothenemus hampei（Ferrari）
咖啡果小蠹

Incisitermes minor（Hagen）
小楹白蚁

Ips spp.（non-Chinese）
齿小蠹（非中国种）

Ischnaspis longirostris（Signoret）
黑丝盾蚧

Lepidosaphes tapleyi Williams
芒果蛎蚧

Lepidosaphes tokionis（Kuwana）
东京蛎蚧

Lepidosaphes ulmi（L.）
榆蛎蚧

Leptinotarsa decemlineata（Say）
马铃薯甲虫

Leucoptera coffeella（Guérin-Méneville）
咖啡潜叶蛾

Liriomyza trifolii（Burgess）
三叶斑潜蝇

Lissorhoptrus oryzophilus Kuschel
稻水象甲

Listronotus bonariensis（Kuschel）
阿根廷茎象甲

Lobesia botrana（Denis et Schiffermuller）
葡萄花翅小卷蛾

Mayetiola destructor（Say）
黑森瘿蚊

Mercetaspis halli（Green）
霍氏长盾蚧

Monacrostichus citricola Bezzi
桔实锤腹实蝇

Monochamus spp.（non-Chinese）
墨天牛（非中国种）

Myiopardalis pardalina（Bigot）
甜瓜迷实蝇

Naupactus leucoloma（Boheman）
白缘象甲

Neoclytus acuminatus（Fabricius）
黑腹尼虎天牛

Opogona sacchari（Bojer）
蔗扁蛾

Pantomorus cervinus（Boheman）
玫瑰短喙象

Parlatoria crypta Mckenzie
灰白片盾蚧

Pharaxonotha kirschi Reither
谷拟叩甲

Phenacoccus manihoti Matile-Ferrero
木薯绵粉蚧（2011年6月20日新增）

Phenacoccus solenopsis Tinsley
扶桑绵粉蚧（2009年2月3日新增）

Phloeosinus cupressi Hopkins
美柏肤小蠹

Phoracantha semipunctata（Fabricius）
桉天牛

Pissodes Germar
木蠹象属

Planococcus lilacius Cockerell
南洋臀纹粉蚧

Planococcus minor（Maskell）
大洋臀纹粉蚧

Platypus spp.（non-Chinese）
长小蠹（非中国种）

Popillia japonica Newman
日本金龟子

Prays citri Milliere
桔花巢蛾

Promecotheca cumingi Baly
椰子缢胸叶甲

Prostephanus truncatus（Horn）
大谷蠹

Ptinus tectus Boieldieu
澳洲蛛甲

Quadrastichus erythrinae Kim
刺桐姬小蜂

Reticulitermes lucifugus（Rossi）
欧洲散白蚁

Rhabdoscelus lineaticollis（Heller）
褐纹甘蔗象

Rhabdoscelus obscurus（Boisduval）
几内亚甘蔗象

Rhagoletis spp.（non-Chinese）
绕实蝇（非中国种）

Rhynchites aequatus（L.）
苹虎象

Rhynchites bacchus L.
欧洲苹虎象

Rhynchites cupreus L.
李虎象

Rhynchites heros Roelofs
日本苹虎象

Rhynchophorus ferrugineus（Olivier）
红棕象甲

Rhynchophorus palmarum（L.）
棕榈象甲

Rhynchophorus phoenicis（Fabricius）
紫棕象甲

Rhynchophorus vulneratus（Panzer）
亚棕象甲

Sahlbergella singularis Haglund
可可盲蝽象

Saperda spp.（non-Chinese）
楔天牛（非中国种）

Scolytus multistriatus（Marsham）
欧洲榆小蠹

Scolytus scolytus（Fabricius）
欧洲大榆小蠹

Scyphophorus acupunctatus Gyllenhal
剑麻象甲

Selenaspidus articulatus Morgan
刺盾蚧

Sinoxylon spp.（non-Chinese）
双棘长蠹（非中国种）

Sirex noctilio Fabricius
云杉树蜂

Solenopsis invicta Buren
入侵红火蚁

Spodoptera littoralis（Boisduval）
海灰翅夜蛾

Stathmopoda skelloni Butler
猕猴桃举肢蛾

Sternochetus Pierce
芒果象属

Taeniothrips inconsequens（Uzel）
梨蓟马

Tetropium spp.（non-Chinese）
断眼天牛（非中国种）

Thaumetopoea pityocampa（Denis et Schiffermuller）
松异带蛾

Toxotrypana curvicauda Gerstaecker
番木瓜长尾实蝇

Tribolium destructor Uyttenboogaart
褐拟谷盗

Trogoderma spp.（non-Chinese）
斑皮蠹（非中国种）

Vesperus Latreile
暗天牛属

Vinsonia stellifera（Westwood）
七角星蜡蚧

Viteus vitifoliae（Fitch）
葡萄根瘤蚜

Xyleborus spp.（non-Chinese）
材小蠹（非中国种）

Xylotrechus rusticus L.
青杨脊虎天牛

Zabrotes subfasciatus（Boheman）
巴西豆象

软体动物

Achatina fulica Bowdich
非洲大蜗牛

Acusta despecta Gray
琉球球壳蜗牛

Cepaea hortensis Müller
花园葱蜗牛

Cernuella virgata Da Costa
地中海白蜗牛（2012年9月17日新增）

Helix aspersa Müller
散大蜗牛

Helix pomatia Linnaeus
盖罩大蜗牛

Theba pisana Müller
比萨茶蜗牛

真　菌

Albugo tragopogi（Persoon）Schröter var. *helianthi* Novotelnova
向日葵白锈病菌

Alternaria triticina Prasada et Prabhu
小麦叶疫病菌

Anisogramma anomala（Peck）E. Muller
榛子东部枯萎病菌

Apiosporina morbosa（Schweinitz）von Arx
李黑节病菌

Atropellis pinicola Zaller et Goodding
松生枝干溃疡病菌

Atropellis piniphila（Weir）Lohman et Cash
嗜松枝干溃疡病菌

Botryosphaeria laricina（K.Sawada）Y. Zhong
落叶松枯梢病菌

Botryosphaeria stevensii Shoemaker
苹果壳色单隔孢溃疡病菌

Cephalosporium gramineum Nisikado et Ikata
麦类条斑病菌

Cephalosporium maydis Samra，Sabet et Hingorani
玉米晚枯病菌

Cephalosporium sacchari E. J. Butler et Hafiz Khan
甘蔗凋萎病菌

Ceratocystis fagacearum（Bretz）Hunt
栎枯萎病菌

Chalara fraxinea T. Kowalski
白蜡鞘孢菌（2013年3月6日新增）

Chrysomyxa arctostaphyli Dietel
云杉帚锈病菌

Ciborinia camelliae Kohn
山茶花腐病菌

Cladosporium cucumerinum Ellis et Arthur
黄瓜黑星病菌

Colletotrichum kahawae J. M. Waller et Bridge
咖啡浆果炭疽病菌

Crinipellis perniciosa（Stahel）Singer
可可丛枝病菌

Cronartium coleosporioides J. C. Arthur
油松疱锈病菌

Cronartium comandrae Peck
北美松疱锈病菌

Cronartium conigenum Hedgcock et Hunt
松球果锈病菌

Cronartium fusiforme Hedgcock et Hunt ex Cummins
松纺锤瘤锈病菌

Cronartium ribicola J. C. Fisch.
松疱锈病菌

Cryphonectria cubensis（Bruner）Hodges
桉树溃疡病菌

Cylindrocladium parasiticum Crous，Wingfield et Alfenas
花生黑腐病菌

Diaporthe helianthi Muntanola-Cvetkovic Mihaljcevic et Petrov
向日葵茎溃疡病菌

Diaporthe perniciosa É. J. Marchal
苹果果腐病菌

Diaporthe phaseolorum（Cooke et Ell.）Sacc. var. *caulivora* Athow et Caldwell
大豆北方茎溃疡病菌

Diaporthe phaseolorum（Cooke et Ell.）Sacc. var. *meridionalis* F. A. Fernandez
大豆南方茎溃疡病菌

Diaporthe vaccinii Shear
蓝莓果腐病菌

Didymella ligulicola（K. F. Baker，Dimock et L. H. Davis）von Arx
菊花花枯病菌

Didymella lycopersici Klebahn
番茄亚隔孢壳茎腐病菌

Endocronartium harknessii（J. P. Moore）Y. Hiratsuka
松瘤锈病菌

Eutypa lata（Pers.）Tul. et C. Tul.
葡萄藤猝倒病菌

Fusarium circinatum Nirenberg et O'Donnell
松树脂溃疡病菌

Fusarium oxysporum Schlecht. f. sp. *apii* Snyd. et Hans
芹菜枯萎病菌

Fusarium oxysporum Schlecht. f.sp. *asparagi* Cohen et Heald
芦笋枯萎病菌

Fusarium oxysporum Schlecht. f.sp. *cubense*（E. F. Sm.）Snyd. et Hans（Race 4 non-Chinese races）
香蕉枯萎病菌（4号小种和非中国小种）

Fusarium oxysporum Schlecht. f. sp. *elaeidis* Toovey
油棕枯萎病菌

Fusarium oxysporum Schlecht. f. sp. *fragariae* Winks et Williams
草莓枯萎病菌

Fusarium tucumaniae T. Aoki，O'Donnell，Yos. Homma et Lattanzi
南美大豆猝死综合症病菌

Fusarium virguliforme O'Donnell et T. Aoki
北美大豆猝死综合症病菌

Gaeumannomyces graminis（Sacc.）Arx et D. Olivier var. *avenae*（E. M. Turner）Dennis
燕麦全蚀病菌

Greeneria uvicola（Berk. et M. A. Curtis）Punithalingam
葡萄苦腐病菌

Gremmeniella abietina（Lagerberg）Morelet
冷杉枯梢病菌

Gymnosporangium clavipes（Cooke et Peck）Cooke et Peck
榅桲锈病菌

Gymnosporangium fuscum R. Hedw.
欧洲梨锈病菌

Gymnosporangium globosum（Farlow）Farlow
美洲山楂锈病菌

Gymnosporangium juniperi-virginianae Schwein
美洲苹果锈病菌

Helminthosporium solani Durieu et Mont.
马铃薯银屑病菌

Hypoxylon mammatum（Wahlenberg）J. Miller
杨树炭团溃疡病菌

Inonotus weirii（Murrill）Kotlaba et Pouzar
松干基褐腐病菌

Leptosphaeria libanotis（Fuckel）Sacc.
胡萝卜褐腐病菌

Leptosphaeria lindquistii Frezzi，无性态：*Phoma macdonaldii* Boerma 向日葵黑茎病（2010年10月20日新增）

Leptosphaeria maculans（Desm.）Ces. et De Not.
十字花科蔬菜黑胫病菌

Leucostoma cincta（Fr.:Fr.）Hohn.
苹果溃疡病菌

Melampsora farlowii（J. C. Arthur）J. J. Davis
铁杉叶锈病菌

Melampsora medusae Thumen
杨树叶锈病菌

Microcyclus ulei（P.Henn.）von Arx
橡胶南美叶疫病菌

Monilinia fructicola（Winter）Honey
美澳型核果褐腐病菌

Moniliophthora roreri（Ciferri et Parodi）Evans
可可链疫孢荚腐病菌

Monosporascus cannonballus Pollack et Uecker
甜瓜黑点根腐病菌

Mycena citricolor（Berk. et Curt.）Sacc.
咖啡美洲叶斑病菌

Mycocentrospora acerina（Hartig）Deighton
香菜腐烂病菌

Mycosphaerella dearnessii M. E. Barr
松针褐斑病菌

Mycosphaerella fijiensis Morelet
香蕉黑条叶斑病菌

Mycosphaerella gibsonii H. C. Evans
松针褐枯病菌

Mycosphaerella linicola Naumov
亚麻褐斑病菌

Mycosphaerella musicola J. L. Mulder
香蕉黄条叶斑病菌

Mycosphaerella pini E. Rostrup
松针红斑病菌

Nectria rigidiuscula Berk. et Broome
可可花瘿病菌

Ophiostoma novo-ulmi Brasier
新榆枯萎病菌

Ophiostoma ulmi（Buisman）Nannf.
榆枯萎病菌

Ophiostoma wageneri（Goheen et Cobb）Harrington
针叶松黑根病菌

Ovulinia azaleae Weiss
杜鹃花枯萎病菌

Periconia circinata（M. Mangin）Sacc.
高粱根腐病菌

Peronosclerospora spp.（non-Chinese）
玉米霜霉病菌（非中国种）

Peronospora farinosa（Fries: Fries）Fries f. sp. *betae* Byford
甜菜霜霉病菌

Peronospora hyoscyami de Bary f. sp. *tabacina*（Adam）Skalicky
烟草霜霉病菌

Pezicula malicorticis（Jacks.）Nannfeld
苹果树炭疽病菌

Phaeoramularia angolensis（T. Carvalho et O. Mendes）P. M. Kirk
柑橘斑点病菌

Phellinus noxius（Corner）G. H. Cunn.
木层孔褐根腐病菌

Phialophora gregata（Allington et Chamberlain）W. Gams
大豆茎褐腐病菌

Phialophora malorum（Kidd et Beaum.）McColloch
苹果边腐病菌

Phoma exigua Desmazières f. sp. *foveata*（Foister）Boerema
马铃薯坏疽病菌

Phoma glomerata（Corda）Wollenweber et Hochapfel
葡萄茎枯病菌

Phoma pinodella（L. K. Jones）Morgan-Jones et K. B. Burch
豌豆脚腐病菌

Phoma tracheiphila（Petri）L. A. Kantsch. et Gikaschvili
柠檬干枯病菌

Phomopsis sclerotioides van Kesteren
黄瓜黑色根腐病菌

Phymatotrichopsis omnivora（Duggar）Hennebert
棉根腐病菌

Phytophthora cambivora（Petri）Buisman
栗疫霉黑水病菌

Phytophthora erythroseptica Pethybridge
马铃薯疫霉绯腐病菌

Phytophthora fragariae Hickman
草莓疫霉红心病菌

Phytophthora fragariae Hickman var. *rubi* W. F. Wilcox et J. M. Duncan
树莓疫霉根腐病菌

Phytophthora hibernalis Carne
柑橘冬生疫霉褐腐病菌

Phytophthora lateralis Tucker et Milbrath
雪松疫霉根腐病菌

Phytophthora medicaginis E. M. Hans. et D. P. Maxwell
苜蓿疫霉根腐病菌

Phytophthora phaseoli Thaxter
菜豆疫霉病菌

Phytophthora ramorum Werres，De Cock et Man in't Veld
栎树猝死病菌

Phytophthora sojae Kaufmann et Gerdemann
大豆疫霉病菌

Phytophthora syringae（Klebahn）Klebahn
丁香疫霉病菌

Polyscytalum pustulans（M. N. Owen et Wakef.）M. B. Ellis
马铃薯皮斑病菌

Protomyces macrosporus Unger
香菜茎瘿病菌

Pseudocercosporella herpotrichoides（Fron）Deighton
小麦基腐病菌

Pseudopezicula tracheiphila（Müller-Thurgau）Korf et Zhuang
葡萄角斑叶焦病菌

Puccinia pelargonii-zonalis Doidge
天竺葵锈病菌

Pycnostysanus azaleae（Peck）Mason
杜鹃芽枯病菌

Pyrenochaeta terrestris（Hansen）Gorenz，Walker et Larson
洋葱粉色根腐病菌

Pythium splendens Braun
油棕猝倒病菌

Ramularia beticola Fautr. et Lambotte
甜菜叶斑病菌

Rhizoctonia fragariae Husain et W. E. McKeen
草莓花枯病菌

Rigidoporus lignosus（Klotzsch）Imaz.
橡胶白根病菌

Sclerophthora rayssiae Kenneth，Kaltin et Wahl var. *zeae* Payak et Renfro
玉米褐条霜霉病菌

Septoria petroselini（Lib.）Desm.
欧芹壳针孢叶斑病菌

Sphaeropsis pyriputrescens Xiao et J. D. Rogers
苹果球壳孢腐烂病菌

Sphaeropsis tumefaciens Hedges
柑橘枝瘤病菌

Stagonospora avenae Bissett f. sp. *triticea* T. Johnson
麦类壳多胞斑点病菌

Stagonospora sacchari Lo et Ling
甘蔗壳多胞叶枯病菌

Synchytrium endobioticum（Schilberszky）Percival
马铃薯癌肿病菌

Thecaphora solani（Thirumalachar et M. J. O'Brien）Mordue
马铃薯黑粉病菌

Tilletia controversa Kühn
小麦矮腥黑穗病菌

Tilletia indica Mitra
小麦印度腥黑穗病菌

Urocystis cepulae Frost
葱类黑粉病菌

Uromyces transversalis（Thümen）Winter
唐菖蒲横点锈病菌

Venturia inaequalis（Cooke）Winter
苹果黑星病菌

Verticillium albo-atrum Reinke et Berthold
苜蓿黄萎病菌

Verticillium dahliae Kleb.
棉花黄萎病菌

原核生物

Acidovorax avenae subsp. *cattleyae*（Pavarino）Willems et al.
兰花褐斑病菌

Acidovorax avenae subsp. *citrulli*（Schaad et al.）Willems et al.
瓜类果斑病菌

Acidovorax konjaci（Goto）Willems et al.
魔芋细菌性叶斑病菌

Alder yellows phytoplasma
桤树黄化植原体

Apple proliferation phytoplasma
苹果丛生植原体

Apricot chlorotic leafroll phtoplasma
杏褪绿卷叶植原体

Ash yellows phytoplasma
白蜡树黄化植原体

Blueberry stunt phytoplasma
蓝莓矮化植原体

Burkholderia caryophylli（Burkholder）Yabuuchi et al.
香石竹细菌性萎蔫病菌

Burkholderia gladioli pv. *alliicola*（Burkholder）Urakami et al.
洋葱腐烂病菌

Burkholderia glumae（Kurita et Tabei）Urakami et al.
水稻细菌性谷枯病菌

Candidatus Liberobacter africanum Jagoueix et al.
非洲柑橘黄龙病菌

Candidatus Liberobacter asiaticum Jagoueix et al.
亚洲柑橘黄龙病菌

Candidatus Phytoplasma australiense
澳大利亚植原体候选种

Clavibacter michiganensis subsp. *insidiosus*（McCulloch）Davis et al.
苜蓿细菌性萎蔫病菌

Clavibacter michiganensis subsp. *michiganensis*（Smith）Davis et al.
番茄溃疡病菌

Clavibacter michiganensis subsp. *nebraskensis*（Vidaver et al.）Davis et al.
玉米内州萎蔫病菌

Clavibacter michiganensis subsp. *sepedonicus*（Spieckermann et al.）Davis et al.
马铃薯环腐病菌

Coconut lethal yellowing phytoplasma
椰子致死黄化植原体

Curtobacterium flaccumfaciens pv. *flaccumfaciens*（Hedges）Collins et Jones
菜豆细菌性萎蔫病菌

Curtobacterium flaccumfaciens pv. *oortii*（Saaltink et al.）Collins et Jones
郁金香黄色疱斑病菌

Elm phloem necrosis phytoplasma
榆韧皮部坏死植原体

Enterobacter cancerogenus（Urosevi）Dickey et Zumoff
杨树枯萎病菌

Erwinia amylovora（Burrill）Winslow et al.
梨火疫病菌

Erwinia chrysanthemi Burkhodler et al.
菊基腐病菌

Erwinia pyrifoliae Kim, Gardan, Rhim et Geider
亚洲梨火疫病菌

Grapevine flavescence dorée phytoplasma
葡萄金黄化植原体

Lime witches' broom phytoplasma
来檬丛枝植原体

Pantoea stewartii subsp. *stewartii*（Smith）Mergaert et al.
玉米细菌性枯萎病菌

Peach X-disease phytoplasma
桃X病植原体

Pear decline phytoplasma
梨衰退植原体

Potato witches' broom phytoplasma
马铃薯丛枝植原体

Pseudomonas savastanoi pv. *phaseolicola*（Burkholder）Gardan et al.
菜豆晕疫病菌

Pseudomonas syringae pv. *morsprunorum*（Wormald）Young et al.
核果树溃疡病菌

Pseudomonas syringae pv. *persicae*（Prunier et al.）Young et al.
桃树溃疡病菌

Pseudomonas syringae pv. *pisi*（Sackett）Young et al.
豌豆细菌性疫病菌

Pseudomonas syringae pv. *maculicola*（McCulloch）Young et al
十字花科黑斑病菌

Pseudomonas syringae pv. *tomato*（Okabe）Young et al.
番茄细菌性叶斑病菌

Ralstonia solanacearum（Smith）Yabuuchi et al.（race 2）
香蕉细菌性枯萎病菌（2号小种）

Rathayibacter rathayi（Smith）Zgurskaya et al.
鸭茅蜜穗病菌

Spiroplasma citri Saglio et al.
柑橘顽固病螺原体

Strawberry multiplier phytoplasma
草莓簇生植原体

Xanthomonas albilineans（Ashby）Dowson
甘蔗白色条纹病菌

Xanthomonas arboricola pv. *celebensis*（Gaumann）Vauterin et al.
香蕉坏死条纹病菌

Xanthomonas axonopodis pv. *betlicola*（Patel et al.）Vauterin et al.
胡椒叶斑病菌

Xanthomonas axonopodis pv. *citri*（Hasse）Vauterin et al.
柑橘溃疡病菌

Xanthomonas axonopodis pv. *manihotis*（Bondar）Vauterin et al.
木薯细菌性萎蔫病菌

Xanthomonas axonopodis pv. *vasculorum*（Cobb）Vauterin et al.
甘蔗流胶病菌

Xanthomonas campestris pv. *mangiferaeindicae*（Patel et al.）Robbs et al.
芒果黑斑病菌

Xanthomonas campestris pv. *musacearum*（Yirgou et Bradbury）Dye
香蕉细菌性萎蔫病菌

Xanthomonas cassavae（ex Wiehe et Dowson）Vauterin et al.
木薯细菌性叶斑病菌

Xanthomonas fragariae Kennedy et King
草莓角斑病菌

Xanthomonas hyacinthi（Wakker）Vauterin et al.
风信子黄腐病菌

Xanthomonas oryzae pv. *oryzae*（Ishiyama）Swings et al.
水稻白叶枯病菌

Xanthomonas oryzae pv. *oryzicola*（Fang et al.）Swings et al.
水稻细菌性条斑病菌

Xanthomonas populi（ex Ride）Ride et Ride
杨树细菌性溃疡病菌

Xylella fastidiosa Wells et al.
木质部难养细菌

Xylophilus ampelinus（Panagopoulos）Willems et al.
葡萄细菌性疫病菌

线 虫

Anguina agrostis（Steinbuch）Filipjev
剪股颖粒线虫

Aphelenchoides fragariae（Ritzema Bos）Christie
草莓滑刃线虫

Aphelenchoides ritzemabosi（Schwartz）Steiner et Bührer
菊花滑刃线虫

Bursaphelenchus cocophilus（Cobb）Baujard
椰子红环腐线虫

Bursaphelenchus xylophilus（Steiner et Bührer）Nickle
松材线虫

Ditylenchus angustus（Butler）Filipjev
水稻茎线虫

Ditylenchus destructor Thorne
腐烂茎线虫

Ditylenchus dipsaci（Kühn）Filipjev
鳞球茎茎线虫

Globodera pallida（Stone）Behrens
马铃薯白线虫

Globodera rostochiensis（Wollenweber）Behrens
马铃薯金线虫

Heterodera schachtii Schmidt
甜菜胞囊线虫

Longidorus（Filipjev）Micoletzky（The species transmit viruses）
长针线虫属（传毒种类）

Meloidogyne Goeldi（non-Chinese species）
根结线虫属（非中国种）

Nacobbus abberans（Thorne）Thorne et Allen
异常珍珠线虫

Paralongidorus maximus（Bütschli）Siddiqi
最大拟长针线虫

Paratrichodorus Siddiqi（The species transmit viruses）
拟毛刺线虫属（传毒种类）

Pratylenchus Filipjev（non-Chinese species）
短体线虫（非中国种）

Radopholus similis（Cobb）Thorne
香蕉穿孔线虫

Trichodorus Cobb（The species transmit viruses）
毛刺线虫属（传毒种类）

Xiphinema Cobb（The species transmit viruses）
剑线虫属（传毒种类）

病毒及类病毒

African cassava mosaic virus，ACMV
非洲木薯花叶病毒（类）

Apple stem grooving virus，ASPV
苹果茎沟病毒

Arabis mosaic virus，ArMV
南芥菜花叶病毒

Banana bract mosaic virus，BBrMV
香蕉苞片花叶病毒

Bean pod mottle virus，BPMV
菜豆荚斑驳病毒

Broad bean stain virus，BBSV
蚕豆染色病毒

Cacao swollen shoot virus，CSSV
可可肿枝病毒

Carnation ringspot virus，CRSV
香石竹环斑病毒

Cotton leaf crumple virus，CLCrV
棉花皱叶病毒

Cotton leaf curl virus，CLCuV
棉花曲叶病毒

Cowpea severe mosaic virus，CPSMV
豇豆重花叶病毒

Cucumber green mottle mosaic virus，CGMMV
黄瓜绿斑驳花叶病毒

Maize chlorotic dwarf virus，MCDV
玉米褪绿矮缩病毒

Maize chlorotic mottle virus，MCMV
玉米褪绿斑驳病毒

Oat mosaic virus，OMV
燕麦花叶病毒

Peach rosette mosaic virus，PRMV
桃丛簇花叶病毒

Peanut stunt virus, PSV
花生矮化病毒

Plum pox virus, PPV
李痘病毒

Potato mop-top virus, PMTV
马铃薯帚顶病毒

Potato virus A, PVA
马铃薯A病毒

Potato virus V, PVV
马铃薯V病毒

Potato yellow dwarf virus, PYDV
马铃薯黄矮病毒

Prunus necrotic ringspot virus, PNRSV
李属坏死环斑病毒

Southern bean mosaic virus, SBMV
南方菜豆花叶病毒

Sowbane mosaic virus, SoMV
藜草花叶病毒

Strawberry latent ringspot virus, SLRSV
草莓潜隐环斑病毒

Sugarcane streak virus, SSV
甘蔗线条病毒

Tobacco ringspot virus, TRSV
烟草环斑病毒

Tomato black ring virus, TBRV
番茄黑环病毒

Tomato ringspot virus, ToRSV
番茄环斑病毒

Tomato spotted wilt virus, TSWV
番茄斑萎病毒

Wheat streak mosaic virus, WSMV
小麦线条花叶病毒

Apple fruit crinkle viroid, AFCVd
苹果皱果类病毒

Avocado sunblotch viroid, ASBVd
鳄梨日斑类病毒

Coconut cadang-cadang viroid, CCCVd
椰子死亡类病毒

Coconut tinangaja viroid, CTiVd
椰子败生类病毒

Hop latent viroid, HLVd
啤酒花潜隐类病毒

Pear blister canker viroid, PBCVd
梨疱症溃疡类病毒

Potato spindle tuber viroid, PSTVd
马铃薯纺锤块茎类病毒

杂 草

Aegilops cylindrica Horst
具节山羊草

Aegilops squarrosa L.
节节麦

Ambrosia spp.
豚草

Ammi majus L.
大阿米芹

Avena barbata Brot.
细茎野燕麦

Avena ludoviciana Durien
法国野燕麦

Avena sterilis L.
不实野燕麦

Bromus rigidus Roth
硬雀麦

Bunias orientalis L.
疣果匙荠

Caucalis latifolia L.
宽叶高加利

Cenchrus spp.（non-Chinese species）
蒺藜草（非中国种）

Centaurea diffusa Lamarck
铺散矢车菊

Centaurea repens L.
匍匐矢车菊

Crotalaria spectabilis Roth
美丽猪屎豆

Cuscuta spp.
菟丝子

Emex australis Steinh.
南方三棘果

Emex spinosa（L.）Campd.
刺亦模

Eupatorium adenophorum Spreng.（见 *Ageratina adenophorum*
（Spreng.）King & H. Rob.）
紫茎泽兰

Eupatorium odoratum L.（见 *Chromolaena odorata* L.）
飞机草

Euphorbia dentata Michx.
齿裂大戟

Flaveria bidentis（L.）Kuntze
黄顶菊

Ipomoea pandurata（L.）G. F. W. Mey.
提琴叶牵牛花

Iva axillaris Pursh
小花假苍耳

Iva xanthifolia Nutt.
假苍耳

Knautia arvensis（L.）Coulter
欧洲山萝卜

Lactuca pulchella（Pursh）DC.
野莴苣

Lactuca serriola L.
毒莴苣

Lolium temulentum L.
毒麦

Mikania micrantha Kunth
薇甘菊

Orobanche spp.
列当

Oxalis latifolia Kubth
宽叶酢浆草

Senecio jacobaea L.
臭千里光

Solanum carolinense L.
北美刺龙葵

Solanum elaeagnifolium Cav.
银毛龙葵

Solanum rostratum Dunal.
刺萼龙葵

Solanum torvum Swartz
刺茄

Sorghum almum Parodi.
黑高粱

Sorghum halepense（L.）Pers.（Johnsongrass and its cross breeds）
假高粱（及其杂交种）

Striga spp.（non-Chinese species）
独脚金（非中国种）

Subgen *Acnida* L.
异株苋亚属

Tribulus alatus Delile
翅蒺藜

Xanthium spp.（non-Chinese species）
苍耳（非中国种）

备注：

1. 非中国种，是指中国未有发生的种；
2. 非中国小种，是指中国未有发生的小种；
3. 传毒种类，是指可以作为植物病毒传播介体的线虫种类。